Invading Ecological Networks

Until now, biological invasions have been conceptualised and studied mainly as a linear process: from introduction to establishment to spread. This volume charts a new course for the field, drawing on key developments in network ecology and complexity science. It defines an agenda for Invasion Science 2.0 by providing new framings and classification of research topics and by offering tentative solutions to vexing problems. In particular, it conceptualises a transformative ecosystem as an open adaptive network with critical transitions and turnover, with resident species learning heuristically and fine-tuning their niches and roles in a multiplayer eco-evolutionary game. It erects signposts pertaining to network interactions, structures, stability, dynamics, scaling and invasibility. It is not a recipe book or a road map, but an atlas of possibilities: a 'hitchhiker's guide'.

CANG HUI is a professor of mathematical biology and holds the South African Research Chair in Mathematical and Theoretical Physical Biosciences at Stellenbosch University. He is a trustee of the International Initiative for Theoretical Ecology. He has published widely on biological invasions and ecological networks.

DAVID M. RICHARDSON is Director of the Centre for Invasion Biology at Stellenbosch University. He is a member of the Species Survival Specialist Group on Invasive Organisms for the International Union for Conservation of Nature. His main expertise is in invasion ecology, and particularly alien tree invasions. He has published extensively on invasive species and restoration ecology.

ECOLOGY, BIODIVERSITY AND CONSERVATION

The world's biological diversity faces unprecedented threats. The urgent challenge facing the concerned biologist is to understand ecological processes well enough to maintain their functioning in the face of the pressures resulting from human population growth. Those concerned with the conservation of biodiversity and with restoration also need to be acquainted with the political, social, historical, economic and legal frameworks within which ecological and conservation practice must be developed. The new Ecology, Biodiversity, and Conservation series will present balanced, comprehensive, up-to-date, and critical reviews of selected topics within the sciences of ecology and conservation biology, both botanical and zoological, and both 'pure' and 'applied'. It is aimed at advanced final-year undergraduates, graduate students, researchers, and university teachers, as well as ecologists and conservationists in industry, government and the voluntary sectors. The series encompasses a wide range of approaches and scales (spatial, temporal, and taxonomic), including quantitative, theoretical, population, community, ecosystem, landscape, historical, experimental, behavioural and evolutionary studies. The emphasis is on science related to the real world of plants and animals rather than on purely theoretical abstractions and mathematical models. Books in this series will, wherever possible, consider issues from a broad perspective. Some books will challenge existing paradigms and present new ecological concepts, empirical or theoretical models, and testable hypotheses. Other books will explore new approaches and present syntheses on topics of ecological importance.

Ecology and Control of Introduced Plants
Judith H. Myers and Dawn Bazely

Invertebrate Conservation and Agricultural Ecosystems
T. R. New

Birds and Climate Change: Impacts and Conservation Responses
James W. Pearce-Higgins and Rhys E. Green

Marine Ecosystems: Human Impacts on Biodiversity, Functioning and Services
Tasman P. Crowe and Christopher L. J. Frid

Wood Ant Ecology and Conservation
Jenni A. Stockan and Elva J. H. Robinson

Detecting and Responding to Alien Plant Incursions
John R. Wilson, F. Dane Panetta and Cory Lindgren

Conserving Africa's Mega-Diversity in the Anthropocene: The Hluhluwe-iMfolozi Park Story
Joris P. G. M. Cromsigt, Sally Archibald and Norman Owen-Smith

National Park Science: A Century of Research in South Africa
Jane Carruthers

Plant Conservation Science and Practice: The Role of Botanic Gardens
Stephen Blackmore and Sara Oldfield

Habitat Suitability and Distribution Models: With Applications in R
Antoine Guisan, Wilfried Thuiller and Niklaus E. Zimmermann

Ecology and Conservation of Forest Birds
Grzegorz Mikusiński, Jean-Michel Roberge and Robert J. Fuller

Species Conservation: Lessons from Islands
Jamieson A. Copsey, Simon A. Black, Jim J. Groombridge and Carl G. Jones

Soil Fauna Assemblages: Global to Local Scales
Uffe N. Nielsen

Curious About Nature
Tim Burt and Des Thompson

Comparative Plant Succession Among Terrestrial Biomes of the World
Karel Prach and Lawrence R. Walker

Ecological-Economic Modelling for Biodiversity Conservation
Martin Drechsler

Freshwater Biodiversity: Status, Threats and Conservation
David Dudgeon

Joint Species Distribution Modelling: With Applications in R
Otso Ovaskainen and Nerea Abrego

Natural Resource Management Reimagined: Using the Systems Ecology Paradigm
Robert G. Woodmansee, John C. Moore, Dennis S. Ojima and Laurie Richards

The Species–Area Relationship: Theory and Application
Thomas J. Matthews, Kostas A. Triantis and Robert J. Whittaker

Ecosystem Collapse and Recovery
Adrian C. Newton

Animal Population Ecology: An Analytical Approach
T. Royama

Why Conserve Nature? Perspectives on Meanings and Motivations
Stephen Trudgill

Invading Ecological Networks

CANG HUI

Stellenbosch University

DAVID M. RICHARDSON

Stellenbosch University

CAMBRIDGE
UNIVERSITY PRESS

CAMBRIDGE
UNIVERSITY PRESS

University Printing House, Cambridge CB2 8BS, United Kingdom

One Liberty Plaza, 20th Floor, New York, NY 10006, USA

477 Williamstown Road, Port Melbourne, VIC 3207, Australia

314–321, 3rd Floor, Plot 3, Splendor Forum, Jasola District Centre, New Delhi – 110025, India

103 Penang Road, #05–06/07, Visioncrest Commercial, Singapore 238467

Cambridge University Press is part of the University of Cambridge.

It furthers the University's mission by disseminating knowledge in the pursuit of
education, learning, and research at the highest international levels of excellence.

www.cambridge.org
Information on this title: www.cambridge.org/9781108478618
DOI: 10.1017/9781108778374

First published 2022

Printed in United Kingdom by TJ Books Limited, Padstow Cornwall

A catalogue record for this publication is available from the British Library.

ISBN 978-1-108-47861-8 Hardback
ISBN 978-1-108-74596-3 Paperback

Imagin'd rather oft than elsewhere seen,
That stone, or like to that which here below
Philosophers in vain so long have sought,
In vain, though by their powerful Art they bind
Volatile Hermes, *and call up unbound*
In various shapes old Proteus *from the Sea,*
Drain'd through a Limbec to his native form.
                John Milton, *Paradise Lost*, III, 599–605

Contents

Color plates are to be found between pages 206–207

Preface

The *I Ching* or *Book of Changes*, published about three thousand years ago, is among the oldest of the Chinese classics. It posits that all things are interlinked and changing, and therefore that the state of any system is in constant flux and transition, while the rules of change are observable in the midst of the diverse patterns of flux exhibited in the interlinked natural world. Indeed, our planet's biosphere, comprising all living things and their interactions, is experiencing unprecedented change, most of it attributable to humans. This rapid change has created a wave of selection, transforming global ecosystems and moulding their structures and functions. Within a few decades we could have entirely new climates and radically altered species compositions and distributions – the emergence of novel networks of biotic interactions extending across multiple scales. The agenda in ecology, however, is largely mired in thoughts of restoring balance and equilibrium. We hope that this book will stimulate readers to think deeply about the concept and ramifications of constant change in our natural world and to ponder the rules of change in our rapidly changing ecosystems.

Alien species, those that have been moved to new areas through human actions, are both drivers and the passengers of the current global change. Although many introduced species fail to establish footholds in new regions, many flourish in their new homes to such an extent that they transform recipient ecosystems. The novel experiences and non-equilibrial dynamics associated with biological invasions, and the complex responses of recipient ecosystems, provide us with a natural experiment to gain an understanding of the rules of change in this emerging transformative ecology. We envisage that the transformative structure and function of an open adaptive ecosystem, driven by eco-evolutionary processes, biological invasions and other drivers of change, will become the mainstay of ecology.

Species do not live alone but share their spaces with many others, both long-time residents and newcomers, the latter often dumped in their new homes without co-evolved partners by human activities. Introduced species and the changes they cause to biotic interactions are both key drivers of community succession and assembly. Along the conceptual spectrum of an ecological community, Frederic Clements (1916) proposed the metaphor of a superorganism evolving towards a stable 'climax state'. He further argued that 'climax' formations could arise, grow, mature, die and reproduce. In opposing the ideas of Clements, Henry Gleason (1926) depicted ecological communities as artefacts shaped by these residing species which happened as merely a coincidence and behave in an individualistic manner. Ecological communities are, of course, neither a superorganism nor an artefact, but an entangled web of biotic interactions. As Charles Darwin (1859) so eloquently portrayed at the end of *On the Origin of Species*,

It is interesting to contemplate a tangled bank, clothed with many plants of many kinds, with birds singing on the bushes, with various insects flitting about, and with worms crawling through the damp earth, and to reflect that these elaborately constructed forms, so different from each other, and dependent upon each other in so complex a manner, have all been produced by laws acting around us.

We view 'ecological networks' as entangled webs of distinct, interdependent and complex biotic interactions among co-occurring species in a landscape; such webs are also called ecological interaction networks. They are the veins of an ecological community, with the emphasis on the strength and structure of biotic interactions rather than the identity of its residing species. To describe an ecological network, we need to delineate its boundaries, enumerate the resident species and define and elucidate the entangled interactions. When asked to define the boundary of an ecological community, Samuel Scheiner quipped, 'That's where an ecologist stops the car.' Although dispersal barriers, such as rivers, gorges or fences, do create clear edge effects on the distribution of biodiversity, their boundaries seldom coincide neatly with those of ecological networks. As is the case in the concepts of meta-population and meta-community, smaller ecological networks can be connected to form, or are embedded within, much bigger spatial networks – meta-networks. Importantly, delimiting the boundary of ecological networks not only affects how we label a species, resident or alien, but also defines the complexity and impact of the entangled biotic interactions.

Network ecology focusses on understanding the functions that emerge from the complexity of biotic interactions; each residing species is thus conscripted into a community by virtue of its functional traits (typically invoking Eltonian niches), not by its identity per se. Mark Davis and colleagues (2011) have argued that we should judge a species not by its origin but rather by its functional roles in an ecosystem. Such views, when considered from the perspective of network ecology and conservation science, represent a paradigm shift from the classic species-centric view towards one focussed on the safeguarding ecosystem functioning and services in the Anthropocene. All crises beget opportunities. Faced with massive ongoing extinctions and failed conservation attempts, such as the Aichi Biodiversity Targets, Chris Thomas (2017) called for ecologists to embrace opportunities created by biological invasions and other facets of global change to deal with anticipated species losses. Such challenges force us to confront critical questions. Can we replace a community of native species with one comprising a mixture of native and alien species while ensuring the same or even more desirable functioning and services? What makes an alien species different from a resident native species?

Ecological networks are far more than just ensembles of co-occurring species. Species interact with each other in a network not simply through trait matching; Dan Janzen (1985) coined the term 'ecological fitting' for this. Species also co-adapt and co-evolve relentlessly, as J. N. Thompson puts it (2013). This is not a jigsaw puzzle, but a melting pot. It is messy and unpredictable. We cannot simplify the ways of change in an ecological network as actions and reactions. As Robert Holt opined, parsimony is the rule of reasoning, but not necessarily the way of nature. Ecological networks, to us, are therefore multi-agent, complex and adaptive systems. Ecosystems facing biological invasions are, consequently, ideal models for studying the structure and function of an open, complex, adaptive and dynamic network.

The young and vibrant discipline of invasion science has a wide range of focal research agendas and uses diverse approaches to address theoretical and practical issues. The field is, however, converging on a few unifying frameworks that serve to unite and clarify concepts and define the most pressing questions (Wilson et al. 2020). One of the most widely accepted frameworks in invasion science is a species-centric linear scheme that conceptualises barriers as mediators of progression of the species through different invasion stages (Richardson et al. 2000). Building on the mantra of this barrier construct, a unified framework for invasion biology was proposed to elucidate the factors that drive this progression along a linear introduction–naturalisation–invasion continuum (Blackburn et al. 2011).

This has served as a roadmap for both theoretical and applied studies. Management-focussed work has sought effective intervention strategies to curb or slow progression along the invasion continuum. In adopting this invasion construct, substantial effort has gone into applying correlative statistics to identify the functional traits that allow species, or groups of species, to overcome different barriers. Macroecological approaches are often applied to compare, for example, the niches of alien species in their invaded versus their native ranges; the traits of aliens and those of ecologically similar native species; or to identify traits and environmental factors associated with alien species of contrasting invasion stages. Such research has, to some extent, unified management practices and has yielded reasonable, albeit often unsatisfying, insights into invasion dynamics. This invader-centric phase is unpacked when we describe what we term Invasion Science 1.0 in Chapter 1.

The other chapters of this book explore the potential of a new agenda for invasion science, and perhaps also for global change biology more broadly. In particular, we try to build on tentative steps taken to address key questions pertaining to biological invasions using concepts and tools developed in the fields of network science for complex systems. In a complex system, feedback loops can push a network towards a state of paradox – although patterns derived from observations are highly structured in retrospective views (e.g., invasive traits can be clearly identified from comparative studies), predictions based on such observed structures often perform poorly and have high levels of uncertainty. This is akin to weather forecasts, earthquake predictions or shares in the stock market. Retrospectively, clear patterns emerge from interpolation, but forward extrapolation normally fails. In this new paradigm, we explore such failures, expose their potential causes and propose a way forward.

In tackling the state of paradox facing Invasion Science 1.0, in line with rapid advances in community, network, evolutionary and systems ecology, we tentatively suggested in our earlier book *Invasion Dynamics* the need to shift the metaphor of biological invasions from one based on the linear invasion continuum scheme to one that invokes complex adaptive networks (Hui and Richardson 2017). In the section of our book titled 'Invasion Science 2050' we suggested that such a restructuring is 'not just a call for next-generation quantitative methodologies to improve detection and measurement of feedback loops in ecology' but also 'an appeal for a paradigm shift in ecology and invasion science to embrace adaptive cycles and network thinking'. As the first step in this direction we offered a tentative blueprint of Invasion Science 2.0 in a paper titled 'How to Invade an Ecological Network' (Hui and Richardson 2019). There, we unpacked the practical difficulties and challenges that emerge when seeking to marry concepts and tools from network ecology and invasion ecology. We also demonstrated the potential of this network–invasion synergy by providing a simple generic model. This book expands substantially on this thesis.

We plan to expand our vision for Invasion Science 2050 (Hui and Richardson 2017, 2019) and elucidate the core conceptual issues that need to be considered in the transition towards Invasion Science 2.0. After cherry-picking the key concepts of Invasion Science 1.0 in Chapter 1, the remaining six chapters of the book expand on six core concepts that are essential for understanding how ecological networks harbour numerous interacting species, adapt and respond to biological invasions, as well as the role of each species, native or alien, in this complex multiplayer game. In a nutshell, we hope to show how biotic interactions operate as the building blocks of ecological networks (Chapter 2); the emergence of network complexity and architecture (Chapter 3); how biological invasions modify and break the network complexity–stability relationship, and push the system regime towards marginal instability (Chapter 4); network dynamics, as well as invasion performance, at marginal instability; (Chapter 5); how the response of network structure and function to biological invasions plays out over ranges of spatial and temporal scales (Chapter 6); and finally to ponder on the concept of network invasibility as the sweet-spots in the functional trait space to achieve elevated invasion performance in an adaptive ecological network (Chapter 7).

In 1889, King Oscar II of Sweden established a prize for the n-body problem in mathematics and physics,

Given a system of arbitrarily many mass points which attract each other according to Newton's laws, try to find, under the assumption that no two points ever collide, a representation of the coordinates of each point as a series in a variable which is some known function of time and for all of whose values the series converges uniformly.

We now face a similar but perhaps even more complex n-species problem in ecology. To paraphrase King Oscar II's letter: given an ecological network of arbitrarily many species which engage with each other according to trait-mediated density-dependent inter- and intraspecific biotic interactions, under the assumptions that new species can invade this ecological network and that the resident species can become extinct or be extirpated, we need to predict, for a meaningful time span, the population dynamics of each species, including both the invaders and those resident species, and the evolutionary dynamics of their functional traits. This is the subject of our book. It is not a recipe book, but a hitchhiker's guide, a cloud atlas, to the adventures of future ecology.

References

Blackburn TM, et al. (2011) A proposed unified framework for biological invasions. *Trends in Ecology & Evolution* **26**, 333–339.

Clements FE (1916) *Plant Succession: An Analysis of the Development of Vegetation.* Washington, DC: Carnegie Institute of Washington Publication Sciences.

Darwin CR (1859) *On the Origin of Species by Means of Natural Selection.* London: John Murray.

Davis M, et al. (2011) Don't judge species on their origins. *Nature* **474**, 153–154.

Gleason HA (1926) The individualistic concept of the plant association. *Bulletin of the Torrey Botanical Club* **53**, 7–26.

Hui C, Richardson DM (2017) *Invasion Dynamics.* Oxford: Oxford University Press.

Hui C, Richardson DM (2019) How to invade an ecological network. *Trends in Ecology & Evolution* **34**, 121–131.

Janzen D (1985) On ecological fitting. *Oikos* **45**, 308–310.

Richardson DM, et al. (2000) Naturalization and invasion of alien plants: concepts and definitions. *Diversity and Distributions* **6**, 93–107.

Thomas CD (2017) *Inheritors of the Earth: How Nature Is Thriving in an Age of Extinction.* London: Penguin Books.

Thompson JN (2013) *Relentless Evolution.* Chicago: University of Chicago Press.

Wilson JRU, et al. (2020) Frameworks used in invasion science: progress and prospects. *NeoBiota* **62**, 1–30.

Acknowledgements

We are grateful to so many colleagues and friends who have over the years discussed and collaborated with us, either casually or formally, on topics that somehow shaped our thinking on issues covered in this book. Many students and postdoctoral fellows have passed through our labs; their sparks of inspiration and inquisitiveness spurred us on to dig deeper into this topic. These include, in alphabetical order, Ashleigh Basel, Cecile Berthouly-Salazar, Oonsie Biggs, Tim Blackburn, William Bond, Luz Boyero, Åke Brännström, Chris Broeckhoven, Marc Cadotte, Jonathan Chase, Susana Clusella-Trullas, Franck Courchamp, Richard Cowling, Helen de Klerk, Ulf Dieckmann, Genevieve Diedericks, Benjamin du Toit, Allen Ellis, Marcela Espinaze, Franz Essl, Gordon Fox, Llewellyn Foxcroft, Mirijam Gaertner, Laure Gallien, Richard Gibbs, Lev Ginzburg, Dominique Gravel, Jessica Gurevitch, Xiaozhuo Han, Fangliang He, Steve Higgins, Marinel Janse van Rensburg, Jonathan Jeschke, Jan-Hendrik Keet, Ingolf Kuhn, Christian Kull, Sabrina Kumschick, Bill Kunin, Vitalis Lagat, Pietro Landi, Gilbert Langat, Guillaume Latombe, Jaco Le Roux, Jingjing Liang, Sandy Liebhold, Sandra MacFadyen, Denise Mager, Anne Magurran, Mario Marial, Sonja Matthee, Melodie McGeoch, John Measey, Guy Midgley, Joe Miller, Ony Minoarivelo, Mozzamil Mohammed, Jane Molofsky, Assumpta Nnakenyi, Ana Novoa, Savannah Nuwagaba, Anton Pauw, David Phair, Luke Potgieter, Petr Pyšek, Andriamihaja Ramanantoanina, Axel Rensberg, Mark Robertson, James Rodger, Mathieu Rouget, Núria Roura-Pascual, Helen Roy, Nate Sanders, Jeff Sanders, Wolf-Christian Saul, Maria Schelling, Ross Shackleton, Min Su, Lawal Suleiman, John Terblanche, Jessica Toms, Anna Traveset, Brian van Wilgen, Nicola van Wilgen, James Wilsenach, Sofia Vaz, Peter Verburg, Joana Vicente, Vernon Visser, Sophie von der Heyden, Klaus von Gadow, Olaf Weyl, John Wilson, Darragh Woodford, Theresa Wossler, Yinghui Yang, Dongxia Yue, Feng Zhang and Zihua Zhao.

We have surely missed some names here. Forgive us.

The opportunities for research and friendly collaboration across traditional disciplinary boundaries provided by sixteen years of funding to the DSI-NRF Centre of Excellence for Invasion Biology (C·I·B) cannot be overstated. It has been an exhilarating ride. We also acknowledge financial support from South Africa's National Research Foundation (grants 85417 and 89967 in particular), the Australian Research Council (grant DP200101680), the UK's Natural Environment Research Council (grant NE/V007548/1) and the Oppenheimer Memorial Trust (grant 18576/03).

We appreciate the great editorial support from Dominic Lewis, Michael Usher, Samuel Fearnley and Aleksandra Serocka at Cambridge University Press. We are also grateful to Jonathan Downs, Vanessa du Plessis, Larisse Bolton and Charndré Williams for their assistance in preparing the reference lists and handling copyright issues.

Lastly, we thank our families, Beverley, Keira and Zachary Hui, and Corlia and Sean Richardson, for their patience and tremendous support during the writing of this book, especially through the extended period of time over the lockdown due to the COVID-19 pandemic.

1 · *Invasion Science 1.0*

This [invasion] velocity is proportional to the square root of the intensity of selective advantage and to the standard deviation of scattering in each generation.

(Fisher 1937)

1.1 Welcome to the Anthropocene

At the time of writing this book, we have witnessed an extreme case of biological invasion. A virus, through an evolutionary leap, has jumped onto a new host species, *Homo sapiens*, and has taken advantage of the new host's ambitions and mobility in the zealous phase of globalisation, causing a worldwide pandemic and economic meltdown. The 2019 coronavirus outbreak (COVID-19) is a showcase of the core of invasion science. A list of questions spring to mind. Why this particular virus, and not others? Why now? How fast can it spread? How is its spread mediated by climatic and other environmental factors? What are its vectors and pathways of transmission? Which regions and populations are most susceptible? How much damage can it cause to public health and economies? What factors cause substantial variation in mortality between human populations in different countries? How can we control it? Can we forecast and prevent future outbreaks of emerging infectious diseases? While the whole world scrambles to make sense of COVID-19 and to combat the biggest crisis for humanity since World War II (WWII), we embark on a journey to address these questions to cover many more taxa and situations – the invasion of any biological organism into novel environments.

All species have the means to shift their progeny, either via direct movement or through vector-mediated dispersal. The incentive to move has driven Earth's biota to cover all possible niches, from the Antarctic to the Arctic, from the Himalayas to the Mariana Trench. Most propagules, however, move slowly and over short distances. On very rare occasions do propagules catch a ride on ocean rafts or hurricanes, or become attached to a seabird. Such propensity and limitation of dispersal are key factors behind the world's distinct biotic zones. This process of natural dispersal and spread of species was altered by early hominids. Hunter-gatherer societies had deep knowledge of the animals and plants around them and started to cultivate many species to ensure a sustainable supply of food and fibre. When humans began colonising the entire planet, cultivated plants and animals moved with them, the result being a growing list of species able to thrive in human-dominated environments, with the capacity to transform landscapes. Not only did humans intentionally move many species with them to supply their needs, but their movements also resulted in the accidental movement of many species. These include species associated with useful organisms, such as yeasts, viruses and other microorganisms, and many other types of pest and weed that simply 'hitched a ride' on diverse means of transport. Human selection has resulted in a rather unique assemblage of species, distinct from those that occur in natural communities and which are filtered by natural selection.

Human-mediated movement of species has accelerated dramatically in the era of globalisation, in terms of quantity, distance and speed. Technological innovations have revolutionised ways in which we transport goods. Stretching from Xi'an to Rome, the Silk Road connected the Eurasian supercontinent as early as the first century BC, carrying goods on the backs of horses and camels. Islamic merchants created the Spice Route in the seventh century, thereby connecting the Mediterranean Sea and the Indian Ocean. Global trade started in earnest in the Age of Discovery, when European explorers connected East and West with the Americas in the fifteenth century. Global trade scaled up after the first Industrial Revolution in the eighteenth and nineteenth centuries when global production chains began compartmentalising (e.g. meat export from South America). The trajectory has been interrupted only by two World Wars and the COVID-19 pandemic. After WWII, globalisation resumed its march with the mainstream transport of cars, ships and planes (global export totalling US$62 billion in 1950), only being slowed temporarily by the Iron Curtain during the Cold War. A milestone of this globalisation was the launch of the World Trade Organisation in 1995, when global exports reached US$5 trillion. Globalisation then soared over the next two decades, with bumps along the way during the 2008 recession and the COVID-19 pandemic,

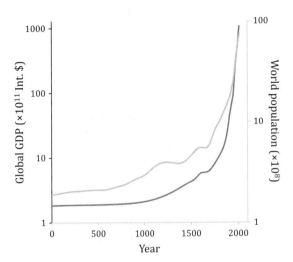

Figure 1.1 Global GDP, in international dollars (2011 price), and world population, in the past two millennia. Based on data from ourworldindata.org under CC-BY Licence.

reaching close to US$19 trillion in 2014. Real gross domestic product (GDP) per capita in the United States in 2014 was four times the size it was in 1950. The human population increased from 2.5 billion in 1950 to 7 billion in 2012 (Figure 1.1), and is projected to reach 10 billion in 2050. Not only has our ecological footprint overshot the planet's carrying capacity, but there are also emerging global crises that are threatening the whole of humanity (e.g. climate change, biodiversity loss and the pandemic).

With the rising dominance of humans in the biosphere, previously characteristic floras and faunas in regional biotic zones have been mixed and reshuffled, resulting in a major homogenisation of the world's biota. The accumulation of non-native species across the globe is continuing with no sign of a slowing of the rate of new records of naturalisation and invasion (Seebens et al. 2017). Putting aside biases in taxonomy and sampling effort, the trend in the global rate of new records of established non-native species is overwhelming (Figure 1.2). Geographic and taxonomic variations in the dynamics and rate of non-native establishment reflect the role and history of regional countries in global trade. With the rise of global trade, the rate of establishment of non-native species has increased steadily, as stowaways, contaminants and pets since 1800, and

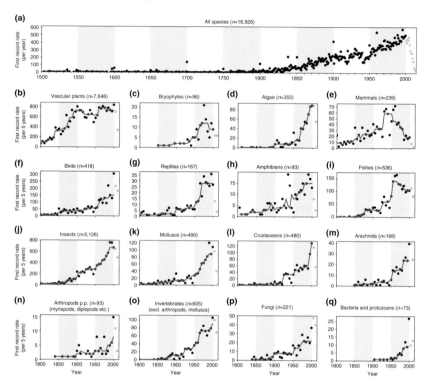

Figure 1.2 Global temporal trends in the rates of first records of the establishment of non-native species. Global temporal trends in first record rates (dots) for all species (a) and taxonomic groups (b–q) with the total number of established non-native species during the respective time periods given in parentheses. Data after 2000 (grey dots) are incomplete because of the delay between sampling and publication, and therefore not included in the analysis. As first record rates were recorded on a regional scale, species may be included multiple times in one plot. (a) First record rates are the number of first records per year during 1500–2014. (b–q) First record rates constitute the number of first records per 5 years during 1800–2014 for various taxonomic groups. The trend is indicated by a running median with a 25-year moving window (red line). For visualisation, 50-year periods are distinguished by white/grey shading. From Seebens et al. (2017) under CC-BY Licence.

accelerated further after 1950 – with the sole exception of mammals and fishes, which exhibit a hump-shaped curve, perhaps due to the regulations on farming for the game and fur industry. The establishment of non-native plant species has maintained a high rate since 1900 (Figure 1.2), coinciding with acclimatisation and colonisation activities in European diasporas. Technology has enabled us to move species around the world in new ways, quickly and in huge numbers; and changing fashions, fads and desires of human societies are continuously modifying and expanding the catalogue of translocated species – not just for essential goods but also for peculiar luxuries and hobbies. We need new ways of categorising and managing the new assemblages of biota that occur in different environments. Not only do we need to understand how many species are moved around the world by humans, but we also need to understand how these species interact with other species and how the added species and the changes that they bring affect the functioning of ecosystems, and thereby influence our well-being, both positively and negatively.

Biological invasions are by no means the only driver of the massive global-scale environmental changes that we are seeing. Invasive species interact in complex ways with other key builders and shapers of novel ecosystems such as agriculture, urbanisation, altered biogeochemical cycles, excessive carbon emission and pollution. For instance, of the documented 291 records of plant species extinction (Le Roux et al. 2019), agriculture, urbanisation, grazing, habitat degradation and destruction, together with biological invasions, are found to be implicated. The exact role of each of these factors is difficult to discern in most cases, but each surely has its own distinct temporal pattern and role to play (Figure 1.3). With these burgeoning factors affecting the planet's biosphere, we are witnessing pervasive alterations to physical systems, disturbance regimes and biogeochemical cycles, leading to a downward spiral in the integrity and health of ecosystems, accompanied by biodiversity loss and ecosystem transformation. In some cases, biological invasions are directly responsible for the decline of native biota, e.g. native plant species in Mediterranean-type ecosystems have been severely affected by non-native plants, particularly by Australian acacias (Figure 1.4; Gaertner et al. 2009). Recent reviews on the role of biological invasions in reducing the biodiversity of recipient ecosystems overwhelmingly support this view of the detrimental role of invasive species, more so at local than regional levels (Figure 1.5; Chase et al. 2018). These forces of change sometimes reinforce each other at different spatial and temporal scales, often with lags, leading to complex and intertwined challenges to the well-being of humanity and ecosystems (Díaz et al. 2015; Essl et al. 2015a). On this wagon of humanity, many hitchhiker species proliferate, creating harmful impacts on human well-being. The huge number of species that have been transported by us in different quantities and rates, intentionally or not, directly or not, define the subject and context of invasion science (Pyšek et al. 2020a).

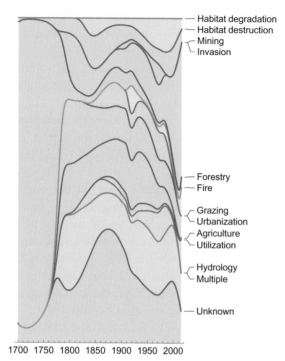

Figure 1.3 Primary drivers of plant extinctions over the last 300 years shown as area graphs to visualise the temporal changes in the relative contribution of the 11 identified primary extinction causes. Data from Le Roux et al. (2019).

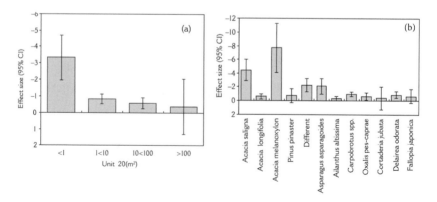

Figure 1.4 Effect size (95% CI) of invasion on species richness for different (a) unit sizes and (b) taxonomical groups in Mediterranean-type ecosystems. Q-test shows significant different effect sizes (heterogeneity) between unit sizes and between species. From Gaertner et al. (2009), reproduced with permission.

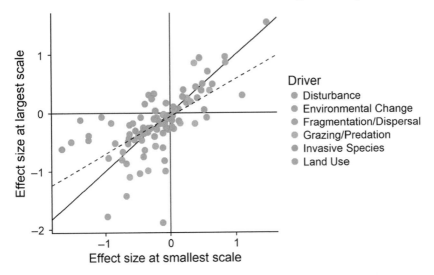

Figure 1.5 Results of a meta-analysis of scale-dependent responses to a number of different ecological drivers. Points represent the log response ratio comparing species richness in control to treatments in a given comparison measured at the smallest (x-value) and largest (y-value) scale. The solid line indicates the 1:1 line expected if effect sizes were not scale dependent. Points above and below this line indicate effect sizes that are larger or smaller, respectively, as scale increases; points in the upper left and lower right quadrants represent cases where the direction of change shifted from positive to negative, or vice versa, with increasing scale. The dashed line indicates the best fit correlation, which is significantly different than the 1:1 line (P < 0.01), indicating that overall, effect sizes tend to be larger at smaller scales than at larger scales. Colours for points indicate categorisations into different ecological drivers. From Chase et al. (2018), reproduced with permission.

1.2 The Making of a Discipline

Although the human-mediated translocation of species has been documented anecdotally since antiquity, the concept of biological invasions is a very recent construct. Many naturalists in the 1800s wrote of non-native species, but it was only in the mid-1900s that the scale of human-mediated movements of species and the growing importance of the implications of such movement became apparent. Pioneers of ecology in the nineteenth century – among them Charles Darwin, Augustin and his son Alphonse de Candolle, Joseph Dalton Hooker and Charles Lyell – explored the role and performance of a small number of non-native species in competition with indigenous ones. Lyell (1832) wrote, every species which has spread itself from a small point over a wide area, must, in like manner, have marked its progress by the diminution, or entire extirpation, of some other, and must maintain its ground by a successful struggle against the encroachments of other plants and animals.

Such appreciation of invasive spread leading to species extinctions predates the rise of global change biology in the late twentieth century (Wilkinson 2002). When writing about the European thistle cardoon, *Cynara cardunculus*, in his journal of research into the geology and natural history of the various countries visited by HMS *Beagle*, Darwin (1839) commented,

I doubt whether any case is on record, of an invasion on so grand a scale of one plant over the aborigines [of South America].

Following these early accounts of non-native species, many ecologists in the early twentieth century began synthesising the scattered knowledge of the ecology of non-natives, unknowingly taking the first tentative steps towards creating a framework for conceptualising biological invasions. Albert Thellung, in his 1912 Habilitation thesis *La Flore Adventice de Montpellier*, offered an early population-based definition of naturalisation which implied the notion of penetration of environmental barriers. He also devised concepts to classify the non-native flora of Montpellier in France according to their degree of naturalisation, introduction pathways and residence time (Kowarik and Pyšek 2012). Unfortunately, such work did not have much, if any, influence on the emerging field of ecology, and the ideas were only rediscovered in the late twentieth century, as the underpinning concepts of invasion science began coming under intense scrutiny.

Charles Elton's (1958) classic book *The Ecology of Invasions by Animals and Plants* is recognised as a milestone in the development of the field now known as invasion science (Richardson and Pyšek 2007, 2008). Already expressed in Elton's (1927) book on *Animal Ecology*, the Eltonian niche is an important concept for formulating a species' position in an ecological network using its functional traits, as will be elaborated in later chapters. Following this line of thinking, Elton (1958) speculated that island assemblages are filtered for a small portion of colonisers, which subsequently cannot fully explore the island's resources and are therefore more susceptible to invasions than those on the mainland. However, the publication of Elton's book was not immediately followed by a significant rallying of research effort. Unlike some other books on environmental topics, Elton's book on invasions had a negligible impact on public perceptions and launched no major actions (Hobbs and Richardson 2010). At about the same time as Elton's book appeared, geneticists began synthesising concepts pertaining to the evolution and genetics of colonising species (Baker and Stebbins 1965). These insights provided crucial stepping stones to the development of the central tenets of invasion science, including the determinants of invasion success, life-history trade-offs, generalist versus specialist strategies, general-purpose genotypes, adaptive phenotypic plasticity, mating systems and the influence of bottlenecks on genetic variation (Barrett 2015). Perhaps the most important linkage of Elton's (1958) classic volume to the theme of our book is his notion that decreased diversity leads to decreased stability. This complexity–stability relationship has stimulated long-lasting debates in ecology with substantial inputs from many figures in the field, including Robert MacArthur, Robert May and G. Evelyn Hutchinson. As will be shown in Chapter 4, ecological networks facing biological invasions typically violate this relationship but simultaneously reveal their trajectory of transition and turnover.

In 1980, the third international conference on mediterranean-type ecosystems, termed MEDECOS, was held in Stellenbosch, South Africa. The invasion of fynbos vegetation by non-native trees, a prominent topic of discussion at this meeting, conflicted with the dominant view of the time, which was that human-induced disturbance was the prerequisite for invasion into pristine ecosystems. A proposal drafted at the Stellenbosch meeting led to an international programme on the ecology of biological invasions under the auspices of the Scientific Committee on Problems of the Environment (SCOPE) (Mooney 1998). Its first five-year plan (1982–1986) revisited Elton's key assumptions and generalisations, reviewed the status of invasions worldwide and addressed three key questions relating to invasiveness, invasibility and management. The SCOPE programme attracted some of the world's top ecologists and comprised national, regional and thematic groups covering all aspects of invasions (Drake et al. 1989). Through the SCOPE programme, invasion science has firmly established itself as an exciting and relevant research field within global change biology (Simberloff 2011). In 1996, an influential conference in Trondheim, Norway, concluded that invasions had become one of the most significant threats to global biodiversity and called for a global strategy to address the problem (Mooney 1999; Sandlund et al. 1999). This led to the launch of the Global Invasive Species Programme (GISP Phase 1) in 1997, with more transdisciplinary goals than the SCOPE programme, acknowledging the need for work on economic valuation, stakeholder participation and pathway analysis and management (Mooney et al. 2005). The Convention on Biological Diversity (CBD), Article 8(h), calls on member governments to control, eradicate or prevent the introduction of those non-native species that threaten ecosystems, habitats or species. In 2000, the IUCN published their guidelines for the prevention of biodiversity loss caused by non-native invasive species. The 1990s saw the blossoming of invasion science, with the number of publications growing rapidly in all related fields (Vaz et al. 2017). In 2018 the Intergovernmental Science-Policy Platform on Biodiversity and Ecosystem Services (IPBES) launched a thematic assessment of invasive non-native species and their control.

Invasion science, as is the case with any emerging discipline, has exhibited different phases. From 1950 to 1990, studies on biological invasions were rather sparse, with fewer than ten publications per year according to the ISI Web of Science. In 1999, the journal *Biological Invasions* was launched, with its founding editor James T. Carlton (1999) stating,

[the aim of] *Biological Invasions* [the journal] ... is to seek the threads that bind for an evolutionary and ecological understanding of invasions across terrestrial, fresh water, and salt water environments. Specifically, we [the journal] offer a portal for research on the patterns and processes of invasions across the broadest menu: the ecological consequences of invasions as they are deduced by experimentation, the factors that influence transport, inoculation, establishment, and persistence of non-native species, the mechanisms that control the abundance and distribution of invasions, and the genetic consequences of invasions.

The period 1990 to 2010 saw the rapid rise of invasion science and its multidisciplinary tentacles (Richardson et al. 2011; Vaz et al. 2017). During this phase, competing concepts, hypotheses, models and knowledge frameworks have been proposed and debated, and consensus has been reached on many fronts; we call this 'Invasion Science 1.0'. Knowledge systems developed during this period accumulated mainly through individual case studies and comparative studies, with the focus being on the invader itself. Developments in the study of invasions at this time must be considered within the context of the intellectual landscape of the day. Indeed, following the Clements–Gleason debate, the Gleasonian individualistic notion that species function independently from the influence of others was implicitly accepted by most researchers as the foundation on which to build frameworks and concepts about ecological communities (Mascaro et al. 2013). As a result, Invasion Science 1.0 sought synergies mainly with population ecology, initially, and macroecology, more recently. The toolbox assimilated via this route served the field reasonably well, until problems began emerging early in the new millennium. As detailed later in this chapter, the growing frustration from contextual complexity and the lack of genuine predictability in invasion science, tentatively due to Gleason's individualistic view, has driven many to search for alternatives, especially with the rise of network science, and call for the synergy between invasion science and community and network ecology; we call this 'Invasion Science 2.0' – the focus of this book. To start off, however, we use the rest of this chapter to delve into Invasion Science 1.0 and discuss its key concepts, achievements and shortcomings.

1.3 A Unified Invasion Framework

With the flourishing of invasion science in the new millennium, a number of frameworks emerged that were grasped by researchers to guide research on the ecology and management of invasions (Wilson et al. 2020a). Several protocols were proposed to synthesise the current understanding of invasiveness and invasibility into simple flow-chart models for assessing the risk of newly introduced species becoming invasive. The Australian Weed Assessment scheme (originally proposed by Pheloung et al. 1999) has been the most widely applied of these protocols. Phase 2 of GISP (2006–2010) set out to improve the scientific basis for decision-making to enhance the ability to manage invasive species; assess the impacts of invasions on major economic sectors; and create a supportive environment for the improved management of invasions. This initiative stimulated many new directions in research to elucidate the multiple dimensions of the impacts of invasive species and to utilise existing knowledge and incorporate new ideas and methodologies to inform options for management (e.g., Richardson et al. 2000; Clout and Williams 2009; Wilson et al. 2017). Substantial progress was made towards deriving general models of invasion, such as the 'unified framework for biological invasions' (Blackburn et al. 2011), which sought to merge insights from many previous attempts to conceptualise key aspects of invasion dynamics for all taxa (Figure 1.6; Table 1.1). This model, reinforcing the utility of conceptualising the many processes implicated in the phenomena of biological invasions using a series of barriers along an introduction–naturalisation–invasion continuum, has

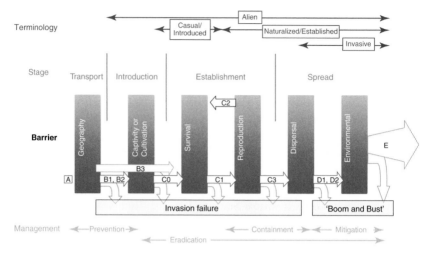

Figure 1.6 The unified framework for biological invasions. The framework recognizes that: the invasion process can be divided into a series of stages; that in each stage there are barriers that need to be overcome for a species or population to pass on to the next stage; that species are referred to by different terms depending on the stage in the invasion process they have reached; and that different management interventions are feasible at different stages. Different parts of this framework emphasise views of invasions that focus on individuals, populations, processes or species. The unfilled block arrows describe the movement of species along the invasion framework with respect to the barriers; alphanumeric codes associated with the arrows relate to the categorisation of species with respect to the invasion pathway given in Table 1.1. From Blackburn et al. (2011); reproduced with permission.

been widely applied (Wilson et al. 2020b). It provided an objective framework for linking theoretical and applied aspects in invasion science, and reinforced the foundations for a standard lexicon of terms for the field of invasion science (Richardson et al. 2011).

1.4 Pathways and Propagules

The premise of any biological invasion is the introduction of non-native propagules via invasion pathways that are required to breach ecological barriers and physical distance. Elton (1958) summarised his view in a chapter called 'The Invasion of the Continents', on pathways and the breakdown of geographic isolation through human-induced movement of organisms around the world. He likened the continents to great tanks of water, connected by narrow tubing blocked by taps. Using this

Table 1.1 *A categorisation scheme for populations in the unified framework of biological invasions. From Blackburn et al. (2011).*

Category	Definition
A	Not transported beyond limits of native range
B1	Individuals transported beyond limits of native range, and in captivity or quarantine (i.e., individuals provided with conditions suitable for them, but explicit measures of containment are in place)
B2	Individuals transported beyond limits of native range, and in cultivation (i.e., individuals provided with conditions suitable for them but explicit measures to prevent dispersal are limited at best)
B3	Individuals transported beyond limits of native range, and directly released into novel environment
C0	Individuals released into the wild (i.e., outside of captivity or cultivation) in location where introduced, but incapable of surviving for a significant period
C1	Individuals surviving in the wild (i.e., outside of captivity or cultivation) in location where introduced, no reproduction
C2	Individuals surviving in the wild in location where introduced, reproduction occurring, but population not self-sustaining
C3	Individuals surviving in the wild in location where introduced, reproduction occurring and population self-sustaining
D1	Self-sustaining population in the wild, with individuals surviving a significant distance from the original point of introduction
D2	Self-sustaining population in the wild, with individuals surviving and reproducing a significant distance from the original point of introduction
E	Fully invasive species, with individuals dispersing, surviving and reproducing at multiple sites across a greater or lesser spectrum of habitats and extent of occurrence

analogy, he conceptualised the processes of biological invasions as interlinked tanks of propagule sources via introduction tubes (Figure 1.7),

Fill these tanks with different mixtures of a hundred thousand different chemical substances in solution... then turn on each tap for a minute each day... the substances would slowly diffuse from one tank to another. If the tubes were narrow and thousands of miles long, the process would be very slow. It might take quite a long time before the system came into final equilibrium, and when this happened a great many of the substances would have been recombined and, as specific compounds, disappeared from the mixture, with new ones from other tanks taking their places. The tanks are the continents, the tubes represent human transport along lines of commerce.

Figure 1.7 Elton's vision of global biological invasions across continents via invasion pathways. Artwork by Lorraine Blumer. Copyright: DSI-NRF Centre of Excellence for Invasion Biology. Reproduced with permission.

Six such introduction tubes, now known as invasion pathways, have been described (Table 1.2; Hulme et al. 2008) and adopted by the CBD. Invasion management targeting surveillance, mitigation and rapid response can efficiently target related invasion pathways and associated non-native taxa, before the species can gain foothold in new territories. Except for the unaided pathway of natural dispersal, the rest reflect clear management negligence and surveillance gaps that should be targeted by various different agencies (Hulme 2015). When combined with the unified framework (Section 1.3), biological invasions via different pathways were found to have different levels of invasion performance (Wilson et al. 2009). For instance, of those non-native plant species introduced to Central Europe, the pathways of release and contamination have resulted in higher than expected proportions of naturalised and invasive species, whereas the pathways of escape and stowaway do not (Pyšek et al. 2011). Each pathway also faces its own cataloguing issues as multiple pathways can be involved in one invasion, while the contribution of each pathway to non-native species introduction varies in terms of geographic, taxonomic and temporal contexts (Essl et al. 2015b).

Table 1.2 *Six invasion pathways and related policy issues. Several key science and policy issues that should be considered if invasion pathways are to be successfully managed to prevent the introduction of invasive non-native species. From Hulme (2015) with modified examples.*

Pathway	Example	Science issues	Policy issues
Release	The Asian Carp was introduced to the southern United States to clean commercial ponds in the 1970s but considered invasive by US FWS in 2006 and threatening the Great Lakes ecosystem.	Improved risk assessment tools to quantify the likelihood of economic, environmental and social impact of species with no prior history of introduction outside their native range.	Legislation that holds parties undertaking assisted colonisation responsible for any costs arising from the impacts and management of such introductions (e.g., assurance bonds).
Escape	Wild pet trade, for example, Burmese python *Python bivittatus* (Kuhl) in Florida.	Identify current commercial activities affecting risk management in the wild pet trade, reasons for non-compliance and ways to change such behaviours.	Promote a whitelist approach to the trade in wild pet species, based on sound risk assessment yet also offering significant commercial benefits.
Contaminant	As a result of adult beetles or larvae hidden in wooden pallets on a ship, the Polyphagous Shot Hole Borer *Euwallacea fornicatus* is rapidly killing trees in South Africa by transmitting the fungus *Fusarium euwallaceae*.	Enhanced next-generation DNA tools and biosystematic data to allow rapid screening of live commodities for cryptic pathogens and parasites.	Extend existing policies on risk prevention of emerging diseases to address the threats posed to biodiversity and ecosystem function.

(*cont.*)

Table 1.2 (*cont.*)

Pathway	Example	Science issues	Policy issues
Stowaway	Tourists might carry SARS–CoV-2 and cause the COVID-19 pandemic.	Analyse the risks from increasing tourist numbers, changes in countries of origin of tourists and shifts in location and types of tourism activities.	Robust codes of practice for tourism operators, aiming to prevent the introduction and movement of invasive non-native species.
Corridor	Canal development, for example, rabbitfish *Siganus luridus* introduced through Suez Canal.	Better prediction of the invaders that might be facilitated by corridors, the potential costs of future impacts and the value of practical measures to mitigate impacts.	International legislation to support environmental risk assessments of major infrastructure projects that include transboundary consequences.
Unaided	Lyme disease outbreak in Canada from bacteria *Borrelia burgdorferi* transmitted through bites by immature blacklegged ticks (the vector has extended its northern range potentially due to climate change).	New modelling tools to predict how wind and sea currents as well as extreme weather events can lead to long-distance dispersal of non-native species.	Polluter pays principle applied where countries fail to contain or eradicate an invasive non-native species with potential to cause detrimental impacts should it spread beyond national borders.

For example, the invasion of marine species in Western Europe is predominantly driven by shipping and aquaculture, whereas in the Eastern Mediterranean countries marine invasions are largely the result of the human–mediated opening of corridors, in this case largely the Suez Canal (Figure 1.8a). After WWII, aquaculture gradually declined as an

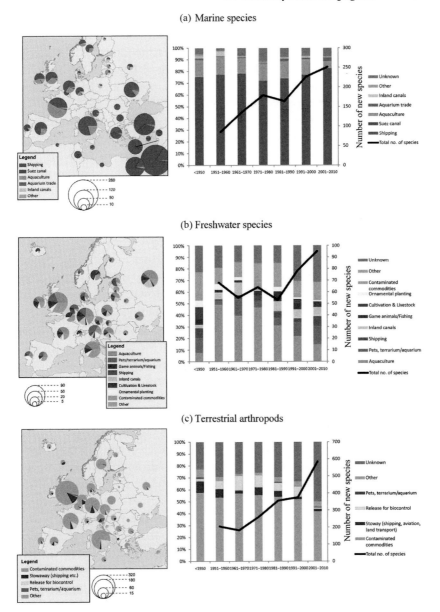

Figure 1.8 Geographic, taxonomic and temporal variation in the importance of the main pathways of introduction for non-native (a) marine species, (b) freshwater species or (c) terrestrial arthropods in Europe. The size of the pie charts indicates the approximate numbers of non-native species per recipient country of first introduction. Temporal trends of new introductions (the right panels) are given as black lines (the right axes). The pathway 'Suez Canal' (a) refers to Red Sea species that moved unaided into the Mediterranean via the Suez Canal. From Essl et al. (2015b), reproduced with permission.

important pathway for freshwater invasions, while the pathway through the pet trade, terrarium and aquarium is on the rise (Figure 1.8b). The release of non-native terrestrial arthropods for biocontrol was a reasonable practice in Europe until 2000 (green belt in Figure 1.8c), probably due to the implementation of the EPPO (European and Mediterranean Plant Protection Organization) standard in 1999.

In reality, multiple pathways can facilitate invasions from different source areas at different stages, while propagules are transferred through these pathways at different rates and times in different contexts. This web of pathways facilitates recurrent bridgehead effects of invasions and gives rise to a pathway network. For instance, three quarters of the non-native ant species intercepted by US air and maritime ports between 1914 and 2013 are from countries into which the species were previously introduced, via similar or other pathways from its native range (Figure 1.9; Bertelsmeier et al. 2018). This further increases the genetic diversity of the introduced species and potentially improves their invasion performance in their new homes. However, a fully-fledged Pathway Network Analysis for most types of biological invasions is still not feasible because of limited data availability and resolution problems. Comprehensive pathway network analyses are undertaken to trace disease-related, protein-coding genes with gene expression data; these analyses use powerful statistical approaches and software developed to identify critical pathways for drug design and disease treatment (Khatri et al. 2012). Such pathway network analyses may be feasible in invasion science when the field has fully embraced data science and informatics, and when new methods are in place to collect and manage large volumes of high-resolution data and geographic and taxonomic coverage at a pace relevant to management.

The total number of non-native propagules introduced to an area through a pathway network, known as the propagule pressure (Lockwood et al. 2005), is the best-supported driver of invasion establishment and success. Although invasion performance is often caused by a chain of demographic actions (Gurevitch et al. 2011), early demographic advantage can provide a long-lasting boost to invasion dynamics, and often leaves an imprint on subsequent invasion dynamics (related to transient dynamics; Stott et al. 2011; Caswell 2019). Propagule pressure typically trumps any niche processes and filters imposed in the recipient ecosystems (Carr et al. 2019). According to a recent meta-analysis (Cassey et al. 2018), the relationship between propagule pressure and non-native population

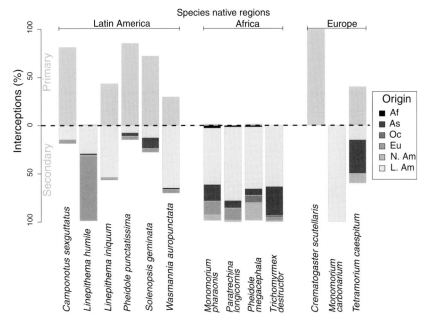

Figure 1.9 Percentage of primary v. secondary introductions of the most frequently intercepted ant species in the United States. The proportion of interceptions from the species' native countries is shown above the x-axis (in grey) and the proportion of interceptions from countries in the non-native range below the x-axis (in colour). The colour code indicates the origin of the secondarily intercepted species. Species were visually grouped on the x-axis according to their native range (Dataset S1). Af, Africa; As, Asia; Eu, Europe; N. Am, North America; L. Am, Latin America; Oc, Oceania. From Bertelsmeier et al. (2018); reproduced with permission.

establishment success is consistently strong, showing a logistic form against the log-transformed propagule size (Figure 1.10). Such a saturation shape resembles the relationship between population viability and initial population size (McGraw and Furedi 2005), suggesting a mirror image between invasion and extinction (Figure 1.11; Colautti et al. 2017). As such, biological invasions are better formulated as an open and non-equilibrial system (Hui and Richardson 2019). Constant propagule influx is crucial to erase the genetic bottleneck anticipated for most invasive species as only a small sample of propagules are actually introduced (Simberloff 2009), although the impact of such obvious genetic bottlenecks on establishment might be weaker than previously thought because of rapid local adaptation (Dlugosch and Parker 2008).

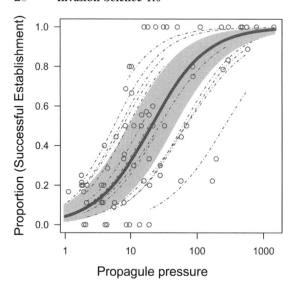

Figure 1.10 Estimated relationship of establishment success with propagule pressure and 95% credible interval (shaded). Dashed lines are individual experimental relationships based on a logistic model with random variation in the intercept and slope among individual experiments. Data points are raw data from 14 relationships from 11 studies that experimentally tested associations between propagule pressure and establishment probability. From Cassey et al. (2018), under CC-BY Licence.

1.5 Invasion Dynamics

With invasion pathways and propagule pressure clarified, many researchers have focused their efforts on investigating the behaviour and mechanisms underlying the diverse forms of invasion dynamics. Invasion dynamics are highly stochastic and context dependent, making attempts to synthesise knowledge and predict particular cases challenging to say the least (Pyšek et al 2020b). This is not only a result of the spatiotemporal complexity of any given ecosystems but also the stochastic nature of invasive spread itself. The two demographic processes involved in spreading – growth and dispersal – both contribute to demographic stochasticity and con-textual dependence (Hui et al. 2011b). Demographic rates such as the population growth rate of an invader can be scale-dependent and often exhibit specific spatial covariance structures in the invaded range (Gurevitch et al. 2016; Hui et al. 2017). The dispersal of an invader, often depicted as the dispersal kernel (probability distribution of dispersal distance), can also be anisotropic and reflect context-dependent movement strategies (e.g., good-stay, bad-disperse dispersal behaviour;

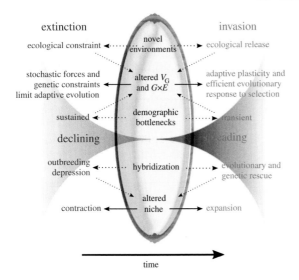

Figure 1.11 Extinctions and invasions conceptualised 'Through the Looking Glass' of evolutionary ecology. Extinctions (left side) represent population decline over time, while biological invasions (right side) represent an increase in abundance. Both invasions and extinctions reflect a common set of elements (central column) because subtle but influential ecological and genetic differences (outer columns) can cause opposite population growth trajectories. From Colautti et al. (2017), reproduced with permission.

Hui et al. 2012). The availability of natural enemies and dispersal barriers, as well as the novel ecological and evolutionary experiences facing each non-native organism (Schittko et al. 2020), make each turn a decision that affects the future possibilities of its invasion performance and dynamics. For instance, in the early 1900s when acclimatisation societies introduced the common starling, *Sturnus vulgaris*, into North America, southeastern Australia and the Western Cape of South Africa, the different propagule pressures, geographies and climates of the three regions resulted in distinct rates and directions of invasive spread across the three regions (Okubo 1980: Phair et al. 2018). Despite such challenges, progress has been made that allows us to grasp the processes and mechanisms behind a plethora of invasion dynamics (Hui and Richardson 2017). Here we highlight only a few aspects that are especially relevant to later chapters.

Once geographic barriers have been breached, via human facilitation, non-native organisms embark on their invasion and spread in recipient ecosystems. Patterns of invasion dynamics are diverse but can be summarised into different types for convenience (Hui and Richardson 2017), although the reality is much fuzzier around these thematic curves. Essentially, invasion dynamics can be divided into transient and asymptotic phases, where the former is highly flexible but the latter more consistent, reflecting the potential within a given habitat (see invasion curves in Figure 1.12a). In the invasion science literature, the curve of a specific invasion is often found to have multiple phases (Figure 1.12b), with a lag phase after introduction

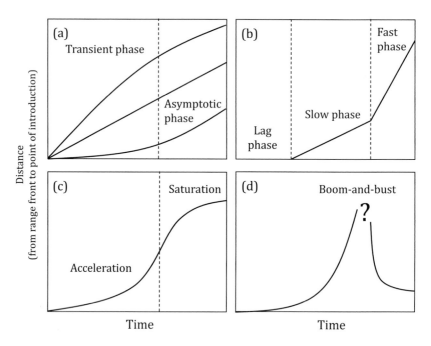

Figure 1.12 A variety of possible invasion dynamics. The invasion expansion of a non-native species is often documented as the distance from the advancing range front to the point of introduction over time. (a) Invasion dynamics are divided into two phases: a transient phase that is highly context-dependent which converges gradually to an asymptotic phase during which spread occurs at constant velocity. (b) Invasion dynamics are divided into a lag phase (no expansion for a period after introduction), and then slow and fast phases during which expansion occurs at a constant velocity. (c) A typical logistic curve for highly mobile species with an acceleration phase and a saturation phase as invasible space is occupied. (d) Boom-and-bust invasions are often caused by the collapse of local demographic processes, or by the encounter of an efficient natural enemy that has switched to target the non-native resource species when the non-native resource reaches an abundance threshold.

and before invasive expansion, and a biphasic range expansion of a slow phase followed by a fast phase. Over a longer period, the trajectory resembles a logistic curve and can be divided into acceleration and saturation phases (Figure 1.12c). For some invaders, success is transient, and its population size may follow a boom-and-bust trajectory, eventually settling at a much lower level, while the invaded range can percolate back from a continuous range to multiple smaller scattered local populations (Figure 1.12d). Such diverse forms of invasion dynamics – the duration, timing and speed of different phases – can be explained by different constraints and limiting factors through the invader constantly experiencing and exploring the novel environment in its invaded range, not only behaviourally by individuals of the non-native species but also ecologically and evolutionarily by its niche dynamics.

In terms of spreading dynamics, physicists and modellers have made great strides in elucidating the phenomenon of particle diffusion and dispersion in a suspension matrix. The Brownian movement of pollen grains in water, driven by the collision of numerous water molecules and the pollen, has enjoyed the attention of many renowned physicists, including Albert Einstein (1905). Collectively, such random movements of particles can be studied using specific models of partial differential equations known as reaction–diffusion models. Adolf Eugen Fick (1829–1901) described two laws of diffusion in 1855. Fick's first law relates the diffusive flux to the concentration, assuming a steady state (Fick 1855). It postulates that the flux goes from regions of high concentration to regions of low concentration, with a magnitude proportional to the concentration gradient (i.e., spatial derivative); in simplistic terms, a solute moves from a region of high concentration to a region of low concentration across a concentration gradient. Fick's second law predicts how diffusion causes the concentration to change with respect to time. Based on Fick's second law, Ronald Fisher (1937) developed the now famous reaction–diffusion model which depicts the advancing wave of advantageous genes in the context of population dynamics

$$\frac{\partial n}{\partial t} = m(1 - n) + D\frac{\partial^2 n}{\partial x^2}, \tag{1.1}$$

where n represents the population density and is a function of time t and location x (thus more explicitly, $n(t, x)$). The left of this equation describes the time derivative of population density. The first term on the right depicts a simple logistic growth, with r reactivity (here, the intrinsic rate of growth). The second term on the right depicts diffusion as a second order derivative of population density over space, known as the Laplacian operator; this term describes how uneven population densities (i.e., density gradients) in a local area are smoothed out. The diffusion cannot even out constant gradients of densities (e.g., at the range front), and consequently propels the population to spread along the direction of the gradient, forming a travelling wave. Fisher (1937) derived the travelling wave solution of the system. He concluded that the spreading velocity 'is proportional to the square root of the intensity of selective advantage [r] and to the standard deviation of scattering in each generation [σ]', as $v = \sigma\sqrt{2r}$, or 'equivalently to the square root of the diffusion coefficient when time is measured in generation', as $v = 2\sqrt{rD}$. This milestone not only provides a way to estimate the diffusion rate based on movement records (e.g., from ringing and mark-recapture data) as $D = \sigma^2/2$ but also derives a commonly used estimate of spreading velocity that has been the backbone of many models of spread ever since. A classic example is by Okubo (1980) who used this model to explore the invasive spread of common starlings, *Sturnus vulgaris*, in North America. This has resulted in such developments as velocity estimates under biotic interactions, density dependence (e.g., Allee effect), heterogeneous habitats, drift/convection in environments and biotic interactions (see review by Hui et al. 2011b).

A major challenge facing diffusion-type models is related to Reid's (1899) paradox of rapid northward plant migration after the last glacial maximum. In Reid's words, 'the oak, to gain its present most northerly position in northern Britain after being driven out by the cold, probably had to travel fully six hundred miles, and this, without external aid, would take something like a million years' (see updated review by Davis and Shaw 2001; Figure 1.13). This is probabilistically impossible when estimated based on r and D measured for today's oak populations, while the unrealistically high levels of population growth rate and diffusion rate suggest alternatives (Skellam 1951). The inability of reaction–diffusion models to accommodate Reid's paradox has led to a systematic shift in paleoecology away from the use of population ecology models that partition observed velocity into demographic factors of population growth and dispersal in driving such range shifts (Clark et al. 1998).

Two mechanisms have received substantial support in invasion science for explaining such an augmented spreading velocity, as anticipated by Reid's paradox. First, the forms of dispersal kernels can be diverse. While the standard deviation of dispersal distances is only a representative shape

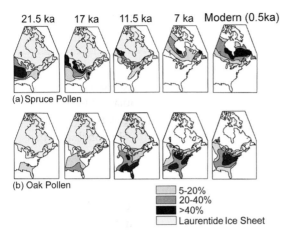

| 21.5 ka | 17 ka | 11.5 ka | 7 ka | Modern (0.5ka) |

(a) Spruce Pollen

(b) Oak Pollen

- 5-20%
- 20-40%
- >40%
- Laurentide Ice Sheet

Figure 1.13 Ranges, as indicated by pollen percentages in sediment, of spruce (*Picea*) and oak (*Quercus*) in eastern North America at intervals during the late Quaternary. The continental ice sheet is shown in blue; pollen proportions are shown in shades of green. The shoreline is not drawn to reflect changes in sea level. (a) Spruce pollen representing three extant species plus the extinct *P. critchfieldii*. More recent data show that spruce was abundant further south in the Mississippi valley during the Last Glacial Maximum than shown here. Both southern and northern range boundaries of spruce shifted northward. (b) Oak pollen representing some or all of the 27 extant *Quercus* species in eastern North America. Oak expanded from the southeast but continued to grow near locations of full-glacial refuges. From Davis and Shaw (2001); reproduced with permission.

metric for Gaussian-type kernels, it fails to capture the 'average' for highly skewed, often fat-tailed, dispersal kernels. For instance, extremely rare dispersal events of non-native species over large distances can occur, often mediated by human activities that skew the dispersal kernel (e.g., Suarez et al. 2001). To accommodate rare long-distance dispersal events, models with stratified dispersal kernels (i.e., a combination of different modes of dispersal) have been proposed, especially to explain biphasic range expansion (Figure 1.12b; Shigesada and Kawasaki 1995). More explicit modelling, however, needs to implement realistic dispersal kernels directly. This can be handled flexibly by using integral equations. Integral equations have a long history in mathematics (Fredholm 1903); they were first applied in invasion ecology by Van den Bosch (1992) and Kot et al. (1996). A generic form of an integrodifference equation is

$$n(x, t+1) = \int k(x, y) f(n(y, t)) n(y, t) dy. \tag{1.2}$$

where the population density $n(x, t+1)$ at time $t+1$ in location x can be calculated as the integral (or the summation) of propagules from all possible locations in the previous time step $n(y, t)$ multiplied by the per-capita population growth rate during one time step $f(n(y, t))$, and weighted by the dispersal kernel $k(x, y)$ that depicts the chance of a propagule moving from location y to location x within one time step. A vector format of such equations is easily extended to accommodate population structures of species with complex life-cycles and life stage-dependent dispersal modes. In such models, dispersal kernels are typically implemented as only distance dependent $k(d)$, with the distance $d = \|x - y\|$, while more realistic movement can be captured directly by considering explicitly the beginning and end locations, y and x respectively. One important way to boost up the rate of spread comes with the realisation that the movement of most organisms, especially those that disperse via means other than gravity, follows the pattern of Lévy flights (e.g., $k(d) \sim d^{-3/2}$ in Figure 1.14 for the common starling). Unlike Gaussian-type kernels with finite mean and variance, such a power-law dispersal kernel is of infinite variance, meaning that the movement can have diverse characteristic length, and is highly flexible

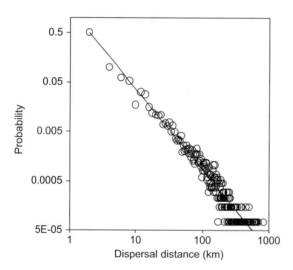

Figure 1.14 The inverse power function of the dispersal kernel for all movements of European starling *Sturnus vulgaris* within Britain during the breeding season. Dispersal kernels are produced using a 2 km distance class; that is, all records are binned to the dispersal distance class of < 2 km, 2–4 km, 4–6 km, and so on. From Hui et al. (2012); reproduced with permission.

to handle complex and novel context. A power law dispersal kernel could reflect that the movements of individuals are self-organised near criticality that enables the species to cover novel and heterogeneous resource landscapes with maximum flexibility and minimum energy input (Muñoz 2018).

Second, spatial sorting has emerged as a clear pattern in many invasive species when comparing individuals from the core/introduction versus the periphery/front populations (Shine et al. 2011). The invasion of cane toads (*Rhinella marina*) in Australia provides the textbook example of spatial sorting in invasion ecology (Figure 1.15). This species was

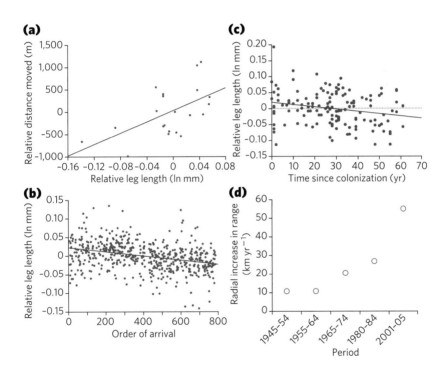

Figure 1.15 Morphology of cane toads (*Rhinella marina*) in relation to their speed and invasion history. (**a** and **b**) Compared with their shorter-legged conspecifics, cane toads with longer hind limbs move further over 3-day periods ($r^2 = 0.34$) (a), and are in the vanguard of the invasion front (based on order of arrival at the study site; $r^2 = 0.11$) (b). (**c**) Cane toads are relatively long-legged in recent populations, and show a significant decline in relative leg length with time in older populations ($r^2 = 0.05$). (**d**) The rate at which the toad invasion has progressed through tropical Australia has increased substantially since toads were first introduced in 1935 ($r^2 = 0.92$). From Phillips et al. (2006); reproduced with permission.

(imprudently, with the benefit of hindsight) introduced to Australia to control insect pests in sugar-cane fields. The annual rate of progress of the toad invasion front has increased fivefold since its introduction, with toads at the front having longer legs and moving faster (Phillips et al. 2006). The invasion of the common mynah (*Acridotheris tristis*) in South Africa also exhibited a clear pattern of spatial sorting. Dispersal-related traits such as wing-loading and length were clearly sorted along the distance to the introduction point, while foraging traits such as the ratio of bill length to width corresponded more to local habitat quality (Berthouly-Salazar et al. 2012). Spatial sorting is driven by both the non-equilibrium dynamics (ongoing advancing front) that continuously sieve out the front runners, and the declining density gradient (and thus the diminishing intensity of resource competition and increasing difficulty of finding a mate) from the core to the expansion front. These processes create two selection forces: (i) 'first come first served' that favours individuals with high dispersal ability, as well as those that are more aggressive and competitive; (ii) the low density at the range front further allows these individuals to possess higher fitness, thus creating the phenomenon of spatial sorting and spatial selection when range expansion is ongoing (Shine et al. 2011). The forces combined can create synergies to obviate the trade-off between dispersal and the cost of reproduction remedies to low densities (e.g., selfing), giving rise to the Good Coloniser Syndrome (Cheptou and Massol 2009; Rodger et al. 2018), as postulated by Baker (1955).

When we consider a skewed dispersal kernel and spatial sorting of propagules with different dispersal capacities, we can derive an estimate of spreading velocity using integrodifference equations (Ramanantoanina et al. 2014)

$$v \approx \sqrt{2rd_{\max}^2(1 + \gamma \cdot r/12)}, \tag{1.3}$$

where $d_{\max}^2 = \exp\left(\mu + \sqrt{2\sigma^2}\,\mathrm{erf}^{-1}\left(2^{1-1/n_0} - 1\right)\right)$ represents the maximal dispersal ability in the non-native population, assuming the dispersal ability follows a lognormal distribution with μ and σ^2 its logarithmic mean and variance, and $\mathrm{erf}^{-1}(\cdot)$ the inverse Gaussian error function; note, n_0 represents the initial propagule pressure, while γ is the kurtosis of the dispersal kernel (a measure of second-order skewness). In short, the intrinsic population growth rate (r), the initial propagule pressure (n_0) and diversity (σ), as well as the degree of skewness of the dispersal kernel (γ), can all contribute positively to the spreading velocity of a non-native species.

Besides these overall explanations of the largely accelerated speed of invasion dynamics, it is important to consider context dependence. Not only do different non-native species exhibit different dynamics, but the same species can also have drastically different invasion velocities and follow different archetypes of invasion dynamics at different localities. Even in controlled environments, arguably under identical settings, the same species do not necessarily replicate their own invasion dynamics (Melbourne and Hastings 2012). Invasive species subject to Allee effects can exhibit invasion dynamics that are completely dependent on initial conditions; they can exhibit a whole spectrum of performance levels, from range pinning with stalled range expansion to high-speed range expansion (Keitt et al. 2001; Hui and Li 2006). We need to change our perspectives so to make sense of such a plethora of invasion dynamics scenarios. Think of a car. It can be driven in many ways: at high speed or low speed, with the engine revving fast, idling or stalling; it is impossible to summarise or predict how a car can be driven. However, to achieve a high speed, the way the clutch changes in a gearbox follows the same procedure, to balance the force (impact) and the speed by sequentially moving up the gears. Too high or too low a gear for a given speed can stall or burn the engine, respectively (Figure 1.16). The population structure of any species can be complex. Such complex, multi-stage life-cycles create a demographic transition matrix that propels the population dynamics. Demographic and dispersal rates of this 'population engine' are not constant, but depend on how the gear of population structures and densities of different life stages fits the engine of the demographic transition matrix, while driving along a heterogeneous terrain. During the functioning of this engine and to complete the life-cycle efficiently, particular vital rates become the limiting factor and control the gear. This can normally be revealed by exploring the dominant eigenvector(s) of the transition matrix, and the abundance vector of different life stages. When the abundance vector and the dominant eigenvector are aligned, they function like a perfect gear to maintain and boost speed. Shifting the dominant eigenvector will require a change of gear (abundance vector) to match up. So, to merge the diverse forms of invasion dynamics, we need to examine the engine and gearbox of biological invasions, and pay less attention to the dash-cam footages of particular journeys.

1.6 Invasiveness and Invasion Syndromes

The invasiveness of a non-native species reflects directly its demographic performance (growth and dispersal), while multiple ecological and

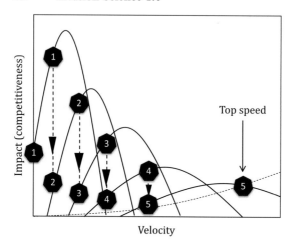

Figure 1.16 A schematic procedure of clutch changes in a gearbox to reach top speed. Each solid polynomial curve represents the performance of a particular gear (performance is depicted here as impact and velocity). A single gear cannot cover all the impact and velocity ranges. To speed up from standing still, we need to sequentially move up through the gears to avoid engine failure. The system can fail in two ways: high gear and low speed, or low gear and high speed. In species with complex life-cycles, different demographic rates serve as different gears (limiting factors) during invasion. Diverse invasion dynamics are therefore anticipated (see, for species with complex life-cycles, Figure 1.12).

evolutionary processes are at play, forming a causal pathway network in determining the demographic performance of an invader (Figure 1.17; Gurevitch et al. 2011). In this causal pathway network, each arrow can be switched on and off, or set flickering to become stronger or weaker in a specific context, giving rise to a contextually realised causal pathway network behind the demographic performance of a particular non-native species. Some of these processes are directly related to invasiveness, while others indirectly affect invasiveness via mediators. We select a few factors that have been considered proxies of invasiveness: invasive traits, range size, dispersal strategies, spatial covariance and a trio platform of trait–site–pathway, which can explain invasion performance of a non-native species.

There has been a long tradition of using comparative methods to identify invasive traits that can help pinpoint the propensity of a non-native species for invasion success (Van Kleunen et al. 2010a). A meta-analysis of trait differences between 125 invasive plant species and 196 paired non-invasive species in the invaded range revealed a number

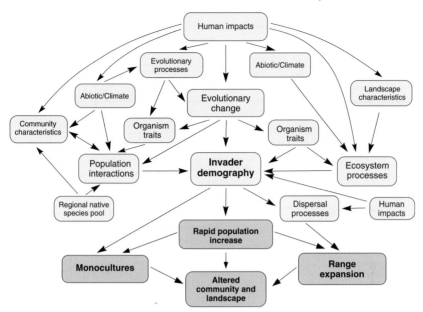

Figure 1.17 A conceptual synthetic invasion meta-framework based on fundamental ecological and evolutionary processes and states. The three different characteristics of invasions and their effect on altering communities and landscapes are in bold capital letters. Transitions between the processes and states are indicated by arrows. Components found in more than one position affect or are affected by more than one set of other processes. From Gurevitch et al. (2011); reproduced with permission.

of consistent results (Van Kleunen et al. 2010b): after traits were grouped by physiology, leaf-area allocation, shoot allocation, growth rate, size and fitness, invasive species were found to have significantly higher values than non-invasive species for all six trait categories (Figure 1.18a). Although trait differences were more pronounced between invasive and native species than between invasive and non-invasive alien species (Figure 1.18b), comparisons between invasive species and native species which were invasive elsewhere yielded no significant trait differences (Figure 1.18c). Differences in physiology and growth rate were larger in tropical regions than in temperate regions (Figure 1.18d–f). Trait differences did not depend on whether the invasive non-native species originated from Europe, nor did they depend on the test environment (Figure 1.18g and h). More importantly, several studies have suggested that successful non-native species possess more distinct traits when compared to those native species in the community (e.g., Divíšek et al. 2018;

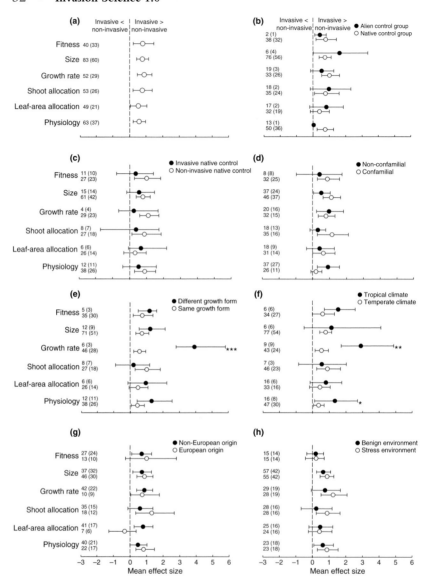

Figure 1.18 Mean effect sizes of differences between invasive non-native plant species and non-invasive plant species for: (a) the six trait categories, and the dependency of these mean effect sizes on (b), whether the control species was a non-invasive non-native species or a native species; (c) whether the native control species is known to be invasive elsewhere; (d) whether the invasive non-native species and non-invasive species belong to the same family; (e) whether the invasive non-native species and non-invasive species have the same

Mathakutha et al. 2019), and such elevated invasion success could correspond to biotic novelty (Schittko et al. 2020). Trait distinctiveness and novelty relate more to how a non-native species takes advantage of empty niches available in an ecological network; we dive deeper into this topic in subsequent chapters.

Can invasion performance mirror how a non-native species 'behaves' in its native range? If the home and away performances mirror each other, one could predict the non-native performance by monitoring performance in the native range (and perhaps also in other invaded ranges). This is exactly what a number of studies have suggested: they show the strong effects of a non-native species' native geographic range on its non-native performance. One obviously needs to consider the habitat and disturbance similarity/overlap between the native and non-native ranges, which limits the performance due to obvious physiological constraints. In a cross-taxon study (Hayes and Barry 2008), when comparing the traits of introduced species vs. those of invasive species, only two species-level characteristics – taxon and geographic range size – were consistently associated with establishment success. However, when comparing the traits of introduced and native species, three species-level characteristics – geographic range size, leaf surface area and fertilisation system (monoecious, hermaphroditic or dioecious) – were consistently found to be significantly different, but only for plants. However, whether geographic range size reflects inherent invasiveness or just human preference in selecting species to move is debatable. For instance, unlike invasive Australian acacias, the native range size of invasive Australian eucalypts is not significantly greater than that of naturalised species (Figure 1.20). Intriguingly, the human preference for introducing species with larger ranges was much greater for acacias than for eucalypts,

---

Figure 1.18 (cont.) growth form; (f) whether the study was performed in a temperate region or in a (sub)tropical region; (g) whether the invasive non-native species is native to Europe; and (h) whether the species were compared under benign environmental conditions or under more stressful environmental conditions. The bars around the means denote bias-corrected 95% bootstrap-confidence intervals. A mean effect size is significantly different from zero when its 95% confidence interval does not include zero. The sample sizes (i.e., number of species comparisons) and, in parentheses, the numbers of studies are given on the left-hand side of each graph. Positive mean effect sizes indicate that the invasive non-native species had larger trait values than the non-invasive species. Significant differences between factor levels (see Appendix S3): ★ $P < 0.05$; ★★ $P < 0.01$; ★★★ $P < 0.001$. From van Kleunen et al. (2010b); reproduced with permission.

as the geometric mean range sizes of introduced, naturalised and invasive acacias are 2.04, 1.88 and 3.59 times those of eucalypts that had attained the same invasion status (introduced/naturalised/invasive) (Hui et al. 2011a; 2014). The selection preference of acacias during introduction is thus for species that can rapidly expand their range; in contrast, slow spreading eucalypts have been selected for dissemination. In other words, humans appear to have selected for highly invasive acacias but against introducing highly invasive eucalypts (Figure 1.19).

Besides particular functional traits and species-level proxies that can tentatively explain species invasiveness, invasive performance has emerged as being highly context dependent. The results of invasive–native trait contrasts can be highly context dependent. When comparing resource-use traits in native and invasive plant species across eight diverse vegetation communities distributed across the world's five Mediterranean climate regions, invasive traits differed strongly across regions (Funk et al. 2016). However, in regions with functional differences between native and invasive species groups, invasive species displayed traits consistent with high resource acquisition. The impacts of invasive species can also be highly context dependent. For instance, invasive plants exert consistently

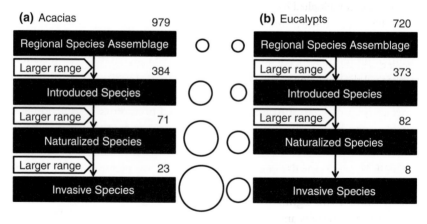

Figure 1.19 A schematic illustration of the selection bias of Australian (a) *Acacia* species and (b) eucalypts (genera *Angophora*, *Eucalyptus* and *Corymbia*) at different invasion stages for four cascaded lists. Only significant differences of native range sizes between two species lists are presented; differences in percolation intercept and exponent are not shown for conciseness. The number of species in each stage is indicated above each box. The circles between the boxes are proportional in area to the geometric mean range sizes of species in that stage. From Hui et al. (2014); reproduced with permission.

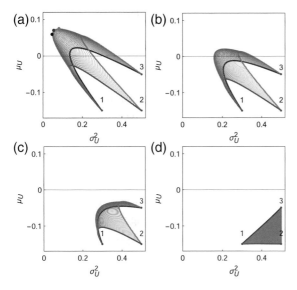

Figure 1.20 Expectation and variance of relative growth rate for an ensemble of three populations. Red, green and black curves: attainable sets for combinations of two populations. Blue mesh: attainable sets for combinations of three populations. Covariances between populations are the same in each plot and are calculated as the correlation multiplied by the standard deviations of the two populations. Correlation = -0.5 in (a), 0 in (b) representing independent; $= 0.5$ in (c); and $= 1$ in (d) representing perfect synchrony. The efficient frontier is calculated for the maximum growth portfolio (green dot in a), for the minimum variance portfolio (black dot in a) and for the portfolio on the efficient frontier where the attainable set is stretched the most along the ensemble shifting direction (purple dot in a). From Hui et al. (2017); reproduced with permission.

significant impacts on some outcomes (survival of resident biota, activity of resident animals, resident community productivity, mineral and nutrient content in plant tissues, and fire frequency and intensity), whereas for outcomes at the community level, such as species richness, diversity and soil resources, the significance of impacts is determined by interactions between species traits and the invaded biome (Pyšek et al. 2012). There is no universal measure of impact and the pattern observed depends on the ecological feature of interest. In short, the dynamics, performance and impact of an invasion are highly context dependent.

A potential explanation for this widely acknowledged context dependence is that successful invaders need to learn about their new environment and adapt to it, and therefore need to be flexible in heuristic learning. This generic learning can be achieved through a number of features – plasticity, rapid evolution or power-law dispersal (behaviour flexibility at criticality). For instance, invasive species are nearly always more plastic in their response to greater resource availability than non-invasive species, although this plasticity is not always correlated with a fitness gain. In other words, invasive species are more plastic in a variety of traits but non-invasive species respond just as well, if not better, when resources are limited (Davidson et al. 2011). Second, such a high level of context dependence is perhaps due to the rapid evolution anticipated in many introduced species (Whitney and Gabler 2008). Rapid evolutionary changes are common during invasions, perhaps as a result of new environments, effects of hybridisation and coevolution in recipient communities (Prentis et al. 2008). Typical invasion traits (e.g., growth rate, dispersal ability, generation time) can undergo evolutionary change following introduction over very short periods (e.g., within a decade). Rapid evolutionary change in many invasive species calls for added caution during risk assessment that often assumes a fixed life-history strategy in invaders. Finally, there could be a flexible learning strategy such as the 'win-stay, lose-shift' game strategy and the 'good-stay, bad-disperse' behaviour that could give rise to the power law dispersal kernel (Figure 1.14). A power-law dispersal kernel can emerge from self-organised movement of foragers when exploring a new heterogeneous resource landscape (Muñoz 2018). It can also give rise to specific spatial covariance between local population demographics, minimising population volatility but maximising population growth, such as those exhibited in the landscape demographics of the gypsy moth in the northeastern United States (Figure 1.20). Predicting invasive dynamics, performance and impacts under such extreme contextual dependence becomes 'Mission: Impossible'.

Dealing with such highly context-dependent invasion performance, risk assessment and management has edged forwards by following a simplified three-pronged approach. Catford et al. (2009) summarised existing invasion hypotheses into three groups: those that involve propagule pressure, those that consider traits associated with invasiveness and those that deal with features of the environment that affect invasibility. Subsequently, invasion management has also been tasked with seeking balanced prioritisation by considering pathways, species traits and site characteristics (Figure 1.21a; McGeoch et al. 2016). The idea of a three-pronged approach has been further expanded by the concept of 'invasion syndromes' (Figure 1.21b; Novoa et al. 2020), where each of the trio – pathways, traits and sites – can be fine-tuned by invasion symptoms and

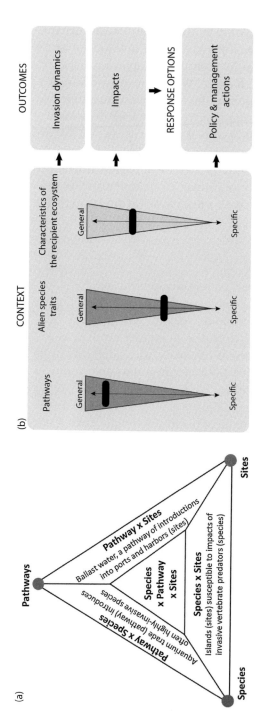

Figure 1.21 The three components of invasion syndromes (a) The three foci for a comprehensive approach to prioritising investment in management of biological invasions. Examples of combined prioritised risks associated with these focus areas, with the example in the centre being ornamental species in gardens as escapees (pathway) into adjacent protected areas (sites). From McGeoch et al. (2016); reproduced with permission. (b) An invasion syndrome is defined as a combination of pathways, non-native species traits and characteristics of the recipient ecosystem which collectively result in predictable dynamics and impacts, and which can be managed effectively using specific policy and management actions. For it to be coherent, the shared characteristics (pathways, non-native species traits and characteristics of the recipient ecosystem) must result in predictable outcomes (regarding invasion dynamics and impacts) which in turn can be best managed using similar management or policy responses. This means that invasion syndromes should be created from generalisations that are as broad as possible but which are still robust and useful. The invasion context is displayed here on three vertical axes (i.e., vertical black bars) that range from general (at the top) to specific (at the bottom). For example, the non-native species traits axis could vary (top/general to bottom/specific) from all aquatic species to aquatic species within a specific genus and to congeneric freshwater species within a specific body size range. The positions along the axes (i.e., black boxes) are adjusted so that all invasion events within the selected context result in similar outcomes and response options. A change in any one of the axes, or a change in the outcomes or response options that are to be encompassed by the invasion syndrome, will likely affect all other aspects of the framework, which means that circumscribing a syndrome is an iterative process. From Novoa et al. (2020), under CC-BY Licence.

contexts, and by our knowledge, from both the specific and generic. A specific invasion syndrome defines an archetype of invasion dynamics and impacts which can potentially be managed by applying a specific suite of policy and management actions. While the trio idea makes conceptual sense, the practicality of using this method is still questionable as most of the supporting cases are retrospective. Only with the accumulation of a large number of well-described cases will we be able to construct robust syndromes using a bottom-up approach with vigorous cross-validation. Well-described invasion cases and their management actions and consequences are only beginning to be systematically collected; it could require decades for such robust invasion syndromes with solid support to emerge.

1.7 From Trees to Networks

Invasion Science 1.0 has followed an invader-centric view and has sought synergies from population ecology on the role of habitat quality and from demography in determining species performance. This species-centric view packages all factors that an invader needs to negotiate as eco-environmental barriers. What an invader confronts in its recipient ecosystem is not a passive pool of antagonists and resources that it has to contend with. Rather, all resident species and other socioeconomic components of the recipient ecosystem respond to the invasion simultaneously. The outcome of the invasion thus depends not only on the invader's own strategies (behavioural, demographic, ecological, evolutionary) but also on the actions and strategies of all resident species and embedded components. This is a typical setting of game theory, where one player's payoff not only depends on his/her own play but also that of its opponents. As Elton (1958) anticipated in his classic book

It is a very long haul from handling a small group of four species like the lemon tree, the nightshade, the black scale, and a chalcid parasite, to the contemplation of the most inconceivable and profuse richness of a tropical rain forest, or even to the several thousand species living in Wytham Woods, Berkshire. It is a question for future research, but an urgent one, how far one has to carry complexity in order to achieve any sort of equilibrium.

Indeed, we need to formulate the complexity of these 'thousand species' to achieve better understanding and predictability in invasion science. We think this can be done by nudging Invasion Science 1.0 and its synergies with population ecology towards Invasion Science 2.0, which embraces advances in community ecology, network ecology and ecosystem science. This does not require us to shift focus away from species but to expand it to allow for the critical inclusion of biotic interactions between and among species. It requires us to consider biological invasions in the context of ecological networks.

If we look again at the unified framework of invasion (Figure 1.6), the mother lode of Invasion Science 1.0, its linear and tree-like structure is obvious. Trees have been the metaphor for orderly and structured knowledge since antiquity, from the Tree of Life, genealogy, Aristotle's taxonomic classification, to Darwin's tentative sketch of a phylogeny. The barriers and stages of the invasion framework allow for structured response and prioritisations for targeting early stages for maximum interventions. However, these processes and barriers at different stages are intertwined and are species and context dependent. If we look at the current knowledge landscape in invasion science (Figure 1.22), we are faced with a complex network, comprising arguably five clusters (Enders et al. 2020). This landscape, of course, represents only the knowledge of invasion performance that has accumulated over the last few decades, mainly since 2000. More hypotheses and links, especially hyperlinks that simultaneously connect existing hypotheses, could emerge to drastically complicate or, oddly, simplify this congested landscape. Looking closely, we can see that the pathway–trait–site troika has been expanded here into pathway (propagule cluster), trait (trait cluster and Darwin's cluster) and site (biotic interaction cluster and resource availability cluster). Although the resource availability cluster was positioned at the centre of this knowledge network, we put biotic interactions, more precisely ecological networks of biotic interactions, at the centre of subsequent chapters to highlight what we think is the key to understanding the complexity of biological invasions.

How many invasive species can our ecosystems tolerate? Human activities are intentionally and unintentionally introducing more and more species to new regions of the world – for example, via commodity transport or tourism. Some of these non-native species have negative consequences for biodiversity and human well-being, e.g., by displacing native species or transmitting diseases. However, while we have relatively good information on the historical spread of non-native species, there is still little knowledge on which to base predictions of future trends on invasion impacts. A poll of expert opinion suggests that an increase of 20 to 30 per cent in the number of newly introduced non-native species

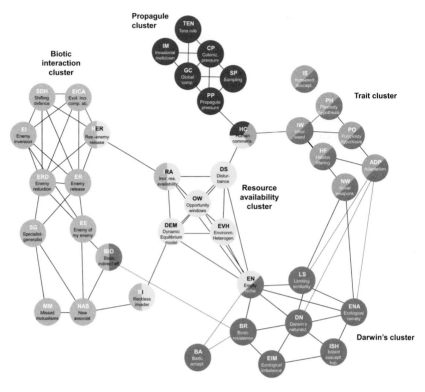

Figure 1.22 Network of 39 common hypotheses in invasion biology. Clusters were calculated using the local link-clustering algorithm. Colours indicate membership of hypotheses to concept clusters. The representation is simplified in that, for example, the node empty niche (EN) appears to be split into two equal parts, whereas it actually belongs slightly more to Darwin's cluster (6/11 = 55%) than in the resource availability cluster (5/11 = 45%). From Enders et al. (2020), under CC-BY Licence.

is considered sufficient to cause massive global biodiversity loss (Figure 1.23; Essl et al. 2020). Such an increase in the magnitude of invasions is likely to be reached soon, as the number of introduced species is constantly increasing. Furthermore, humans are the main drivers of the future spread of non-native species. The experts identified three main reasons: primarily the increase in the global transport of goods, followed by climate change and the extent of economic growth. The study also showed that the spread of non-native species can be greatly slowed down by ambitious countermeasures. Politicians have the responsibility to act.

This daunting scientific and management challenge has led to authors labelling the management of some biological invasions a 'wicked problem' (Woodford et al. 2016). Indeed, the number of major success stories in the management of biological invasions is alarmingly small. The Aichi Biodiversity Target 9 states,

By 2020, invasive non-native species and pathways are identified and prioritised, priority species are controlled or eradicated, and measures are in place to manage pathways to prevent their introduction and establishment.

During the writing of this book in 2020, during the most dramatic phase of the invasion dynamics and impacts of COVID-19, it was clear that we have collectively failed to reach this target. The choice of a warlike word 'invasion' to describe this research field has been met with mixed sentiments. Many have vivid memories of reading stories of non-native invasions and the feelings associated with them as unfamiliar, strange or even scary, while invasion itself has been recorded throughout history as brutal and uncivilised, creating fear and sometimes irrational response. However, faced with global changes with multiple drivers that transform global ecosystems and the biosphere, fear is normal. If there is a way out of each crisis, perhaps we need to look attentively with fear and curiosity and see where the tides (and tsunamis) will take us. This fear and opportunity has driven some to deny, some to restore and some to leave the research field, and others to embrace the change as a whole new world of novel ecosystems. Even without using the warlike metaphor 'invasion', scientists in the field of global change are clearly loaded with fear and a slice of excitement. The pervasiveness of the current unprecedented global changes suggests that we can never turn back the clock to some bygone era. Instead, we need to steer or surf through the Volatility, Uncertainty, Complexity and Ambiguity (VUCA) wave of knowledge landscapes (Figure 1.24), transformed ecosystems and increasingly complex societal changes and responses. Of the four components of the VUCA diagram, the current agenda in invasion science is pushing the veiled line that divides Invasion Science 1.0 from Invasion Science 2.0 towards the top right. In Invasion Science 1.0, we have largely rooted out Ambiguity and are rapidly reducing Uncertainty by embracing informatics and data science, while we envisage embracing Complexity, and experiencing further Volatility in the knowledge enterprise of invasion science when it transitions to Invasion Science 2.0.

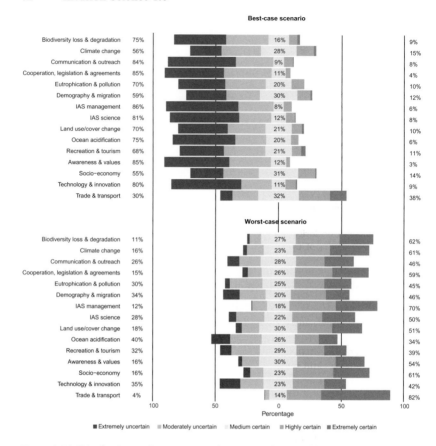

Figure 1.23 Distributions of uncertainty of 15 major drivers of biological invasions exhibit major impacts on the environment by 2050 under a best- and worst-case scenario, based on answers provided by 36 experts. The uncertainty categories follow a five-point Likert scale. The estimates shown include all responses across 14 contexts regarding taxonomic groups, zonobiomes, realms and socio–economic status. The stacked bars represent the uncertainty categories, with the bars and percentage value for the medium certain category centred at 0% on the x-axis. Bars and percentage values on the left refer to the uncertainty categories extremely and moderately uncertain, and bars and percentage values on the right refer to the answers in the categories highly and extremely certain. Categories sum up to 100%. From Essl et al. (2020); reproduced with CC-BY Licence.

In the following chapters we move tentatively into the Complexity component by approaching it as an ecological network facing biological invasions. This resonates with the paradigm shift in human knowledge systems from the classic tree structures to the emerging network metaphor

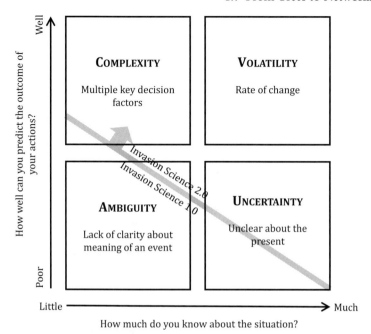

Figure 1.24 Invasion science in a VUCA world. By defining and unifying terminology, metrics and frameworks, invasion science has largely rooted out ambiguity in the field. However, work is still underway to reduce uncertainty associated with the current knowledge of invasion status and the ecology of invasive species. Future work must embrace complexity in invasion science, both in terms of ecosystems and multiple stakeholders, while volatility in invasion science has yet to truly emerge, as the rate of invasions and their impacts are currently increasing unabated, although critics have started sounding alarms.

(Lima 2015). How does a non-native species invade an ecological network? Which aspects and components of the network should concern us? In network science, a complex network normally possesses five features: interdependence (watch out for knock-on effects), interconnectivity (six-degree; all connected eventually), decentralisation (lack of governance), multiplicity (exponential/explosive) and non-linearity (1+1 is not 2; feedbacks). Interdependence describes the interactions between connected nodes in a network; interconnectivity portrays the emerged network structures and architectures; decentralisation depicts the function and regime/stability of a network; multiplicity talks about the dynamics of nodes in a network; non-linearity characterises the response of a network to invasions (i.e., invasibility). For an ecological network affected by biological invasions, we not only dissect the system into these five aspects but also address the issue of scale. We devote full chapters to, in sequence, network interactions, network patterns, network stability, network dynamics, network scaling and network invasibility. Again, this is not a recipe book but an atlas of clues and ideas for the roads ahead. Please fasten your seatbelts. Bon voyage.

References

Baker HG (1955) Self-compatibility and establishment after 'long-distance' dispersal. *Evolution* 9, 347–349.

Baker HG, Stebbins GL (eds.) (1965) *The Genetics of Colonizing Species.* London: Academic Press.

Barrett SCH (2015) Foundations of invasion genetics: the Baker and Stebbins legacy. *Molecular Ecology* 24, 1927–1941.

Bertelsmeier C, et al. (2018) Recurrent bridgehead effects accelerate global alien ant spread. *Proceedings of the National Academy of Sciences USA* 115, 5486–5491.

Berthouly-Salazar C, et al. (2012) Spatial sorting drives morphological variation in the invasive bird, *Acridotheris tristis*. *PLoS ONE* 7, e38145.

Blackburn TM, et al. (2011) A proposed unified framework for biological invasions. *Trends in Ecology & Evolution* 26, 333–339.

Carlton JT (1999) A Journal of Biological Invasions. *Biological Invasions* 1, 1.

Carr AN, et al. (2019) Long-term propagule pressure overwhelms initial community determination of invader success. *Ecosphere* 10, e02826.

Cassey P, et al. (2018) Dissecting the null model for biological invasions: a meta-analysis of the propagule pressure effect. *PLoS Biology* 16, e2005987.

Caswell H (2019) *Sensitivity Analysis: Matrix Methods in Demography and Ecology.* Cham: Springer.

Catford JA, et al. (2009) Reducing redundancy in invasion ecology by integrating hypotheses into a single theoretical framework. *Diversity and Distributions* 15, 22–40.

Chase JM, et al. (2018) Embracing scale-dependence to achieve a deeper understanding of biodiversity and its change across communities. *Ecology Letters* 21, 1737–1751.

Cheptou PO, Massol F (2009) Pollination fluctuations drive evolutionary syndromes linking dispersal and mating system. *The American Naturalist* 174, 46–55.

Clark JS, et al. (1998) Reid's paradox of rapid plant migration: Dispersal theory and interpretation of paleoecological records. *BioScience* 48, 13–24.

Clout MN, Williams PA (2009) *Invasive Species Management: A Handbook of Principles and Techniques.* Oxford: Oxford University Press.

Colautti RI, et al. (2017) Invasions and extinctions through the looking glass of evolutionary ecology. *Philosophical Transactions of the Royal Society B: Biological Sciences* 372, 20160031.

Darwin CR (1839) *The Voyage of the Beagle: Journal of Researches into the Geology and Natural History of the Various Countries visited by HMS Beagle.* London: Henry Colburn.

Davidson AM, et al. (2011) Do invasive species show higher phenotypic plasticity than native species and, if so, is it adaptive? A meta-analysis. *Ecology Letters* 14, 419–431.

Davis MB, Shaw RG (2001) Range shifts and adaptive responses to quaternary climate change. *Science* 292, 673–679.

Díaz S, et al. (2015) A Rosetta stone for nature's benefits to people. *PLoS Biology* 13, e1002040.

Divíšek J, et al. (2018) Similarity of introduced plant species to native ones facilitates naturalization, but differences enhance invasion success. *Nature Communications* 9, 4631.

Dlugosch KM, Parker IM (2008) Founding events in species invasions: Genetic variation, adaptive evolution, and the role of multiple introductions. *Molecular Ecology* 17, 431–449.

Drake JA, et al. (1989) *Biological Invasions: A Global Perspective*. Chichester: John Wiley.

Einstein A (1905) Über die von der molekularkinetischen Theorie der Wärme geforderte Bewegung von in ruhenden Flüssigkeiten suspendierten Teilchen. *Annalen der Physik* 322, 549–560.

Elton CS (1927) *Animal Ecology*. Chicago: The University of Chicago Press.

Elton CS (1958) *The Ecology of Invasions by Animals and Plants*. London: Metheun.

Enders M, et al. (2020) A conceptual map of invasion biology: Integrating hypotheses into a consensus network. *Global Ecology and Biogeography* 29, 978–991.

Essl, F. et al. (2015a) Historical legacies accumulate to shape future biodiversity change. *Diversity and Distributions* 21, 534–547.

Essl F, et al. (2015b) Crossing frontiers in tackling pathways of biological invasions. *BioScience* 65, 769–782.

Essl F, et al. (2020) Drivers of future alien species impacts: An expert-based assessment, *Global Change Biology* 26, 4880–4893.

Fick A (1855) Ueber diffusion. *Annalen der Physik* 170, 59–86.

Fisher RA (1937) The wave of advance of advantageous genes. *Annals of Eugenics* 7, 355–369.

Fredholm I (1903) Sur une classe d'équations fonctionnelles. *Acta Mathematica* 27, 365–390.

Funk JL, et al. (2016) Plant functional traits of dominant native and invasive species in Mediterranean-climate ecosystems. *Ecology* 97, 75–78.

Gaertner M, et al. (2009) Impacts of alien plant invasions on species richness in Mediterranean-type ecosystems: A meta-analysis. *Progress in Physical Geography* 33, 319–338.

Gurevitch J, et al. (2011) Emergent insights from the synthesis of conceptual frameworks for biological invasions. *Ecology Letters* 14, 407–418.

Gurevitch J, et al. (2016) Landscape demography: Population change and its drivers across spatial scales. *The Quarterly Review of Biology* 91, 459–485.

Hayes KR, Barry SC (2008) Are there any consistent predictors of invasion success? *Biological Invasions* 10, 483–506.

Hobbs RJ, Richardson DM (2010) Invasion ecology and restoration ecology: Parallel evolution in two fields of endeavour. In Richardson DM (ed.), *Fifty Years of Invasion Ecology*, pp. 61–69. Oxford: Blackwell Publishing Ltd.

Hui C, Li Z (2006) Distribution patterns of metapopulation determined by Allee effects. *Population Ecology* 46, 55–63.

Hui C, Richardson DM (2017) *Invasion Dynamics*. Oxford: Oxford University Press.

Hui C, Richardson DM (2019) Network invasion as an open dynamical system: Pesponse to Rossberg and Barabás. *Trends in Ecology & Evolution* 34, 386–387.

Hui C, et al. (2011a) Macroecology meets invasion ecology: Linking the native distributions of Australian acacias to invasiveness. *Diversity and Distributions* 17, 872–883.

Hui C, et al. (2011b) Modelling spread in invasion ecology: A synthesis. In Richardson DM (ed.) *Fifty Years of Invasion Ecology: The Legacy of Charles Elton*, pp. 329–343. Oxford: Wiley-Blackwell.

Hui C, et al. (2012) Flexible dispersal strategies in native and non-native ranges: Environmental quality and the 'good-stay bad-disperse' rule. *Ecography* 35, 1024–1032.

Hui C, et al. (2014) Macroecology meets invasion ecology: Performance of Australian acacias and eucalypts around the world foretold by features of their native ranges. *Biological Invasions* 16, 565–576.

Hui C, et al. (2017) Scale-dependent portfolio effects explain growth inflation and volatility reduction in landscape demography. *Proceedings of National Academy of Sciences USA* 114, 12507–12511.

Hulme PE (2015) Invasion pathways at a crossroad: Policy and research challenges for managing alien species introductions. *Journal of Applied Ecology* 52, 1418–1424.

Hulme PE, et al. (2008) Grasping at the routes of biological invasions: A framework for integrating pathways into policy. *Journal of Applied Ecology* 45, 403–414.

Keitt TH, et al. (2001) Allee effects, invasion pinning, and species' borders. *The American Naturalist* 157, 203–216.

Khatri P, et al. (2012) Ten years of pathway analysis: Current approaches and outstanding challenges, *PLoS Computational Biology* 8, e1002375.

Kot M, et al. (1996) Dispersal data and the spread of invading organisms. *Ecology* 77, 2027–2042.

Kowarik I, Pyšek P (2012) The first steps towards unifying concepts in invasion ecology were made one hundred years ago: Revisiting the work of the Swiss botanist Albert Thellung. *Diversity and Distributions* 18, 1243–1252.

Le Roux JL, et al. (2019) Recent anthropogenic plant extinctions differ in biodiversity hotspots and coldspots. *Current Biology* 29, 2912–2918.

Lima M (2015) A visual history of human knowledge. TED Talk. www.ted.com/talks/manuel_lima_a_visual_history_of_human_knowledge

Lockwood JL, et al. (2005) The role of propagule pressure in explaining species invasions. *Trends in Ecology & Evolution* 20, 223–228.

Lyell C (1832) *Principles of Geology*. London: John Murray.

Mascaro J et al. (2013) Origins of the novel ecosystem concept. In Hobbs RJ, Higgs EC, Hall CM (eds.) *Novel Ecosystems: Intervening in the New Ecological World Order*, pp. 45–57. Oxford: Wiley-Blackwell.

Mathakutha R, et al. (2019) Invasive species differ in key functional traits from native and non-invasive alien plant species. *Journal of Vegetation Science* 30, 994–1006.

McGeoch MA, et al. (2016) Prioritizing species, pathways, and sites to achieve conservation targets for biological invasion. *Biological Invasions* 18, 299–314.

McGraw JB, Furedi MA (2005) Deer browsing and population viability of a forest understory plant. *Science* 307, 920–922.

Melbourne BA, Hastings A (2012) Extinction risk depends strongly on factors contributing to stochasticity. *Nature* 454, 100–103.

Mooney HA (1998) *The Globalization of Ecological Thought: Excellence in Ecology 5.* Oldendorf/Luhe: Ecology Institute.

Mooney HA (1999) The global invasive species program (GISP). *Biological Invasions* 1, 97–98.

Mooney HA, et al. (eds.) (2005) *Invasive Alien Species: A New Synthesis.* Washington: Island Press.

Muñoz MA (2018) Criticality and dynamical scaling in living systems. *Reviews of Modern Physics* 90, 031001.

Novoa A, et al. (2020) Invasion syndromes: A systematic approach for predicting biological invasions and facilitating effective management. *Biological Invasions* 22, 1801–1820.

Okubo A (1980) *Diffusion and Ecological Problems: Mathematical Models.* New York: Springer-Verlag.

Phair DJ, et al. (2018) Context-dependent spatial sorting of dispersal-related traits in the invasive starlings (*Sturnus vulgaris*) of South Africa and Australia. BioRxiv, 342451.

Pheloung PC, et al. (1999) A weed risk assessment model for use as a biosecurity tool evaluating plant introductions. *Journal of Environmental Management* 57, 239–251.

Phillips B, et al. (2006) Invasion and the evolution of speed in toads. *Nature* 439, 803.

Prentis PJ, et al. (2008) Adaptive evolution in invasive species. *Trends in Plant Science* 13, 288–294.

Pyšek P, et al. (2011) Alien plants introduced by different pathways differ in invasion success: Unintentional introductions as a threat to natural areas. *PLoS ONE* 6, e24890.

Pyšek P, et al. (2012) A global assessment of invasive plant impacts on resident species, communities and ecosystems: The interaction of impact measures, invading species' traits and environment, *Global Change Biology* 18, 1725–1737.

Pyšek P, et al. (2020a) Scientists' warning on invasive alien species. *Biological Reviews* 95, 1511–1534.

Pyšek P, et al. (2020b) Framework for Invasive Aliens (MAFIA): Disentangling large-scale context dependency in biological invasions. *NeoBiota* 62, 407–461.

Ramanantoanina A, et al. (2014) Spatial assortment of mixed propagules explains the acceleration of range expansion. *PLoS ONE* 9, e103409.

Reid C (1899) *The Origin of the British Flora.* London: Dulau & Co.

Richardson DM (2011) Invasion science: The roads travelled and the roads ahead. In Richardson DM (ed.) *Fifty Years of Invasion Ecology: The Legacy of Charles Elton,* pp. 397–407. Oxford: Wiley-Blackwell.

Richardson DM, Pyšek P (2007) Classics in physical geography revisited: Elton, C.S. 1958: The ecology of invasions by animals and plants. Methuen: London. *Progress in Physical Geography* 31, 659–666.

Richardson DM, Pyšek P (2008) Fifty years of invasion ecology: The legacy of Charles Elton. *Diversity and Distributions* 14, 161–168.

Richardson DM, et al. (2000) Naturalization and invasion of alien plants: Concepts and definitions. *Diversity and Distributions* 6, 93–107.

Richardson DM, et al. (2011) A compendium of essential concepts and terminology in invasion ecology. In Richardson DM (ed.) *Fifty Years of Invasion Ecology: The Legacy of Charles Elton*, pp. 409–420. Oxford: Wiley-Blackwell.

Rodger JG, et al. (2018) Heterogeneity in local density allows a positive evolutionary relationship between self-fertilisation and dispersal. *Evolution* 72, 1784–1800.

Sandlund OT, et al. (1999) Introduction: The many aspects of the invasive alien species problem. In Sandlund OT, Schei PJ, Viken Å (eds.) *Invasive Species and Biodiversity Management*, pp. 1–7. Dordrecht: Kluwer Academic Publishers.

Schittko C, et al. (2020) A multidimensional framework for measuring biotic novelty: How novel is a community? *Global Change Biology* 26, 4401–4417.

Seebens H, et al. (2017) No saturation in the accumulation of alien species worldwide. *Nature Communications* 8, 14435.

Shigesada N, et al. (1995) Modeling stratified diffusion in biological invasions. *The American Naturalist* 146, 229–251.

Shine R, et al. (2011) An evolutionary process that assembles phenotypes through space rather than through time. *Proceedings of the National Academy of Sciences USA* 108, 5708–5711.

Simberloff D (2009) The role of propagule pressure in biological invasions. *Annual Review of Ecology, Evolution, and Systematics* 40, 81–102.

Simberloff D (2011) SCOPE Project. In Simberloff D, Rejmánek M (eds.) *Encyclopedia of Biological Invasions*, pp. 617–619. Berkeley: University of California Press.

Skellam JG (1951) Random dispersal in theoretical populations, *Biometrika* 38, 196–218.

Stott I, et al. (2011) A framework for studying transient dynamics of population projection matrix models. *Ecology Letters* 14, 959–970.

Suarez AV, et al. (2001) Patterns of spread in biological invasions dominated by long-distance jump dispersal: Insights from Argentine ants. *Proceedings of the National Academy of Sciences USA* 98, 1095–1100.

Van den Bosch F, et al. (1992) Analysing the velocity of animal range expansion. *Journal of Biogeography* 19, 135–150.

Van Kleunen M, et al. (2010a), Are invaders different? A conceptual framework of comparative approaches for assessing determinants of invasiveness. *Ecology Letters* 13, 947–958.

Van Kleunen M, Weber E, Fischer M (2010b) A meta-analysis of trait differences between invasive and non-invasive plant species. *Ecology Letters* 13, 235–245.

Vaz AS, et al. (2017) The progress of interdisciplinarity in invasion science. *Ambio* 46, 428–442.

Whitney KD, Gabler CA (2008) Rapid evolution in introduced species, 'invasive traits' and recipient communities: Challenges for predicting invasive potential. *Diversity and Distributions* 14, 569–580.

Wilkinson DM (2002) Ecology before ecology: Biogeography and ecology in Lyell's 'Principles'. *Journal of Biogeography* 29, 1109–1115.

Wilson JRU, Panetta FD, Lindgren C (eds.) (2017) *Detecting and Responding to Alien Plant Incursions*. Cambridge: Cambridge University Press.

Wilson JRU, et al. (2009) Something in the way you move: Dispersal pathways affect invasion success. *Trends in Ecology & Evolution* 24, 136–144.

Wilson JRU, et al. (2020a) Frameworks used in invasion science: progress and prospects. *NeoBiota* 62, 1–30.

Wilson JRU, et al. (2020b) Is invasion science moving towards agreed standards? The influence of selected frameworks. *NeoBiota* 62, 569–590.

Woodford DJ, et al. (2016) Confronting the wicked problem of managing invasive species. *NeoBiota* 31, 63–86.

2 · *Relentless Evolution*

Slow though the process of selection may be, if feeble man can do much by his powers of artificial selection, I can see no limit to the amount of change, to the beauty and infinite complexity of the coadaptations between all organic beings, one with another and with their physical conditions of life, which may be affected in the long course of time by nature's power of selection.

Darwin 1859, p.109

2.1 Learning through Games

This book deals with the roles and impacts of the entangled web of biotic interactions that an alien species partakes in as it infiltrates ecological networks. We partition related issues into six topics (network interactions, structures, stability, dynamics, scaling and invasibility). We start unpacking these issues here and will dive deeper into each in subsequent chapters. To embrace the complexity of ecological networks we need to introduce a few simple mathematical models and associated concepts that are fundamental to network analyses, visualisation and the ideas we develop. We keep the mathematical details to a minimum and provide intuitive explanation of their meaning and rationale; we also discuss, using simple terminology wherever possible, key procedures that lead to deductive conclusions. Most of the models we cite have been elucidated in great detail elsewhere and can be implemented in any computational language. Although we will not provide technical details, readers will be able to design their models and conduct analyses based on what is provided here to suit their own needs. Although we have tried to determine consensus views in the literature, the transdisciplinary nature of this field makes the knowledge landscape rugged and fluid. Answers are often not definite but contextualised. Let our journey begin.

One decade after the publication of his most famous book *The Origin of Species*, Charles Darwin (1868) published another book, this one on

The Variation of Animals and Plants under Domestication. Here, Darwin laid out two types of artificial selection – methodical selection (also called deliberate, conscious or intentional) and unconscious selection – besides his monumental idea of natural selection. Both artificial and natural selection work on heritable variation that is already present to create biases in reproductive success; together these forms of selection can create profound changes in species traits and performance (Gregory 2009). Alien species, like any other species, experience both artificial and natural selection. While the drivers of selection are both abiotic (e.g. habitat suitability and disturbance) and biotic (human intervention and interactions with other residing species in an ecosystem), most attention has been given to elucidating the roles that biotic interactions play in driving evolution (Thompson 2013). An alien species, from the time it is introduced, then established, and potentially when it becomes invasive (i.e., spreads widely) in recipient ecosystems, faces all four forces of evolution – selection, drift, dispersal and mutation. The roles of these forces change along the introduction–naturalisation–invasion continuum, potentially leading to profound changes in the ways that alien species perform and interact with other species.

Biotic interactions, ubiquitous and essential for the survival of any species, play diverse and critical roles for alien species in novel environments (Traveset and Richardson 2020). The inter-dependency of organisms is largely because they (cellular organisms at least) require a constant influx of energy to function. Functioning is achieved through biochemical reactions via catabolic and anabolic pathways within cells to drive the metabolism of an organism. Finding ways to improve means for capturing, storing and utilising energy are part of life's basic operation. Primary producers such as plants can convert abiotic sources of energy directly via photosynthesis with the help of microbes. Most species, however, are heterotrophs and require other organisms to supply meals and services to complete their life cycles. This is perhaps why early studies on ecological networks focused mainly on food chains and food webs (Egerton 2007). If we expand the resolution of species and consider greater scales of space and time, food webs can become extremely large and complex (e.g., Figure 2.1). Increasingly, food webs are rendered even more complex when alien species involved. We now zoom in to consider the building blocks of ecological networks – biotic interactions – and explore how these are dealt with in various hypotheses that have been proposed to explain or conceptualize biological invasions.

Biotic interactions are choreographed concerts involving multiple players. To survive and complete their life-cycles, individual organisms

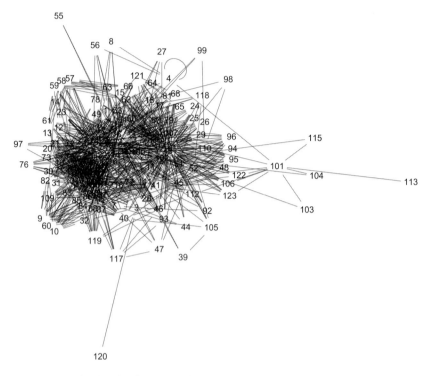

Figure 2.1 A food web of 1205 trophic links between 123 free-living species (incl. six assemblages not identified to species level) of an intertidal mudflat ecosystem of Otago Harbour, New Zealand. Directions of the trophic links are omitted here for simplicity, but largely extend from central to outlaying nodes. Species of selected nodes: 98, green-lipped mussel (*Perna canaliculus*); 99, mudflat snail (*Amphibola crenata*); 101, yellow-eyed mullet (*Aldrichetta forsteri*); 113, white-fronted tern (*Sterna striata*); 114, mallard duck (*Anas platyrhynchos*); 115, Caspian tern (*Hydroprogne caspia*); 120, black swan (*Cygnus atratus*). Data from Mouritsen et al. (2011).

must constantly 'play games', in so doing interacting with co-occurring individuals from the same and other species to extract resources and capitalize on opportunities. Individuals, carrying specific sets of ecological and evolutionary strategies, will likely receive payoffs (positive or negative) from such games. The nature of the payoffs depends not only on their strategies but also those of their opponents, in the context of the prevailing environment and over given spatial and temporal scales. Payoffs can normally be calculated as the benefit minus the cost of each game played (Figure 2.2); eventually, payoffs affect the fitness of a player (a performance measure of survival and reproduction). If the benefit

	Player 2	
	Strategy C	Strategy D
Player 1 Strategy A	P_{AC} P_{CA}	P_{AD} P_{DA}
Player 1 Strategy B	P_{BC} P_{CB}	P_{BD} P_{DB}

Figure 2.2 A payoff matrix of a two-player game where each player has two alternative strategies. The payoffs (p) to player 1 are located on the bottom left of each entry, while those to player 2 are on the top right.

exceeds the cost (i.e., a positive payoff), the game is considered beneficial to the focal player; otherwise, it is detrimental. Players in a game can choose strategies that improve, or at least conserve, their expected payoffs, often without considering the group-level average payoff – a source for individual–group conflicts. As Herbert Gintis (2009) argued in *The Bounds of Reason*,

Game theory is a general lexicon that applies to all life forms. Strategic interaction neatly separates living from non-living entities and defines life itself.

Each individual in an ecological community struggles for its survival and the chance to pass on its genetic information, while its payoff in this multiplayer game depends crucially on its strategies (traits) in relation to those of others. As will be clarified in the book, the relationship between the strategies of the alien and resident individuals in an ecological network defines invasiveness and invasibility (Hui et al. 2016). In the next section we expand on some key concepts of game theory before examining biotic interactions between alien and resident species in ecological networks.

A typical game involves two rational players, each having several alternative strategies that aim to maximise their own payoff only. Although Darwin's theory invokes largely antagonistic interactions such as competition and predation, mutualistic and altruistic interactions that are pervasive in nature are generally ignored. Many well-known games have been designed to fill this gap by exploring the evolution of cooperation and altruism in antagonistic systems. For instance, in a Prisoner's Dilemma game, both players have the same set of alternative strategies: to defect or to cooperate. Each receives a payoff of R (reward) if they cooperate with each other, but a payoff of P (punishment) if they defect each other. If one chooses to cooperate but the other player chooses to defect, the cooperator receives a payoff of S (sucker), while the defector receives a payoff of T (temptation). The Prisoner's Dilemma game assumes that the temptation to defect is greater than the reward associated with cooperation, while the punishment to defect is also greater than the sucker's toll: $T > R > P > S$. The Nash equilibrium is the set of dominant strategies that, once taken by the two players, ensures that any change of strategies from one side will result in the reduction of its payoff. In other words, no other rare strategies can invade the system (i.e., perform better than the existing strategy pair) when the system resides at the Nash equilibrium. The verb *invade* is commonly used in game theory to describe both the process of a rare strategy entering a system where it is previously absent as well as the outcome of the process. The system at the time of invasion is normally, but not necessarily, considered to be at equilibrium (i.e., payoffs have been equalised among strategies). Invasions of rare strategies were introduced mainly to explore whether a system equilibrium is evolutionarily stable (Maynard Smith and Price 1973), in other words whether a rare novel strategy can break the balance between/among existing strategies and establish in the game. In the Prisoner's Dilemma game, the Nash equilibrium occurs when both players choose to defect (D, D), while any change of strategies in a player makes it a sucker while the opponent gets the reward. Clearly, the Nash equilibrium in this game is not the Pareto optimal (strategies to maximise the total payoffs for both players, which can be achieved when both choose to cooperate (C, C)). Although the Pareto strategy (C, C) is optimal for both players (and thus for the entire network of players), it can be easily invaded by the defective strategy. This explains the so-called Tragedy of the Commons, where even though cooperative behaviours can lead to greater societal and eventually personal gain, individual players will still follow the Nash equilibrium and behave selfishly, leading to the depletion of public resources. This means that being rational does not ensure being optimal. In a Hawk–Dove game (also called a snowdrift game), where $T > R > S > P$, there are two Nash equilibriums (C, D) and (D, C); this is called a set of mixed strategies. This is because any pure strategies can be easily *invaded* by a rare alternative strategy; i.e., a strategy anticipates a greater payoff when rare, a case of negative frequency-dependent selection. In other words, in a population with all defectors, a rare alternative strategy (a cooperator) cannot invade the population if the game follows the code of a Prisoner's Dilemma, but it can invade if the rules of a Hawk-Dove game apply. Negative frequency-dependent selection is, therefore, the necessary condition for a successful invasion.

Further elaboration of the role of biotic interactions in an ecological network requires us to examine an important extension of game theory – evolutionary game theory (Maynard Smith and Price 1973). Before the proposal of evolutionary game theory, trait evolution was typically explored by assuming that such traits are adaptive and should therefore follow the optimality model. For instance, in classic clutch-size evolution (Parker and Begon 1986), the trade-off in reproductive investment between clutch size and egg size can result in a simple trait-dependent fitness function: $F(x) = (f_0 + c_f \cdot x)(s_0 + c_s(1 - x))$, where x is the proportion of reproductive investment allocated to increasing clutch size, and thus $1 - x$ is the proportion to increasing egg size; f_0 and s_0 the basal fertility and survival rate; c_f and c_s are the conversion rates from the reproductive investment to fertility and survival, respectively. Optimal clutch size can be easily sought by setting the derivative of fitness $F(x)$ with respect to the adaptive strategy x to zero, $dF(x)/dx = 0$, which gives $x^* = (c_f c_s + c_f s_0 - f_0 c_s)/(2c_f c_s)$. This is a typical approach in mathematics as any maximum (or minimum) of a curve or surface actually has a locally flat tangent (the derivative). Evidently, in an optimality model different traits beget different fitness consequences, and thus fitness is trait dependent, while evolution maximises fitness and thus selects for the optimal trait. However, a critical weakness of such optimality models is that they ignore how interactions between individuals with different traits and their consequences can potentially rewrite the fitness function; trait evolution therefore becomes a hill-climbing process in a static fitness landscape (*sensu* Wright 1932). This flaw has been effectively addressed in evolutionary game theory.

In an evolutionary game, a dynamic framework of interactions between players is added to the standard game theory. Under such conditions, an individual does not choose its strategy but inherits its strategy (functional traits) from its parents, while the average payoff of a population following a particular strategy will receive a trait-dependent payoff that further affects its frequency or population size in the next generation. This then imposes frequency-dependent selection on individuals with trait-mediated fitness. A typical formulation of the evolutionary game theory follows the so-called Replicator Equations (Hofbauer and Sigmund 1998; Gintis 2000). Let us still consider the two-player game with each inheriting its predecessor's strategy (a cooperator or a defector). Let the proportion of cooperative individuals be p_C and the proportion of defective individuals be p_D; note, $p_C + p_D = 1$. We can estimate the expected fitness (payoff) for a cooperator, $F_C = R \cdot p_C + S \cdot p_D$, and a defector, $F_D = T \cdot p_C + P \cdot p_D$, as well as the fitness of an average individual, $F = F_C \cdot p_C + F_D \cdot p_D$, in a well-mixed population. Compared to the fitness formulation in the optimality model, we can see that the fitness now depends on the frequency of different strategies in the evolutionary game. The Replicator Equation then simply considers the dynamics of a strategy being proportional to its relative fitness, $\dot{p}_C = p_C(F_C - F)$ and $\dot{p}_D = p_D(F_D - F)$. With this dynamic formulation, we can then use the well-developed stability analysis in dynamical systems to explore the stability and dynamics of evolutionary equilibriums. For instance, the Nash equilibrium then gives rise to the concept of Evolutionarily Stable Strategy (ESS; Cressman and Tao 2014). It is important to note that in a complex system with players having adaptive strategies, a system equilibrium needs to be evolutionarily unstable to be open to invasion.

Before we move on to discuss trait-dependent invasion performance and invasive traits, we should first consider the winning strategies in game contests. In 1980, political scientist Robert Axelrod convened two tournaments involving game theorists to solicit strategies into paired contests for 200 iterations of a Prisoner's Dilemma game, looking at three separate questions (Axelrod and Hamilton 1981),

(1) Robustness. What type of strategy can thrive in a variegated environment inhabited by others using a wide variety of more or less sophisticated strategies?
(2) Stability. Under what conditions can such a strategy, once fully established, resist invasion by mutant strategies?
(3) Initial viability. Even if a strategy is robust and stable, how can it ever get a foothold in an environment which is predominantly non-cooperative?

We can translate these three criteria into questions that are clearly pertinent to considerations about an alien species infiltrating an ecological network. Robustness seeks a generic invasive trait that is not overly context dependent. Stability is the condition of dominance and invasive impacts. Initial viability is about establishment success and overcoming initial bottleneck effects from being rare in a novel and often hostile environment. The reciprocal strategy Tit For Tat (TFT), proposed by Anatol Rapoport, passed all three criteria and came on top among many proposed strategies (Axelrod 1984). A TFT player is to cooperate in the first move but then to follow whatever the opponent did in the previous round. It is surprising that such a simple heuristic learning strategy wins over many crafty and tricky strategies, to say the least. Although 'Always Defect' is both the only ESS and the only strict Nash equilibrium of the repeated Prisoner's Dilemma game (Maynard Smith and Price 1973), it did not emerge from the tournaments as the winner (Axelrod and Hamilton 1981). This is because evolution in finite-sized populations need not reach the strict Nash equilibrium; instead, evolution favours a robust winner that can withstand stochasticity and uncertainty (Imhof et al. 2005). Axelrod (1984) summarised the winning strategies of a robust winner in his tournaments,

Won, not by doing better than the other player, but by eliciting cooperation by promoting the mutual interest rather than by exploiting the other's weakness.

A decade later, a similar reciprocal strategy, known as Pavlov depicting the heuristic learning behaviour of Win-Stay, Lose-Shift (WSLS), outperformed TFT (Nowak and Sigmund 1993). Pavlov can outcompete TFT by correcting occasional mistakes and exploiting unconditional co-operators; it is described as a trusting avenger who is exploiting yet repentant (Kraines and Kraines 2000). Interestingly, in games with more than two strategies, such as the rock-paper-scissors game that theoretically does not possess a Nash equilibrium, winning players in real contests seem to follow the same WSLS or TFT reciprocal strategy (Wang et al. 2014).

There has been little discussion around the fact that these winning reciprocal strategies (e.g. TFT and WSLS) are actually part of heuristic learning models in decision-making and cognitive psychology, akin to Bayesian inference (Bonawitz et al. 2014). Such a heuristic learning model can be combined with other reinforcing learning models to capture the tendency of staying with the same option or switching to alternatives after trial and error (Worthy and Maddox 2014). Therefore, we could argue that a winning strategy or an invasive trait might not be simply the strategy that outcompetes resident species in a community, but is rather a better learning strategy that enables the player to handle the novel setting in an uncertain environment. For instance, a similar Recent Experience Driven foraging behavioural strategy was also found to be capable of quickly learning the resource context in a novel environment by being picky and conservative in diet choice when the forager's energy reserve is close to full capacity, while being opportunistic when it is hungry (Zhang and Hui 2014). Along this line, biotic interactions not only can provide an individual a way of exploiting and outcompeting other players in the evolutionary game, but also the means to know one's Eltonian niche by learning from other players through its game experience. In an ecological network, a species can assert its functional and niche positions by imposing pressures on others through biotic interactions, while simultaneously experiencing pressures from others to learn and adjust its functional and niche position in the network. In this regard, biotic interactions resemble handshakes that serve to assess the strength of your opponents and to position one's position or ranking in a community.

To explain improved success in hostile and defective environments, strategies need to follow the so-called Hamilton's (1964) rules (e.g., Nowak 2006; Zhang et al. 2010). Simply put, the benefit (b) or the fitness bestowed through adoption of a strategy needs to be greater than its cost (c) to ensure successful establishment and invasion of the strategy, $rb > c$, where r represents a mechanism that can boost the strategy's benefit. In the case of kin selection, the genetic relatedness (r) between individuals can augment fitness b to become inclusive fitness rb and thus boost the chance of success. In the case of direct reciprocity, such as TFT and WSLS, the relative chance to meet the same opponent again in the next round (r) can boost the strategy's chance of establishment. In the case of indirect reciprocity, this chance of interacting with the same or similar opponents is assured by social acquaintanceship. In the case of network reciprocity, this is achieved by reducing network connectivity (i.e., positions with fewer linked nodes on graphs). In the case of group selection, smaller but more groups make the establishment of a strategy easier. In the spatial Prisoner's Dilemma game, parameter r is the difference between the probability of a cooperator being adjacent to a fellow cooperator and the probability being adjacent to a random player, while this difference increases in a harsh environment but decreases in a conducive environment. Transferring such thinking to ecological networks, we suggest that an alien species can similarly use some of these mechanisms to gain a foothold in recipient ecosystems.

Let us consider biotic interactions as ensembles of games between individuals of different species, each having its own traits. This allows us to explore the dimensions and dynamics of biotic interactions. First, the payoff of a game play depends on the strategies of all involved players; in other words, the fitness and performance of a species in a biotic interaction depends on its traits and on those of other players in the ecological network.

Second, the average payoff of a group of players depends on how we place these players into groups. Such groupings in ecological networks are typically done at the species level, but such categorization can also be done at higher or lower taxonomic levels by denoting nodes in the network based on genera, subspecies, Operational Taxonomic Units (OTUs) or localised population. The groupings can also be based on the roles and functions of involved species in the ecosystem, such as functional guilds (e.g., insect parasites, avian pollinators, nitrogen-fixing plants). Such groupings can be further relaxed by considering labels, as is typical in social network analyses. For instance, friendship and club membership will affect how each player in the broader social network behaves and interacts with others. In ecological networks, we typically define labels on the basis of whether a species is alien, native, endemic and rare. Such labelling may (in the case of ecosystem engineers), but does not necessarily, affect the nature of biotic interactions between players (species), but influences human perception and management response towards either the mitigation or conservation of certain labelled species and their interactions with others. Finally, due to the heterogeneity and size of each group (node) in a network, performance and interaction strength between groups are inevitably scale dependent and suffer the Modifiable Areal Unit Problem (Openshaw 1984; Fotheringham and Wong 1991). An invasion manager can choose to report the success of management of local pest populations without reporting collective failures at larger scales when the invasion dynamics of local populations are pooled (e.g., Zhao et al. 2019).

Considering biotic interactions as ensembles of scale-dependent and intertwined dynamic games embedded in a bigger system opens exciting opportunities for gaining new insights on the invasion performance and invasiveness of alien species in an ecosystem. This can explain a plethora of invasion theories and hypotheses that invoke the role of biotic interactions in various processes that mediate biological invasions (Table 2.1; Catford et al. 2009; Enders et al. 2018, 2020; Hui et al. 2020). These invasion hypotheses are typically erected to explain the success or failure of alien species in overcoming the different barriers along the introduction–naturalisation–invasion continuum (Richardson and Pyšek 2006; Blackburn et al. 2011). Most of these hypotheses invoke the role of biotic interactions, with an undertone of why some introduced species perform better than others or than resident native species with similar traits. Most also implicitly choose the default adaptive view; for instance, species lacking predators will evolve to become stronger competitors. These are intuitive but not necessarily evolutionarily feasible

Table 2.1 *Hypotheses invoking biotic interactions to explain aspects of biological invasions. From Hui et al. (2020) based on Catford et al. (2009) and Enders et al. (2018)*

Processes	Hypothesis	Definition
Based on cross-guild biotic interactions only	Enemy of my enemy aka accumulation-of-local-pathogens hypothesis (EE, Eppinga et al. 2006)	Introduced enemies of a non-native species are less harmful to the non-native as compared to the native species
	Enemy reduction (ERD, Colautti et al. 2004)	The partial release of enemies in the non-native range is a cause of invasion success
	Enemy release (ER, Keane and Crawley 2002)	The absence of enemies in the non-native range is a cause of invasion success
	Increased susceptibility (IS, Colautti et al. 2004)	If a non-native species has a lower genetic diversity than the native species, there will be a low probability that the non-native species establishes itself due to increased pressure from enemies
	Missed mutualisms (MM, Colautti et al. 2004; Mitchell et al. 2006)	In their non-native ranges, non-native species suffer from missing mutualists
	Specialist-generalist (SG, Callaway et al. 2004)	Non-native species are more successful in a new region if the local predators are specialists and local mutualists are generalists
	New associations (NA, Colautti et al. 2004)	New relationships between non-native and native species can positively or negatively influence the establishment of the non-native species
	Invasion meltdown (IM, Simberloff and Von Holle 1999; Sax et al. 2007)	The presence of non-native species in an ecosystem facilitates invasion by additional species, increasing their likelihood of survival and/or their ecological impact
Based on within-guild biotic interactions	Novel weapons (NW, Callaway and Ridenour 2004)	In the non-native range, non-native species can have a competitive advantage against native species because they possess a novel weapon, i.e. a trait that is new to the resident community of native species and therefore affects them negatively

Table 2.1 (*cont.*)

Processes	Hypothesis	Definition
	Biotic indirect effects (BID, Callaway et al. 2004)	Non-native species benefit from different indirect effects triggered by native species
Based on biotic interactions and abiotic selection	Empty niche (EN, MacArthur 1970)	The invasion success of non-native species increases with the availability of empty niches in the non-native range
	Enemy inversion (EI, Colautti et al. 2004)	Introduced enemies of non-native species are less harmful for them in the non-native than the native range, due to altered biotic and abiotic conditions
	Resource-enemy release (RER, Blumenthal 2006)	The non-native species is released from its natural enemies and can spend more energy in its reproduction, and invasion success increases with the availability of resources
	Darwin's naturalization (DN, Darwin 1859)	The invasion success of non-native species is higher in areas that are have few closely-related species than in areas that house many closely-related species
	Limiting similarity (LS, MacArthur and Levins 1967)	The invasion success of non-native species is high if they highly differ from native species, and it is low if they are similar to native species
	Sampling (SP, Crawley et al. 1999)	A large number of different non-native species is more likely to become invasive than a small number due to interspecific competition. Also the species identity of the locals is more important than the richness in terms of the invasion of an area
	Adaptation (ADP, Duncan and Williams 2002)	The invasion success of non-native species depends on the adaptation to the conditions in the non-native range before and/or after the introduction. Non-native species that are related to native species are more successful in this adaptation

(cont.)

Table 2.1 (*cont.*)

Processes	Hypothesis	Definition
Based on biotic interactions, abiotic selection and drift	Opportunity windows (OW, Johnstone 1986)	The invasion success of non-native species increases with the availability of empty niches in the non-native range, and the availability of these niches fluctuates spatio-temporally
	Dynamic equilibrium (DEM, Huston 1979)	The establishment of a non-native species depends on natural fluctuations of the ecosystem, which influences the competition of local species
Based on biotic interactions and evolution	Evolution of increased competitive ability (EICA, Blossey and Nötzgold 1995)	After having been released from natural enemies, non-native species will allocate more energy in growth and/or reproduction (this re-allocation is due to genetic changes), which makes them more competitive

explanations. Consequently, most hypotheses are often indefinite and there is little in the way of consensus in the literature. We therefore discuss them within a standard theoretical framework to explore which one is ecologically feasible, while which one is evolutionarily feasible but irrelevant for invasion management over timescales considered by managers. Although most invasion hypotheses have value in some context, our focus is on how the strength and effect of biotic interactions can be measured. First, we need to establish the concept of fitness as demographic performance that is both trait mediated and density dependent in an ecological network. To do so, we present a list of concepts relating to biotic interactions. These concepts are generic, to afford explicit investigation of their structures and dynamics using tractable mathematics. They are building blocks for designing more complex and realistic models that will be discussed in later chapters.

2.2 Interaction Strength

Once individuals have been sorted into groups (nodes), we can calculate the average or overall payoff for a group when its members are playing

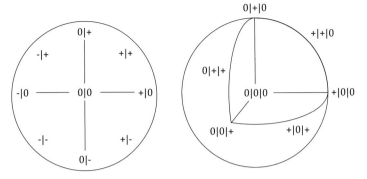

Figure 2.3 A compass of pairwise interactions (left) and a sphere of three-way interactions (right). Pairwise interactions can be sorted on a compass by the strength and sign of an interaction, while three-way interactions extend the possible types of biotic interactions in a three-dimensional sphere. The sign of an interaction is divided to indicate its contribution to a particular species in the interaction; for instance, +|− represents an interaction that is beneficial to the first species but detrimental to the second.

games with individuals of another group (inter-specific) or of the same group (intra-specific). Biotic interactions can thus be defined as the ensembles of such one-on-one games after grouping individuals according to specific labels. Typical biotic interactions include those that are mutually beneficial (e.g., pollination) or detrimental (e.g., resource competition), as well as those that are beneficial to one group but detrimental to another (e.g., herbivory) (Figure 2.3 left for pairwise interactions). Any such group-level classification of interactions inevitably overlooks the complexity of individual strategies within and between groups. As each individual has a unique set of strategies/traits, the performance of its embedded group depends not only on the performance of the average strategy but also the plasticity, diversity and distribution of individual strategies within the group. Moreover, conceptualizing interactions as being 'pairwise' (involving two players) is usually a gross simplification of the complex interactions involved; games can be played simultaneously by more than two players, and higher-order interactions involving more than three players have also been highlighted in the recent literature (Figure 2.3 right; e.g., Grilli et al. 2017; Mayfield and Stouffer 2017). However, a prerequisite for an interaction is that individuals need to first encounter each other to initiate the game and exchange payoffs; consequently, higher-order interactions can only occur at a rate one magnitude of order lower than the encounter rate of pairwise interactions (Wootton 1994; Terry et al. 2017). As such, although they are important for system stability, higher-order interactions are often neglected in surveys and in modelling.

With each species resuming its functional role from past evolutionary games by carrying a set of traits and abundance, biotic interactions can affect the demographic performance and fitness of involved species within an ecological network. Consequently, the dynamic behaviour of involved species should reflect this game play of heuristically learning from each other. To portray the building blocks of an ecological network and to assess the invasiveness of an introduced species into such a network, we need to elaborate on the trait-mediated density-dependent interactions between species. This can be achieved in two steps. First, we explore the concept and metrics of *interaction strength* that quantify the effect and impact of an interaction on the demographic performance of involved species. Second, we relate interaction strength to the traits of involved species by formulating its value to niche overlap and differentiation, or more precisely to trait distance and dissimilarity, according to functions of *interaction kernel*. As shown later, the strength and kernel of interactions are not necessarily static, but they can be dynamic, depending on population densities over ecological timescales and on traits over evolutionary timescales.

In a generic setting, let us consider species i interacting with species j in an ecological network,

$$\begin{bmatrix} \dot{n}_i \\ \dot{n}_j \end{bmatrix} = \begin{bmatrix} F_i \\ F_j \end{bmatrix}, \tag{2.1}$$

where $\dot{n}_i$ is simply the shorthand for the derivative of population density (or population size in a well-mixed scenario) of species i with respect to time, dn_i/dt, while F_i is a generic function of population change rate that needs to be specified for a system. In vector notion, we can simplify these equations to $\dot{n} = F$, where the lower case bold letter represents a vertical vector, and a bold capital letter represents a matrix. In 1715, Brook Taylor discovered that a smooth function can be decomposed at a given reference point into the sum of polynomial series with the coefficients related to the derivatives (steepness) of the function at the reference point. For instance, function $\exp(n)$ can be approximated as $1 + n + n^2/2 + \cdots$ around $n = 0$. Using Taylor expansion we can approximate system (2.1) as a linear function at the point (n_i', n_j'),

$$\begin{bmatrix} \dot{n}_i \\ \dot{n}_j \end{bmatrix} \approx \begin{bmatrix} F_i(n'_i, n'_j) \\ F_j(n'_i, n'_j) \end{bmatrix} + \begin{bmatrix} \dfrac{\partial F_i}{\partial n_i} & \dfrac{\partial F_i}{\partial n_j} \\ \dfrac{\partial F_j}{\partial n_i} & \dfrac{\partial F_j}{\partial n_j} \end{bmatrix} \begin{bmatrix} \Delta n_i \\ \Delta n_j \end{bmatrix}, \tag{2.2}$$

where $\Delta n_i = n_i - n'_i$. When the point (n'_i, n'_j) is chosen to be a system equilibrium (i.e., the solution to zero tangent, $\dot{\boldsymbol{n}} = \boldsymbol{0}$), we have $F_i(n'_i, n'_j) = 0$ and therefore Eq. (2.2) becomes $\dot{\boldsymbol{n}} \approx \boldsymbol{J} \Delta \boldsymbol{n}$ at the equilibrium. Matrix $\boldsymbol{J}$ is known as the Jacobian matrix of the dynamical system, Eq. (2.1), after Carl G.J. Jacobi (1804–1851). This approximation then allows us to define a widely-used metric of interaction strength,

$$s_{ij}^{(J)} = \frac{\partial F_i}{\partial n_j}, \tag{2.3}$$

where the partial derivative is evaluated at the population equilibrium, forming the so-called community matrix (see review by Novak et al. 2016). Interaction strength here measures the impact of changes in the abundance of species j on the population change rate of species i; in other words, the sensitivity of species i's population change rate to species j's population size. If we consider F_i a terrain (Figure 2.4), then this interaction strength metric depicts the steepness of the terrain at this equilibrium point along the direction of increasing the population size of species j while species i's population size stays still. When interaction

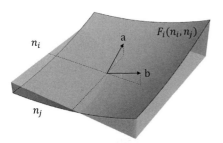

Figure 2.4 An illustration of population change rate of species i as a function of the population sizes of species i and j. Interaction strength can be defined as partial derivative of this function at a reference point, normally at a population equilibrium ($F_i = 0$), along the direction of a particular species. Vectors a and b are tangent lines on the surface of F_i along the direction of n_i and n_j, respectively; the partial derivative (interaction strength) of Eq. (2.3) measures the steepness (i.e., the tangent) of vector b against its projection on the bottom plane.

strengths are measured in such a way, we could then visualise an eco-logical network using its Jacobian matrix at an equilibrium. Jacobian matrices offer exciting opportunities for exploring aspects of system dynamics and stability. (We revisit the features of Jacobian matrices in Chapters 4 and 5.)

Following the reasoning in game theory, this function F_i can be typically simplified into the product of the current population size, n_i, and the average payoff received by an individual of species i, f_i, in the previous round of games with both conspecific and interspecific individuals; that is, $F_i = n_i \cdot f_i$. Note, it is not necessary here to consider the relative pay off as in the Replicator Equations because we are interested in the dynamics of each species' population size, not the proportion or the frequency of each species' population size in the community. The average payoff from the last round, f_i, can be dynamic (typically written as $f_{i,t}$ or $f_i(t)$ to show its dependence on time, t) and is typically known in population ecology as the Relative Growth Rate (RGR). Another typical measure of population growth rate is the multiplicative growth rate, $\lambda_{i,t} = n_{i,t+1}/n_{i,t}$. Consequently, to simulate a species' population dynamics in an ecological network we could choose to either solve the differential equation $\dot{n}_i = n_i \cdot f_i$ or calculate it iteratively as a difference equation, $n_{i,t+1} = n_{i,t}\lambda_{i,t}$, with $\lambda_{i,t} = \exp(f_{i,t})$. In practice, the last-mentioned discrete-time approach is often preferred for computational efficiency. In addition, to consider demographic stochasticity from birth and death events, we can randomly draw $\lambda_{i,t}$ from a Poisson distribution with $\exp(f_{i,t})$ as its mean, or a negative binomial distribution for non-independent birth and death events.

With the relative growth rate defined, we can measure the per-capita interaction strength,

$$s_{ij}^{(I)} = \frac{\partial f_i}{\partial n_j}, \tag{2.4}$$

which is often used as entry for the so-called interaction matrix. Note, when evaluated at the system equilibrium, we simply have $s_{ij}^{(J)} = n_i \cdot s_{ij}^{(I)}$. The exact form of the per-capita demographic fitness, f_i, being both trait mediated and density dependent, is crucial to our understanding of biotic interactions and trait evolution, and as such we embark on a short detour here to lay out some mathematic concepts for later use.

Again, with Taylor expansion, we could write out any forms of f_i into the sum of polynomials of population size of species i, n_i and those of another species, n_j, as the following,

$$f_i(n_i, n_j) \approx f_i(n_i', n_j') + \begin{bmatrix} \dfrac{\partial f_i}{\partial n_i} & \dfrac{\partial f_i}{\partial n_j} \end{bmatrix} \cdot \Delta n + \frac{1}{2} \Delta n^{\mathrm{T}} \begin{bmatrix} \dfrac{\partial^2 f_i}{\partial n_i^2} & \dfrac{\partial^2 f_i}{\partial n_i \partial n_j} \\[2ex] \dfrac{\partial^2 f_i}{\partial n_i \partial n_j} & \dfrac{\partial^2 f_i}{\partial n_j^2} \end{bmatrix} \Delta n.$$

(2.5)

In vector notion, this is

$$f_i(n) \approx f_i(n') + \nabla f(n')\Delta n + \frac{1}{2}\Delta n^{\mathrm{T}} \mathbf{H}(n')\Delta n,$$

where $\nabla f(n')$ represents the gradient of the vector field at the reference point and the i-th row of the Jacobian matrix of per-capita demographic fitness. The last matrix $\mathbf{H}$ is named the Hessian matrix after Ludwig O. Hesse (1811–1874). The Hessian matrix is important for us to understand trait dynamics and eco-evolutionary feedbacks, as will be shown.

As Taylor expansion, we typically ignore higher-order terms, which, however, means that the performance of this approximation is only guaranteed within a close range to the reference point (n_i', n_j'). When exploring system dynamics beyond the neighbourhood of the reference point, we need to consider the density dependence of interaction strength, which is typically formulated as functional response (Holling 1959; Jeschke et al. 2002; Arditi and Ginzburg 2012; Morozov and Petrovskii 2013) due to density-dependent optimal or adaptive foraging (Abrams 1990; Egas et al. 2005; Zhang and Hui 2014). This means that the type and strength of biotic interactions can change with density (e.g., herbivores can switch to preferred food according to resource palatability and abundances; van Baalen et al. 2001) and, conceivably, the quality of environment (e.g., a harsh environment favours facilitative interaction while a conducive environment promotes antagonism). Some strategies can be modified instantaneously by the individual itself (e.g., a predator can change its foraging behaviour through learning; Zhang and Hui 2014), while most are encoded in its genome and can only be naturally selected via relatively slow evolutionary processes over generations. Consequently, an alien species could therefore adopt different strategies at different invasion stages and under different contexts. For instance, annual plant invaders can adapt to novel environments faster than perennial plant invaders, resulting in different life-history strategies being targeted by selection at different invasion stages (Gioria et al. 2018).

If we ignore the Hessian term, rename $f_i(n_i', n_j') - n_i' \partial f_i/\partial n_i - n_j' \partial f_i/\partial n_j$ as r_i and define $\partial f_i/\partial n_i$ as α_{ij}/k_i in Eq. (2.5), population dynamics then become the well-known Lotka–Volterra model,

$$\dot{n}_i = r_i n_i \left(1 + \sum_i \frac{\alpha_{ij} n_j}{k_i} \right), \tag{2.6}$$

where r_i and k_i represent the intrinsic rate of growth and carrying capacity of species i in an ecological network; α_{ij} the interaction coefficient of species j on species i can be considered as interaction strength, forming the alpha matrix,

$$s_{ij}^{(\alpha)} = \alpha_{ij} = \frac{k_i}{r_i} s_{ij}^{(I)}. \tag{2.7}$$

In practice, we could assume that f_i is a linear function of n_j and thus can parameterise interaction strengths by regressing the time series of f_i against the time series of n_j, with r_i being the intercept and α_{ij} the regression coefficient (e.g., $\alpha_{ij} = \text{Cov}(f_i, n_j)/\text{Var}(n_j)$).

There are other proxies of interaction strength that are proportional to these metrics based on system Jacobian (and Hessian). For instance, the total and per capita impact from species j on species i,

$$p_{ij}^{(T)} = s_{ij}^{(\alpha)} n_i n_j = p_{ij}^{(P)} n_i, \tag{2.8}$$

where the mass-action law is assumed for well-mixed interactions, and thus the encounter rate of individuals of species i and j is proportional to the product of their population sizes, $n_i n_j$, and the chance for an individual of species i to encounter individuals of species j is proportional to the population size of species j, n_j. Spatial structure and interaction preference/avoidance can break the assumption of mass action and greatly affect the encounter rate and therefore the realised interaction strength and impact. So far, most empirical studies have not directly measured the interaction strength between introduced species and resident species of an ecosystem. Instead, most such studies have treated biological invasions as a natural experiment and considered it a binary factor (invaded vs. uninvaded). On some rare occasions, consideration is given to more than two invasion stages (e.g., uninvaded vs. lightly invaded vs. heavily invaded). For instance, by comparing invaded vs. uninvaded systems, many cases report on how biological invasions (usually from the perspective of a particular invasive species) have modified the demographic and fitness components of a focal resident species.

The three metrics of interaction strength and other strength proxies are all related to elements of the Jacobian matrix and are tailored to capture only pairwise interactions. To describe higher-order interactions among more than two species simultaneously (Figure 2.3 right), we need to re-insert those higher-order terms in the Taylor series, such as the entries of Hessian matrix and other even higher-order terms ignored in Eq. (2.5) (Mayfield and Stouffer 2017). Together, the Jacobian and the Hessian of the dynamical system that describe pairwise and triplet biotic interactions within a focal ecological network can largely unveil the structure and the fate of each species. Methods for topological analyses of complex networks with higher-order interactions such as simplicial and clique complexes with persistent homology are only now starting to emerge (Rieck et al. 2018; Aktas et al. 2019). Ecological networks with multiple orders of interactions fully resolved are currently unavailable, while such complex networks in co-occurrence networks (e.g., based on zeta diversity similarity of different orders; Hui and McGeoch 2014; Latombe et al. 2018) and collaborative networks are rather common (see Chapter 6).

2.3 Coexistence and Invasion

Biotic interactions can profoundly affect payoffs received by involved species, and a myriad of invasion hypotheses have been proposed to address issues pertaining to alterations to interactions caused by introduced species. Introduced species in their invaded ranges are typically separated from key co-evolved mutualists or antagonists and are faced with new assemblages of species that could initiate direct interactions. Enders et al. (2018) listed 33 invasion hypotheses, of which 19 relate explicitly to biotic interactions (Table 2.1). For instance, the enemy of my enemy (Eppinga et al. 2006), the enemy reduction (Colautti et al. 2004) and the enemy release (Keane and Crawley 2002) hypotheses all assume that a cross-guild antagonistic interaction with an enemy is lower in the introduced than in the native range. This could be the case either because the enemy has a higher negative effect on native species (in which case the absolute effect of the enemy on the alien species is not necessarily lower in the introduced compared to the historical range but is lower relative to other species of the same guild). It could also be because the enemy is less abundant or absent. Either way, this should result in an alien species experiencing less damage and enhanced performance in the introduced range compared with the native range.

On the other hand, the novel weapon hypothesis posits that alien species can have traits that negatively affect the native species of the resident community (Callaway and Ridenour 2004). Some alien plants release phytotoxic biochemical compounds that directly harm native plant species (Callaway and Ridenour 2004) or adopt indirect interactions to harm fungi involved in mutualistic interactions with the native plant species (Callaway et al. 2008). Nevertheless, the premises of these invasion hypotheses (and testing their validity) demand a solid understanding of species coexistence and invasion dynamics.

The population dynamics of two interacting species, say between a resident species (with subscript R) of the recipient ecosystem and an alien species (with subscript A), can be described by two differential equations,

$$\dot{n}_R = F_R = n_R \cdot f_R$$
$$\dot{n}_A = F_A = n_A \cdot f_A \qquad (2.9)$$

As mentioned previously, Taylor expansion of Eq. (2.9) leads to a standard two-species Lotka–Volterra model, described by Eq. (2.6). Without the alien invasion, the basic single-species model with unlimited resources is the simple Malthusian model, $\dot{n}_R = n_R \cdot r_R$, which represents a constant per-capita demographic performance, $\dot{n}_R / n_R$, and leads to an exponential growth of population size, $n_R[t] = n_R[0] \exp(r_R t)$. With limited resources, the per-capita population change rate r_R becomes density dependent and is adjusted to $r_R(1 - n_R / k_R)$. Considering interspecific biotic interactions can be considered as further modification to the per-capita demographic performance and thus the population change rate.

The conditions required to allow the two species to coexist stably can be explored by so-called zero isoclines, while the potential trajectories of the population sizes can be more explicitly visualised as flows in a vector field (Figure 2.5). A standard procedure includes first solving the system equations to identify possible equilibriums by setting the population change rates of all species to zero ($\dot{n}_R = \dot{n}_A = 0$). Second, we can compute the system Jacobian matrix at each equilibrium; see definition in Eq. (2.2). Finally, we can explore the Lyaponov stability of the system at each equilibrium, named after Russian mathematician Aleksandr Lyapunov (1857–1918), based on the spectral distributions (eigenvalues) of the Jacobian matrix. Eigenvalues can be complex numbers and have both real parts and imaginary parts, with each part revealing important information of a dynamic regime. Briefly, species can coexist stably with

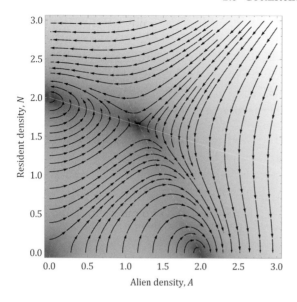

Figure 2.5 An illustration of zero isoclines (white and dotted lines) and the flow trajectories in the vector field. The background indicates the logarithm of field magnitude. Parameters, $r_R = r_A = k_R/2 = k_A/2 = \alpha_{RR} = \alpha_{AA} = 1$; $\alpha_{RA} = 0.3$; $\alpha_{AR} = 0.5$.

each other at an equilibrium if all eigenvalues of its Jacobian matrix at this equilibrium have negative real parts; this means that small perturbations to the system near this equilibrium will diminish exponentially with time. In contrast, any eigenvalue having a positive real part will entail the amplification of the perturbation along its associated eigenvector direction, indicating the equilibrium unstable. We revisit the spectral distribution of system Jacobian of an ecological network in Chapter 4.

Lyapunov stability provides a strong condition for species coexistence near a specific equilibrium. This is not necessary when discussing the coexistence of two species. First, species coexistence is not necessarily around a specific equilibrium, as the system can settle at multiple alternative equilibriums. Second, an unstable equilibrium does not necessarily entail species exclusion or extinction; the two species can still coexist with each other in a local community (e.g., the fluctuating dynamics of lynxes and hares can be unstable even though they enjoy a robust coexistence). The expansion on the condition for species coexistence is a milestone in theoretical ecology and has crucial relevance in invasion ecology.

For two species to coexist, it turns out that we only need to ensure that they fail to expel each other. That is, a species can recover and rebound, or invade, when becoming rare. Specifically, species A needs to be able to invade the system when species R stays at its equilibrium. To allow coexistence, moreover, species R can also invade the system when species A is at its equilibrium. Again, the term *invade* means that a species enters a system from a small initial population size, theoretically negligible and thus close to zero. This bears similarity to the case of biological invasions as many biological invasions result from small founder populations. This condition is termed *mutual invasibility* (also called pairwise invadability or invasibility), which is a more relaxed and robust condition for species coexistence. The condition of mutual invasibility for two species to coexist is composed of two: the *invasion criterion* that one species can successfully invade (recover from being rare) when the other resides at its equilibrium, and the invasion criterion of the other species to invade the focal species; thus, the condition is mutually invasible.

In the context of biological invasions, we are typically only interested in exploring the invasion criterion of the introduced species. For instance, for the alien species in Eq. (2.9), if we look at its per-capita population change rate $(\dot{n}_A/n_A)$ when $n_A \rightarrow 0$ and $n_R \rightarrow k_R/\alpha_{RR}$, also known as the *invasion growth rate* (Gallien et al. 2017), $\dot{n}_A/n_A = r_A(k_A - \alpha_{AR}k_R/\alpha_{RR})/k_A$, we see that a positive value to ensure invasion will require,

$$k_A/k_R > \alpha_{AR}/\alpha_{RR}. \qquad (2.10)$$

This means that the impact of the native species on the alien species (α_{AR}) should be less than the native species' intraspecific competition weighted by the two species' carrying capacities, $\alpha_{RR}k_A/k_R$. Similarly, the invasion criterion for the native species is $k_R/k_A > \alpha_{RA}/\alpha_{AA}$. For simplicity setting $k_R = k_A$, we see that the coexistence condition according to the mutual invadability is, $\alpha_{RR} > \alpha_{AR}$ and $\alpha_{AA} > \alpha_{RA}$; that is, the necessary condition for coexistence is that intraspecific interaction strength needs to be greater than interspecific interaction strength. For the case of biological invasion, we only need to satisfy the condition for the alien species to invade while the native species stands still, $\alpha_{RR} > \alpha_{AR}$, meaning that the self-regulation of the native species needs to be stronger than its resistance against the alien species to allow the alien to invade successfully.

To tackle the role of initial propagule pressure in invasion success we could consider the native setting at its equilibrium while the alien species starts from an initially low density ($n_A = \varepsilon$), the native population size declines at a rate of $r_R\varepsilon\alpha_{RA}/\alpha_{RR}$ and the alien species grows at a rate of $r_A\varepsilon(k_A - \alpha_{AA}\varepsilon - \alpha_{AR}k_R/\alpha_{RR})/k_A$. Evidently, once the invasion criterion for the alien is fulfilled ($\alpha_{RR} > \alpha_{AR}$), the alien species with a higher intrinsic growth rate (large r_A), a larger initial propagule pressure (large ε), less self-regulation (small α_{AA}) and less resistance from the native species (small α_{AR}) will expand its population faster. On the other hand, the native species with a higher intrinsic growth rate (large r_R), less self-regulation (smaller α_{RR}) and a stronger negative impact from the alien species (larger α_{RA}) will suffer a more rapid population decline. For simplicity, after setting the intraspecific effects to $\alpha_{RR} = \alpha_{AA} = -1$, we could summarise the dynamic regimes of this simple model (Figure 2.6).

Mechanisms that foster species coexistence and thus mutual invasibility have been synthesised in community ecology (e.g., Broekman et al. 2019; Grainger et al. 2019). To see the drivers of coexistence, we could write the *relative growth ratio* of the alien to the native species when the population size of the alien is trivial, while the native species sits at its equilibrium,

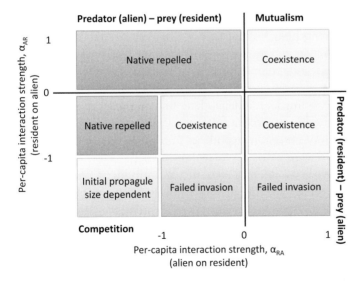

Figure 2.6 The division of biotic interactions based on per-capita interaction strengths. The qualitative outcomes of the biotic interaction (with intraspecific interaction strength assumed to be -1) are given in the blocks; the zero lines separating the four biotic interactions indicate commensalism and amensalism. Redrawn from Fig. 6.1 in Hui and Richardson (2017).

$$\frac{1}{n_A/n_R}\frac{d(n_A/n_R)}{dt}=\frac{\dot{n}_A}{n_A}-\frac{\dot{n}_R}{n_R}=(r_A-r_R)+r_A\left(\frac{r_R}{r_A}-\frac{\alpha_{AR}}{\alpha_{RR}}\right). \quad (2.11)$$

For successful invasion (i.e., a positive relative growth ratio), this ratio must stay positive. Note that the right-hand side of the equation has two components (*sensu* Chesson 2000): fitness difference $(r_A - r_N)$ and stabilising force ($\alpha_{AR} < \alpha_{RR}$, with $\alpha_{AR} = \alpha_{RR}$ suggesting the case of complete niche overlap, and $\alpha_{AR} = 0$ the case of complete niche segregation). When the alien species possesses greater fitness, $r_A > r_N$, niche-based mechanisms predominate the outcome of biotic interactions. When the two species have the same per-capita growth rate, there is no fitness difference (i.e., the setting of the neutral theory) and the invader relies on the stabilising force to invade successfully. In such a case with equal fitness, niche differentiation can improve the chance of successful invasion. As shown in Chapter 7, both factors can be further mediated by propagule pressure (the influx of alien propagules).

Many invasion hypotheses assume that invasion success depends on a combination of biotic (within- or cross-guild) interactions and abiotic selection (Table 2.1), i.e., taking a broad perspective of niche theory. However, these combinations are often quite loosely defined. The resource-enemy release hypothesis (Blumenthal 2006) is similar to the enemy release hypothesis, except that it assumes that invasion success is maximised when resources are high; it assumes that both cross-guild biotic interactions and potentially abiotic selection (if the resource is inorganic) influence invasion success. It therefore combines, in an additive way, two existing and independent invasion hypotheses: enemy release and increased resource availability (Sher and Hyatt 1999). In contrast, the enemy inversion hypothesis (Colautti et al. 2004) assumes that biotic and abiotic processes may influence the cross-guild biotic interactions between an alien species and its introduced enemy (often used as a biocontrol agent), e.g., by providing alternative resources to the latter (Pearson et al. 2000).

Besides the mechanisms of fitness advantage and niche separation, an ecological network also harbours a great diversity of indirect interactions (White et al. 2006; Allen 2020), which can also greatly affect the dynamics of species coexistence and invasion. In particular, in an interaction chain, in which a donor species affects the abundance of a transmitter and has an effect on a recipient. These are thus combinations of at least two pairwise direct interactions. Negative indirect interactions include exploitation competition and apparent competition. Positive indirect interactions include keystone predation, trophic cascade, competitive mutualism and indirect mutualism. For instance, a trophic cascade (or food chain mutualism) takes place when a plant species is positively affected by the decrease in herbivory caused by a predator reducing herbivore numbers (Bosc et al. 2018). Indirect mutualism (or facilitation) comprises a more complex four-species system with two competitors which do not interact directly but consume different prey species competing with each other; if one consumer increases its consumption, its prey decreases, which in turn reduces competition with the second prey, providing more food for the second consumer species (Wootton 1994). A peculiar type of indirect interaction is intransitive competition, where species are arranged in a loop rather than forming a hierarchy of competitive interactions (Levine et al. 2017). Intransitive competition can result from community assembly through trait evolution (Gallien et al. 2018), posed as either a stabilising or destabilising force. Compared to direct interactions, interaction chains involve more species and consider a more specific context. However, interaction chains have not been widely invoked in invasion hypotheses but should be when the multiple or indirect interactions are involved in invasion hypotheses. Interaction intransitivity (akin to the rock–paper–scissors game) can facilitate species coexistence (Allesina and Levine 2011; Grilli et al. 2017; Wandrag and Catford 2020; Yang and Hui 2021). Alien species that can foster intransitive loops in the recipient ecosystems, either directly or indirectly, can improve their chance of invasion success and of having major invasion impact (Gaertner et al. 2014). Together, we could see the three forces at play that steer species coexistence (Figure 2.7), while an introduced species needs to establish by satisfying its invasion criterion, through possessing fitness advantage over and separated niches from resident species and by fostering interaction intransitivity, to be successfully integrated as part of the invaded ecosystem.

As mentioned previously, higher-order interactions can also have a great effect, largely through facilitation, on community stability and species coexistence (Bairey et al. 2016; Grilli et al. 2017; Mayfield and Stouffer 2017). The presence of higher-order interactions represents the formation of hyper-networks that contain hyper-links connecting more than two nodes (Golubski et al. 2016); including such higher-order interactions can improve our predictive power in explaining species' performance in communities (Letten and Stouffer 2019). Almost all species are involved simultaneously in both mutualistic and antagonistic interactions (Strauss et al. 2006). For instance, herbivory

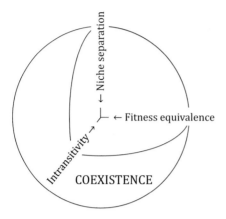

Figure 2.7 Multiple coexistence mechanisms for explaining species coexistence and the invasion performance in a community. Coexistence and invasion can be assured within the sphere along three dimensions each representing one specific mechanism. For species coexistence through mutual invadability, fitness equivalence (or reduced fitness difference) is necessary, while for biological invasions fitness advantage is emphasised over fitness equivalence.

can affect pollination in diverse ways (Lucas-Barbosa 2016): it can reduce pollination through altering floral display, by activating chemical defences and by decreasing the availability and quality of rewards for pollinators (Adler et al. 2006). In contrast, root herbivory has been shown to increase pollinator visits to flowers of the mustard *Sinapis arvensis* (Poveda et al. 2005). In addition, secondary metabolites of plants induced by herbivores can serve as antibiotics for pollinators, thereby enhancing pollination and plant reproduction (Richardson et al. 2016). Reciprocally, pollination can also affect herbivory. Several floral traits for pollination (e.g., corolla size and colour, nectar and floral scent) have been shown to attract herbivores (Adler et al. 2006). Higher-order interactions could result in most invasion hypotheses that rely on direct interactions failing to capture the interaction complexity. However, although the strength of higher-order interactions can have stronger impacts on network stability than the strength of lower-order interactions, as a result of the polynomial nature of the Taylor expansion in Eq. (2.5) we need to consider the encounter rates of interactions for different orders. We foresee that further development in invasion models that include higher-order interactions will improve our ability to predict invasion performance.

2.4 Interaction Rewiring

The recipe for a winning strategy in games touches on issues in a related debate on how natural selection is realised during evolution. As discussed previously, evolution works on inheritable traits and drives traits to change adaptively in diverse ways. Such adaptive changes are in the direction of winning traits, since individuals possessing winning traits are becoming fixed and/or replacing individuals with other traits over generations. The explicit mechanism for evolution was proposed as the *survival of the fittest* by Herbert Spencer (1864) and endorsed by Darwin (1868). In 1888, C. Lloyd Morgan criticised the problem of interpreting natural selection as the survival of the fittest and highlighted the fact that,

in artificial selection it is almost invariably the fittest which are chosen out for survival, it is not so under Nature; the "survival of the fittest" under Nature being in the main the net result of a slow and gradual process of the elimination of the unfit. The well-adapted are not selected; but the ill-adapted are rejected; or rather, the failures are just inevitably eliminated.

Indeed, in a multiplayer game it is clear that rewarding a winning player/ strategy that performs particularly well does not necessarily lead to the same outcome as penalising the losing player/strategy. Smith (2012) revived this history and supported Morgan's (1888) advocacy of using Russel Wallace's (1855) *elimination of the unfit* as a more practical means of evolution. Indeed, in 1877 Wallace wrote explicitly that,

Natural Selection ... does not so much select special variations as exterminate the most unfavourable ones.

As we can see, both WSLS and the elimination of the unfit can be achieved via adaptive switching and rewiring of interaction partners (Figure 2.8). Adaptive interaction switching can occur at three different levels (Hui and Richardson 2017). At the individual level, it is important to choose what to interact with (e.g., habitat or diet selection) or what to avoid (e.g., applying anti-predation strategies). Such preferential interactions could simply arise from optimal or adaptive foraging where individuals aim to maximise their energy intake rate while min- imising the risk of starvation and predation (Zhang and Hui 2014). Consequently, interaction switching can occur rapidly for individuals at a pace even faster than the typical ecological time scale (e.g., host switching in parasites; van Baalen et al. 2001). At the population level, successful interactions often require involved species to possess a certain

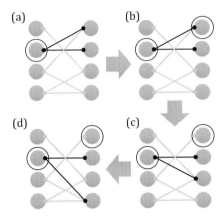

Figure 2.8 Interaction switching via Wallace's elimination of the unfit. At a particular time, a species (node) in a examines its interaction partners and identifies the one contributing the least to its payoff (or fitness) (node with an added circle in c). The species could choose to avoid interacting with this partner and try to interact with a new partner (the newly connected node in c; implemented in Zhang et al. 2011). In other implementations (e.g., Suweis et al. 2013), the focal species will further assess whether its payoff has actually increased after switching to the new partner and will only consolidate the interaction if doing so increases fitness; if not, the species will try out another new partner until fitness increment is assured (new link in d).

level of matching or complementary traits, such as a matching phenology for flowering time in plants, and the foraging activity of insect pollinators (Waser 2015). By adaptively changing the level of trait complementarity and plasticity, populations can adjust interaction strength to their own advantage (Santamaría and Rodríguez-Gironés 2007). At the species level, species can completely lose or regain the possibility of interactions by developing incompatible traits of forbidden links and setting up barriers to exploitation (Jordano et al. 2003).

Adaptive interaction switching occurs when the quantity and quality of available resources changes. Consumers prefer to select highly profitable resources rather than consuming all resources available to them as specified in optimal foraging theory (Stephens and Krebs 1986). Concurrently, consumers also prefer to exploit abundant resources over rare ones to minimize risk (Fossette et al. 2012). Psychologists explain the behaviour of decision making in animals by applying concurrent schedules of reinforcement (e.g., using a Skinner box) and have proposed the momentary maximizing rules of hill climbing and melioration as possible explanations for diet choice (Staddon 2010). However, these rules are sensitive to the specific experimental environment, and their capacity in achieving optimisation has not been verified (Herrnstein and Vaughan 1980; Hinson and Staddon 1983; Vaughan and Herrnstein 1987). Behavioural economists attempt to explain how animals make choices by analysing irrational decisions (e.g., context-dependent choice, state-dependent behaviour and heuristic decision making; Gigerenzer and Gaissmaier 2011; Rosati and Stevens 2009). Neuroscientists seek neural mechanisms of decision making in foraging (Basten et al. 2011; Kolling et al. 2012). Based on these observations, Zhang and Hui (2014) assume that a consumer can (1) recognise the encountered resources (similar to the assumption in optimal diet model); (2) memorise the profitability or utility of resources recently consumed (providing a reference point); and (3) perceive time elapsed since last consumption (an estimate of searching cost or hunger state). This leads to a recent experience-driven (RED) behaviour strategy for decision making during resource consumption and foraging. Importantly, the simple behavioural rule for decision making can cope with changing environment and predict an optimal energy intake rate. Such RED trial-and-error learning underpins interaction rewiring in uncertain and fluctuating environments and again underscores the gaming and learning nature of biotic interactions.

Over short timescales, the nature and strength of cross-trophic interactions depend largely on preferences between prey and on predator behaviour patterns, a phenomenon known as prey switching. In prey switching, the number of attacks on a species is disproportionately large when the species is abundant relative to other prey, and disproportionately small when the species is relatively rare (Murdoch 1969). Jaworski et al. (2013) evaluated the prey switching of the generalist predator *Macrolophus pygmaeus* when it simultaneously encounters the silverleaf whitefly, *Bemisia tabaci*, and the leaf-mining moth, *Tuta absoluta*. Prey switching was observed in both adult and juvenile predators: they over-attacked the most abundant prey when the mixed prey population was biased in favour of either prey species. Some of the prey characteristics that influence predator preference are the nutritional quality of the prey and the ease of attack (Eubanks and Denno 2000). Predation on prey of the highest nutritious value increases the predator's fitness. Capture success generally depends on prey mobility and access to a refuge (enemy-free space) (Fantinou et al. 2009).

Mounting evidence suggests that species often switch their interaction partners not only in antagonistic interactions (i.e., food webs; Murdoch 1969; Staniczenko et al. 2010) but also in mutualistic networks (Basilio et al. 2006; Fortuna and Bascompte 2006; Olesen et al. 2008; Petanidou et al. 2008). For instance, in a pollination network, pollinators continually switch the plant species with which they interact in response to environmental disturbances and the availability of resources (Whittall and Hodges 2007). Plants can also adjust adaptively their phenology (e.g., flowering time) and morphology (e.g., flower heterostyly) to affect their pollinators (Aizen and Vazquez 2006; Kaiser-Bunbury et al. 2010). This implies a dynamic nature of ecological networks; that is, both species abundance and species interaction could affect each other and change over time. Such interaction switches might not have exclusively ecological and environmental reasons (e.g., resource availability) but could also reflect adaptive behaviour of species for enhancing the efficiency of resource utilization (Zhang et al. 2011). Compared with fixed interactions between species, an interaction switch (or alternatively the rewiring of interactions between species) can lead to greater stability in food webs (Staniczenko et al. 2010) and pollination networks (Kaiser-Bunbury et al. 2010).

Without co-evolved mutualists, interaction promiscuity is essential for invasion success, especially for compulsory mutualists (Aizen et al. 2012; Heleno et al. 2013). For instance, alien plants may benefit from forming novel mutualisms with native soil microbes capable of forming associations with a wide range of hosts, thereby facilitating establishment and spread (Reinhart and Callaway 2006), as demonstrated by the similar nitrogen-fixing bacterial communities in invasive legume nodules across invaded and native ranges (Birnbaum et al. 2016). Interaction promiscuity and adaptive rewiring can have a profound influence on the stability of ecological networks, and models implementing adaptive rewiring can explain large variations in many non-random patterns (e.g., modularity and nestedness) in both mutualistic and antagonistic networks (Zhang et al. 2011; Nuwagaba et al. 2015; Nnakenyi et al. 2019).

Several hypotheses on the effect of biotic interaction on invasion performance (e.g., host jump, new associations, enemy release and EICA; see later sections) have been developed based on the behaviour of adaptive interaction switching. This means that successful invaders, especially those requiring high levels of symbiotic specificity (Le Roux et al. 2017), often rely on the co-introduction of co-evolved mutualists to perform well in their invaded ranges (Traveset and Richardson 2014; Mohammed et al. 2018). The missed mutualism hypothesis (Colautti et al. 2004; Mitchell et al. 2006) predicts the opposite of the enemy release hypothesis; it posits that species benefitting from mutualistic interactions in their native range must establish a similar type of positive interaction in the introduced range to invade. In that case, specialised interactions would prevent or limit the ability of an alien species to invade. Missed mutualism is one of the main reasons for invasion failure by alien plants (Traveset and Richardson 2014). However, widespread host generalism appears to make this mechanism relatively rare in nature, for plants and for other taxa (Mitchell et al. 2006).

2.5 Interaction Kernel

Both species coexistence and interaction strength are arguably the functional outcome of biotic interactions, and therefore using one (interaction strength) to explain the other (species coexistence) is phenomenological by nature, and does not reveal the underlying mechanistic forces at work. Although this phenomenological framework is very helpful for clarifying the feasible network structures and the associated relationship between network complexity and stability (May 1972; Allesina and Tang 2012; see Chapter 4), we need to unpack the mechanistic and functional natures of a biotic interaction. In particular, we have only focused on the concept of interaction strength. We have not explained what makes the strength of an interaction high or low. A crucial step in contextualising biotic interactions is to introduce the key concept of niche. In Gause's (1932) competitive exclusion principle, the outcome of two competitively interacting species with similar niches is that one is extirpated. Ecologists battle with the concept of niche. While Chesson (2000) devised a way of measuring niche overlap based on interaction strength coefficients, this approach remains phenomenological in nature and cannot generate metrics of niche overlap for complex networks.

In his seminal paper 'Homage to Santa Rosalia', Hutchinson (1959) pondered the ubiquitous upper bound of 1.3:1 in morphological similarity between closely related species. In 1967, Hutchinson's student Robert MacArthur and Richard Levins, a close colleague of Richard Lewontin and E.O. Wilson, formally addressed this issue and coined the limiting similarity theory, which connected the outcome of biotic interaction to niche similarity (MacArthur and Levins 1967). As demonstrated earlier, the outcome of biotic interactions (species coexistence or exclusion) depends on interaction strength between species, while interaction strength depends on niche similarity. MacArthur and Levins (1967) connected the dots between phenomenon and functionality of biotic interactions by specifying interaction strength as the degree of overlapping between two species' resource utilisation curves. This leads to a normal-shaped niche-based *interaction kernel* of the competition coefficient, $\alpha_{ij} = \exp\left(-d_{ij}^2/(2\sigma^2)\right)$. This opens up a trait paradigm in community ecology. To have a fully mechanistic and functional understanding of biotic interaction, we need to walk a step beyond the Hutchinsonian niche (Holt 2009) to embrace the trait paradigm in community ecology (McGill et al. 2006). In particular, we need to clarify the concept of *interaction kernel* – interaction strength as a function of the traits of involved species.

Invoking interaction kernels opens the door to identifying trait positions in an ecological network to elucidate levels of invasion and demographic performance. Different types of interactions can possibly be synthesised into an overarching formula, with the trait-mediated per-capita interaction strength expressed as the Gaussian-type *interaction kernel* of the trait difference between two interacting species (Doebeli and Dieckmann 2000; Hui et al. 2018a),

$$\alpha\left(x_i, x_j\right) = \pm \exp\left(-\frac{\left(x_i - x_j - \mu\right)^2}{2\sigma^2}\right), \qquad (2.12)$$

for a competitive interaction, the sign of the interaction strength is negative. For mutualism, the sign is positive. For antagonistic interactions (e.g., species i is the predator, while species j the prey), the sign of $\alpha(x_i, x_j)$ is positive for the predator, whereas it is negative for the prey. When $\mu = 0$ (Figure 2.9a), it describes a symmetric interaction ($\alpha(x_i, x_j) = \alpha(x_j, x_i)$) and implies a stronger interaction between species with more similar traits. With $\mu > 0$ there is an optimal trait difference between the two species (e.g., in terms of the optimal body size ratio between a predator and its prey; Figure 2.9b). In this case, the interaction becomes asymmetric ($\alpha(x_i, x_j) \neq \alpha(x_j, x_i)$). The coefficient σ represents the width of the interaction kernel and depicts the reciprocal of interaction specialisation. A small σ implies that interactions only effectively occur between species with similar traits (specialised interactions), whereas a large σ suggests that species with large trait differences can still interact (generalised interactions; compare Figure 2.9a and b).

Other interaction kernels have been proposed, largely for asymmetric interactions. In asymmetric competition, the interaction kernel is defined as (Figure 2.9c; Gallien et al. 2018), $\alpha(x_i, x_j) = \exp\left(-(x_i - x_j)^2/2\sigma^2\right)/\exp\left(\beta(x_i - x_j)\right)$, where σ and β describe the

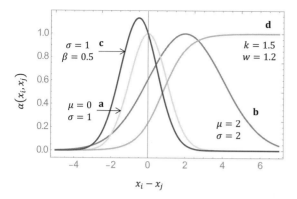

Figure 2.9 A few interaction kernels, describing interaction strength $\alpha(x_i, x_j)$ as a function of trait difference $(x_i - x_j)$. Kernel functions a and b are according to Eq. (2.12), with a portraying a symmetric interaction and b asymmetric interaction. Other two interaction kernels (c and d) are also for asymmetric skewed interactions (see text for formulae).

breadth and asymmetry of competition, respectively. A greater value of σ extends niche overlap and thus increases the competition coefficient α. A greater value of β gives more advantage to the larger competitor over the smaller one, while the intraspecific competition coefficient remains at unit, meaning that the population size reaches its carrying capacity in the absence of interspecific competition.

In mutualistic interactions, such as those between insect pollinators and flowering plants (Pauw et al. 2009), the benefit to a pollinator is normally quantified by the proportion of nectar consumed per visit (assuming the independence of nectar quantity to floral specifics), and the benefit to a flowering plant is measured by the probability of successful pollination. When the proboscis becomes longer in relation to the floral tube, the benefit to the pollinator increases, whereas the benefit to the flower declines. Such a skewed interaction gives the advantage to elongated traits and has been described as a logistic equation (Figure 2.9d; Zhang et al. 2013), $\alpha(x_i, x_j) = \left(1 + \exp(-k(x_i - x_j) + w)\right)^{-1}$, with x_i and x_j denoting the lengths of pollinator proboscis and floral tube, respectively. Interaction kernels have been widely used in evolutionary ecology to document trait-dependent performance and fitness components between co-evolving species (Bergelson et al. 2001; Toju and Sota 2006; Nuismer et al. 2010). Consequently, the range of traits in an interaction kernel normally covers only a portion of the full range of trait values for a species.

Interspecific interaction kernels between all species of an entire ecological network are yet to be reported in literature.

Trait-mediated interaction kernels can be further conceptualised as the evolutionary field over the trait space (Wilsenach et al. 2017). In the Newtonian law of gravitation, two objects interact with each other as a result of the gravitational force that pulls them towards each other, $F = g \cdot m_1 m_2 / d_{12}^2$, where F is the gravitational force acting between two objects; m_1 and m_2 the masses of the two objects; d_{12} the distance between the two objects; and g the gravitational constant. Field is a conjectured concept or tool to simplify calculation of physical forces (i.e., interactions) between multiple objects. For instance, the gravitational field can be measured as the gravitational pull felt by a small object at each point in space, while the small size of this reference object is to ensure that its own gravitational force is negligible (not interfering with the existing field in space); the collective field from multiple objects is simply the sum of each (Weinberg 1977). To see the connection clearly, we can define the functional distance between two species in the evolutionary field as the reciprocal of the per-capital interaction strength $d_{ij}^2 = 1/|\alpha_{ij}|$. For instance, consider that species i is located at (x_i, y_i) in the two-dimensional trait space, the functional distance for symmetric interactions with unit kernel width ($\mu = 0$ and $\sigma = 1$ in Eq. (2.12)) is simply $d_{ij}^2 = \exp(E_{ij})$, where $E_{ij} = (x_i - x_j)^2 + (y_i - y_j)^2$ represents the Euclidean distance of the two species in the trait space. The interaction strength of the evolutionary field in an ecological network can be defined, in a similar way to the gravitational field, as the following (Wilsenach et al. 2017),

$$\Phi_i = \sum_{j=1}^{S} g_{ij} \frac{n_i n_j}{d_{ij}^2}, \tag{2.13}$$

where g_{ij} considers the sign and direction of the field as a result of the interaction between j and i in the trait space. With the field strength defined, its resulted evolutionary dynamics and acceleration of traits can be explored in the trait space (Wilsenach et al. 2017). Interaction kernels are, therefore, explicit and mechanistic tools within the trait paradigm with great potential to explore the dynamics of biotic interactions in an ecological network. This explicit tool, as will be shown in later chapters, further elucidates the role of alien and resident traits in ecological networks and the demographic performance of each species.

2.6 Co-evolution of Traits

While examining the Madagascan orchid *Angraecum sequipedale*, Darwin (1862) speculated on the phenomenon of co-evolution and anticipated a co-evolving species, a moth with a 30 cm long proboscis that could be responsible for the pollination of this orchid species. The two co-evolving species, as Darwin (1862) put it,

might slowly become, either simultaneously or one after the other, modified and adapted in the most perfect manner, by the continued preservation of individuals presenting mutual and slightly favourable deviation of structure.

In 1903, such an animal (the sphinx moth *Xanthopan morganii praedicta*) was discovered and was found to match exactly Darwin's prediction.

Co-evolution can be considered as resulting directly from different types of biotic interactions. For instance, mutualistic interactions between pollinators and flowering plants, between seed plants and seed dispersers, and from antagonistic interactions between predators and their prey, between hosts and their parasites (and pathogens). As portrayed in an evolutionary game, two species are engaged in an arms race through the interaction of their functional traits that affect each other's fitness. Such specific co-evolution typically leads to matched traits through convergence evolution in mutualistic systems and evolutionary cycles known as Red Queen dynamics in antagonistic systems (see examples in this section). In diffuse co-evolution, several species from a functional guild affect each other's fitness by means of their own evolutionary changes. In escape-and-radiate co-evolution, the interaction between species enables one or both species to radiate into a diverse clade.

The numbers of clear examples of adaptive diversification due to co-evolution are increasing (Thompson 2012). One such case has been detected in Darwin's race between a long-proboscid fly *Moegistorhynchus longirostris* of the Nemestinidae family and a long-tubed iris *Lapeirousia anceps* of the family Iridaceae (Figure 2.10). In this arms race, effective feeding occurs when proboscis length exceeds floral tube length because the pollinator is then able to drain all the nectar from the flower. In contrast, effective pollination occurs when floral tube length exceeds proboscis length because this ensures sufficient contact with the stigma and anthers near the entrance of the floral tube (Pauw et al. 2009). These two co-evolving traits thus impose reciprocal directional selection on each other, leading to an escalating arms race. Imbalanced costs to trait elongation, imposed by physiological constraints under environmental variation,

Figure 2.10 Darwin's race between the long-proboscid fly *Moegistorhynchus longirostris* and the long-tubed iris *Lapeirousia anceps* in South Africa. Photo courtesy of A. Pauw.

can trigger divergent selection and trait dimorphism in high-cost species, specifically, in some iris populations (Zhang et al. 2013).

The evolution and co-evolution of quantitative traits or phenotypic characters can be explored using the analytic tool of adaptive dynamics developed in the 1990s by game theorists (e.g., Nowak and Sigmund 1990; Metz et al. 1992; Abrams et al. 1993; Dieckmann and Law 1996). Adaptive dynamics portrays evolutionary changes as being induced by rare and small mutations when fitness is both trait mediated and density dependent (Waxman and Gavrilets 2005). It extends frequency-dependent selection in evolutionary game theory to *density*-dependent selection, and can be considered the evolutionary game theory for continuous traits. A key concept of adaptive dynamics is invasion fitness, defined as the exponential growth rate of a vanishing mutant in a resident population sitting at its demographic equilibrium (Metz et al. 1992), on which the selection gradient that drives trait evolution can be quantified based on underlying ecological interactions, as well as standard mutation and selection processes. This allows the species to move in a dynamic fitness landscape, where diverse evolutionary dynamics can emerge, such as evolutionary branching, evolutionary suicide and traps (Zhang et al. 2013), as well as Red Queen dynamics (Dercole and Rinaldi 2008). To this end, the so-called canonical equation of adaptive dynamics describes the evolution of traits under directional selection through the continuous invasion of rare mutants into resident populations (Metz et al. 1992; Dieckmann and Law 1996). However, the strength of adaptive dynamics is perhaps revealed most clearly by its ability to illuminate how evolution can push traits to evolve towards either fitness maxima or minima, where the latter could trigger disruptive selection and thus sympatric diversification and speciation via evolutionary branching (Geritz et al. 1997, 1998).

To elucidate trait evolution within the framework of adaptive dynamics, we need to accomplish three steps. First, we need to specify interaction kernels (i.e., trait-mediated interaction strengths). Second, we must define the resident–mutant competition model. The final step involves evaluating the per capita mutant growth rate as the invasion fitness, when the mutant population size is negligible and the abundances of resident populations (and other species) are at their equilibria. Using the interaction kernel $\alpha(x_i, x_j)$ defined in Eq. (2.12), and a bell-shaped resource utilisation curve, $k(x_i) = k_0 \exp\left(-(x_i - x_i^*)^2/(2\sigma_i^2)\right)$, we can recast the Lotka–Volterra model (Eq. (2.6)) as follows,

$$\dot{n}(x_i) = r_i n(x_i) \left(1 - \sum_j \frac{\alpha(x_i, x_j) n_j}{k(x_i)} \right). \tag{2.14}$$

Following the trait paradigm, interaction strengths and demographic components of a species depend on its trait in relation to those of other species in the network; for simplicity, we assume that all species have the same intrinsic rate of growth, $r_i = r$. Co-evolution from biotic interactions can drive the trait x_i to adaptively change along the selection gradient.

This adaptive change in species i's trait is achieved through the typical evolutionary process. First, when an ecological network resides at its equilibrium, a small number of mutants emerge in species i with a slightly different trait value, denoted as x_i', normally through mutations in some new-born individuals that did not have identical traits to their predecessors. The fitness of an individual possessing this new mutant trait is the per-capita relative growth rate of these mutants, known as *invasion fitness*.

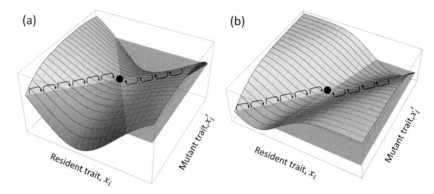

Figure 2.11 Invasion fitness $f(x_i, x_i')$ as a function of resident trait x_i and mutant trait x_i', typically known as a pairwise invadibility plot. Parameters: $\mu = 0$ (symmetric interaction); $\sigma_i = 1$ (width of resource utilisation kernel); $x_i^* = 0$ (optimal resource trait for resource utilisation; indicated by black dots); $\sigma = 1.5$ in a and $= 0.5$ in b (width of interaction kernel). Small arrows indicate successful incremental invasion and replacement of resident trait by mutant trait. For a resident trait value, incremental mutations can occur along a corresponding light grey curve within a close range of the diagonal ($x_i = x_i'$).

For instance, when there is only one species in the system, the population size will settle at the carrying capacity $n_i = k(x_i)$. We can therefore have the following invasion fitness (Figure 2.11),

$$f\left(x_i, x_i'\right) = \frac{1}{n_i'} \frac{dn_i'}{dt}\bigg|_{n_i'=0, n_i=k(x_i)} = r\left(1 - \frac{\alpha\left(x_i', x_i\right)k(x_i)}{k(x_i')}\right). \qquad (2.15)$$

Similarly, we see that the fitness, per-capita relative growth rate, for the resident population is simply $f(x_i, x_i) = 0$ at its equilibrium. Second, if invasion fitness is greater than resident fitness, $f\left(x_i, x_i'\right) > f(x_i, x_i)$, the mutant trait can become competitively dominant and replace the resident trait in the population, changing the trait of the species from x_i to x_i'; otherwise, the mutant cannot establish and will be expelled by the resident trait, with the trait of the species remaining unchanged.

This two-step evolutionary process can be visualised by the so-called pairwise invasibility plot (Figure 2.11). We see that for trait values less than the optimal resource utilisation value ($x_i < x_i^*$), when the mutant happens to be greater than the resident trait ($x_i' > x_i$), the invasion fitness is above zero, $f\left(x_i, x_i'\right) > 0$ and therefore the mutant can invade and replace the resident trait, indicated by a small arrow. Instead, mutants with smaller trait ($x_i' < x_i$) will experience negative invasion fitness and thus fail to invade. Consequently, if the trait value of the resident population is less than the optimal resource utilisation trait value, evolution will push the trait to increase and approach the optimal trait value. Trait evolution, as described here, proceeds with the succession of mutant–resident substitutions along the *selection gradient* (Dieckmann and Law 1996); in our example, it is simply the partial derivative (tangent) of invasion fitness along the mutant trait direction,

$$g(x_i) = \left.\frac{\partial f(x_i, x_i')}{\partial x_i'}\right|_{x_i'=x_i} = -r\left(\frac{x_i - x_i^*}{\sigma_i^2}\right). \tag{2.16}$$

Evidently, when $x_i < x_i^*$, the selection gradient is positive, propelling the trait of the species to increase; when $x_i > x_i^*$, the selection gradient is negative, leading to a reduction of the trait. The steeper the selection gradient is, the faster the speed of trait evolution, described by the canonical equation of adaptive dynamics (Dieckmann and Law 1996),

$$\dot{x}_i = \frac{1}{2} \mu_i s_i^2 \hat{n}_i(x_i) g(x_i), \tag{2.17}$$

where μ_i and s_i^2 are per-capita mutation rate and variance of mutational steps (assumed to be normally distributed with zero; $\hat{n}_i(x_i)$ is the abundance equilibrium of the resident population (in our example $\hat{n}_i(x_i) = k(x_i)$)), and the factor $1/2$ expresses that half of the mutations are lost due to negative fitness (Figure 2.11).

Trait dynamics are driven by the selection gradient. Once the selection gradient diminishes ($g(x_i^*) = 0$), directional selection will halt at the evolutionary singularity. The evolutionary dynamics converge to such a singular trait (black dots in Figure 2.11), since the eigenvalue of the system Jacobian for the trait dynamics is negative, $\partial \dot{x}_i/\partial x_i|_{x_i=x_i^*} = -\frac{1}{2}\mu_i s_i^2 k_0 r/\sigma_i^2 < 0$. This attractive trait of an evolutionary singularity is called convergence stable. In Figure 2.11a, after the trait has reached the convergence stable singularity (black dot), we see that the resident trait now sits at a fitness maximum (the light grey curve through this singularity goes downwards in both directions). This suggests that any mutants in the vicinity of this singularity have negative invasion fitness and therefore cannot invade to replace the resident trait; such an uninvadable trait is called an evolutionarily stable strategy (ESS). A trait singularity that is both convergence stable and evolutionarily stable is called a continuously stable strategy (CSS; Eshel 1983).

A convergence stable singularity does not necessarily indicate the end of trait evolution. In Figure 2.11b, we see that the resident trait traverses and reaches the singularity but finds it sits at a fitness minimum. Any mutants in the vicinity of the singularity can invade (Geritz et al. 1998), creating the precondition for sympatric diversification – disruptive selection (Eshel 1983; Christiansen 1991). This then begets evolutionary branching if the resident and mutant traits can coexist, known as protected dimorphism. That is, in systems with a single adaptive trait, evolutionary branching requires two necessary conditions,

$$\frac{\partial^2 f\left(x_i^*, x_i'\right)}{\partial x_i'^2}\bigg|_{x_i'=x_i^*} > 0 \text{ and } \frac{\partial^2 f\left(x_i, x_i'\right)}{\partial x_i \partial x_i'}\bigg|_{x_i=x_i'=x_i^*} < 0. \qquad (2.18)$$

The first condition ensures the curvature of the light grey curve through the singularity in Figure 2.11b is positive, which implies disruptive selection at this singular trait (i.e., evolutionary instability of a fitness minimum; the green curve tilts upwards). The second condition implies resident–mutant coexistence (i.e., the protection of dimorphism) according to mutual invadability condition (see Section 2.3; Geritz et al. 1998); in other words, the lack of limiting similarity. In a monomorphic case, the sum of the left side of the two conditions equals the derivative of the selection gradient at the singular trait, which means convergence stability plus evolutionary instability imply the protection of dimorphism in monomorphic populations, and thus evolutionary branching. In the example just shown, the only necessary and sufficient condition for evolutionary branching at the monomorphic convergent-stable singular strategy is therefore,

$$\frac{\partial^2 f\left(x_i^*, x_i'\right)}{\partial x_i'^2}\bigg|_{x_i'=x_i^*} = -r\left(\frac{1}{\sigma_i^2} - \frac{1}{\sigma^2}\right) > 0. \qquad (2.19)$$

This means that the kernel width for resource utilisation needs to be greater than the width of interaction kernel, $\sigma_i^2 > \sigma^2$. In such a case, the fitness gain in reducing competition via differentiating traits is greater than the fitness loss from diminishing resource acquisition; consequently, limiting similarity between two competitive species disappears, which allows the resident and the mutant with very similar traits to coexist.

We can classify all possible monomorphic singular strategies according to their convergence and evolutionary stability and resident–mutant coexistence (Figure 2.12). In a polymorphic case, convergence stability is independent from evolutionary stability and dimorphism protection,

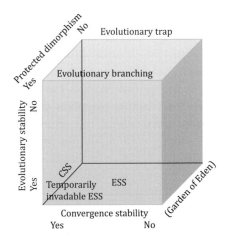

Figure 2.12 Classification of monomorphic singular strategies according to convergence stability, protected dimorphism and evolutionary stability. ESS: Evolutionarily Stable Strategy. CSS: Continuously Stable Strategy (convergence-stable ESS). At a temporarily invadable ESS resident and mutant can coexist but eventually either one of the two will become extinct or their traits will both converge back at the ESS. An evolutionary trap has the potential for diversification under disruptive selection, but resident and mutant cannot coexist. The Garden of Eden represents an unreachable (convergence unstable) fitness maximum. Based on Hui et al. (2018).

and thus all combinations are possible (Hui et al. 2018b; see Chapter 4). Much like the three forces at work in mediating species coexistence, trait-mediated biotic interactions also create three evolutionary mechanisms that steer trait evolution in an ecological network. Thus, a successful invasion of a mutant trait into a resident population implies resident substitution, whereby the mutant traits substitute and expel the resident population, thus becoming the new resident trait; in the fitness landscape this represents an incremental change of the resident trait and, also importantly, a slight change of the fitness landscape. For this reason, the concept of fitness is known as adaptive (Metz et al. 1992), and the formulation of invasion fitness allows us to study evolutionary trait dynamics using the canonical equation (Dieckmann and Law 1996), essentially describing the selection gradient at the current resident trait value.

After the first evolutionary branching, the system can further converge to a new singularity where subsequent evolutionary branching could follow.

As such, co-evolution from biotic interactions is an important source of diversification (Thompson 2012) and the default assumption for the origin and co-adaptation of biotic interactions (Ehrlich and Raven 1964). In principle, trait evolution can be numerically analysed following two iterative steps (Figure 2.13), (1) solving the canonical equations for the period between two evolutionary singularities. and (2) analysing the feasibility of evolutionary branching at a singularity (if yes, replace the current dynamical system with a higher dimensional system taking into account the population and trait dynamics of established mutant populations; if no, continue the simulation with the current system). The procedure of adaptive dynamics can be extended further to consider other higher-order evolutionary phenomena such as evolutionary inertia (Wilsenach et al. 2017).

Typical methods of evolutionary invasion analysis, such as adaptive dynamics, focus on trait means while often ignoring the variation of traits (Dieckmann and Law 1996; Champagnat et al. 2001). However, studying only the evolution of average traits may overlook many important ecological and evolutionary features, as trait diversity can affect the outcome of ecological interactions (Bolnick et al. 2011). For instance, ignoring trait variation can lead to the underestimation of spreading velocity in many invasive species (Ramanantoanina et al. 2014). Moreover, adaptive dynamics of mean traits often rely on the separation of ecological and evolutionary timescales. This assumption seems too strict in view of observations that ecological and evolutionary processes can occur over similar timescales (Yoshida et al. 2003; Jones et al. 2009; Koch et al. 2014). To this end, some recent developments have synthesised a number of theoretical frameworks into the fundamental theorems of evolution based on the Price equation (Queller 2017; Lehtonen 2018) as an internode to connect many evolutionary theories, such as Fisher's equation and the Breeder's equation, as well as the canonical equation of adaptive dynamics.

Let there be individuals with different traits in a population, with $n_{i,t}$ individuals having the trait $x_{i,t}$ at generation (or time) t. Fitness for individuals with trait $x_{i,t}$ can be defined as the ratio between the offspring and parent individuals of this type, $w_{i,t} = n_{i,t+1}/n_{i,t}$. The mean trait value of the population at generation t is $\bar{x}_t = \sum_i x_{i,t} n_{i,t} / \sum_i n_{i,t} \equiv \mathrm{E}(x_{i,t})$, and the mean fitness is $\bar{w}_t = \sum_i w_{i,t} n_{i,t} / \sum_i n_{i,t} = \sum_i n_{i,t+1} / \sum_i n_{i,t} \equiv \mathrm{E}(w_{i,t})$. According to standard statistical probability theory, we can calculate the covariance between the fitness and the trait as follows, $\mathrm{cov}(w_{i,t}, x_{i,t}) = \mathrm{E}(w_{i,t} x_{i,t}) - \mathrm{E}(w_{i,t})\mathrm{E}(x_{i,t})$. We also have the linear operator of

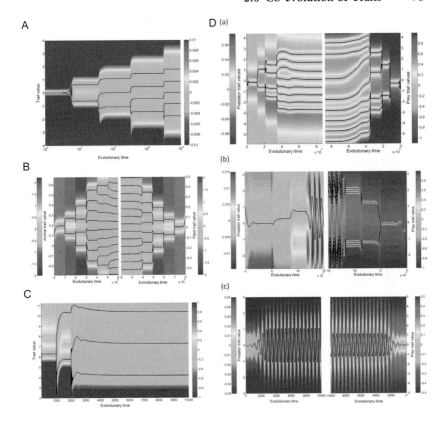

Figure 2.13 Illustrations of trait evolution from co-evolving biotic interactions. Black curves represent traits (vertical axis) in the system over evolutionary time (horizontal axis); background colour represents the invasion fitness. A: Trait diversification under resource competition, starting from a monomorphic population and gradually diversifying into five different morphs. B: Trait diversification under mutualistic co-evolution, starting from a monomorphic population and gradually diversifying into twelve different morphs on each side of the trophic. C: Trait diversification in a food web, starting from a monomorphic heterotroph population and diversifying into different morphs. D: Trait evolution under antagonistic co-evolution, representing trait diversification from antagonistic co-evolution (a) and Red Queen co-evolutionary cycle of prey and predator in polymorphic (b) and monomorphic (c) systems populations. See model details and parameters in Hui et al. (2018), reproduced with permission. *A black and white version of this figure will appear in some formats. For the colour version, please refer to the plate section.*

expectation, $\mathrm{E}(w_{i,t} \cdot \Delta_t x_{i,t}) = \mathrm{E}(w_{i,t} x_{i,t+1}) - \mathrm{E}(w_{i,t} x_{i,t})$. As such, we have $\mathrm{cov}(w_{i,t}, x_{i,t}) + \mathrm{E}(w_{i,t} \Delta_t x_{i,t}) = \mathrm{E}(w_{i,t} x_{i,t+1}) - \mathrm{E}(w_{i,t}) \mathrm{E}(x_{i,t})$. Note, $\mathrm{E}(w_{i,t} x_{i,t+1}) = \mathrm{E}(w_{i,t}) \mathrm{E}(x_{i,t+1})$. Therefore, we have the Price equation,

$$\Delta_t \bar{x}_t = \left(\mathrm{cov}(w_{i,t}, x_{i,t}) + \mathrm{E}(w_{i,t} \Delta_t x_{i,t}) \right) / \bar{w}_t. \tag{2.20}$$

The covariance $\mathrm{cov}(w_{i,t}, x_{i,t})$ captures the direct effect of natural selection: if the trait and the fitness are positively correlated, the mean trait is then expected to increase in the population, and vice-versa. The second term $\mathrm{E}(w_{i,t} \Delta_t x_{i,t})$ represents biased transmission: that is, the fitness advantage of phenotype i can be diluted due to imperfect heritability. Without transmission bias and a focus on fitness as the selected phenotype ($x = w$) one could recover Fisher's fundamental theorem of adaptation, $\Delta \bar{w} = \frac{1}{\bar{w}} \mathrm{Var}(g(w_i))$, where g is the individual breeding value, i.e., the proportion of its trait (in this case, fitness w_i) which is passed on to its offspring. In other words, Fischer's fundamental theorem of adaptation states that the change in the average fitness in a population is proportional to the additive variance in fitness. Using Taylor expansion of the fitness and ignoring transmission bias ($\Delta_t x_{i,t} = 0$), we can further connect the Price equation with gradient dynamics, continuous trait game theory and adaptive dynamics, specifically the selection gradient, singular points, convergence and evolutionary stability, and the canonical equation (Lehtonen 2018). Overall, these sets of models have highlighted the fitness (pay-off) and abundances (frequencies) of different traits as the key variables for depicting evolutionary dynamics.

Evolutionary distribution is another numerical framework for studying the distribution of phenotypic traits in continuous adaptive trait space as the mean phenotypic trait in a population may not be adopted by any individuals (Cohen 2009). Such models commonly take the form of non-local reaction–diffusion equations. The non-local reaction terms are used to incorporate interactions between individuals with different trait values. Diffusion on the other hand captures trait mutation. The evolution of a species undergoing resource competition can be modelled by

$$\frac{\partial n}{\partial t} = m \left(1 - \frac{\int \alpha(x, y) n(y, t) dy}{k(x)} \right) + \mu \frac{\partial^2 n}{\partial x^2}, \tag{2.21}$$

where the competition kernel α denotes the competition strength between individuals of phenotypes x and y; k denotes the resource available to each phenotype; and μ is the trait diffusion rate. Polymorphism occurs when the non-local reaction terms lead to a non-monotonic distribution of the phenotypic traits. A local maximum in the distribution corresponds to a *morph* in the population. Such diversity is sustained by density-dependent selection when the width of the competition kernel is less than that of the resource. In addition, the trait diffusion yields Gaussian-shaped distribution around the local maxima. The emergence of diversity is not limited to the number of morphs but can also include the breadth or the variance of the trait distribution around them (Figure 2.14).

All these theoretical clues suggest that co-evolution can potentially lead to rich evolutionary trajectories via density-dependent selection; in particular, they point to the possibility of diversification and polymorphism via evolutionary branching by disruptive selection in the system. These clues have triggered many theoretical studies that seek to understand how these trait-mediated interactions in co-evolutionary systems trigger disruptive selection and adaptive diversification (Doebeli and Dieckmann 2000). Using phylogenies as a proxy of evolutionary history, studies have shown that co-evolution could explain, to a certain degree, contemporary structures of many ecological networks (Rezende et al. 2007; Minoarivelo et al. 2014). For instance, the mutualistic interaction of seed dispersal by ants (myrmecochory) could have promoted diversification

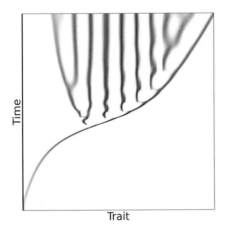

Figure 2.14 Emergence of a food web from a single heterotrophic morph, obtained by using the Evolutionary Distribution method. The level of grey represents relative abundances of a particular trait. Essentially, it is $n(x, t)$ as a function of trait x and time t. Model details and parameters see Hui et al. (2015), reprinted with permission.

in flowering plants (Lengyel et al. 2010). Pollination of flowers by insects could explain why angiosperms are more diverse than gymnosperms. Escape-and-radiate co-evolution could also be common between plants and herbivores, such as between *Blepharida* leaf beetles and their host trees in the genus *Bursera* (Becerra and Venable 1999) and between endosymbiotic bacteria *Buchnera aphidicola* and aphids (Moran and Baumann 1994). Moreover, when multiple species are closely involved in a community, they often form an adaptive co-evolutionary network, with a mixture of mutualistic and antagonistic interactions affecting each other's fitness.

Invasion performance can be profoundly triggered by trait evolution and directional selection from trait-mediated interactions, immediately after the invasion. The 'evolution of increased competitive ability' (EICA; Blossey and Nötzgold 1995) hypothesis assumes that release from native enemies and, therefore, change in biotic interactions, allows an alien species to allocate more energy to growth or reproduction, thereby conferring on it a competitive advantage over native species. It has been observed that post-introduction evolution is indeed common for alien plants, but not necessarily for the purpose hypothesised by EICA (Felker-Quinn et al. 2013). More investigations on the combination of biotic interactions and evolution are therefore needed. In contrast, increased susceptibility suggests that the opposite of enemy reduction, i.e., an increase in susceptibility of alien species to enemies in the introduced range, can occur if the alien species has a lower genetic diversity than the native species, therefore decreasing the chance to have individuals with specific defence mechanisms (Colautti et al. 2004).

Darwin's naturalisation hypothesis (Darwin 1859), the adaption hypothesis (Duncan and Williams 2002) and limiting similarity (MacArthur and Levins 1967) all assume that similarity between functional traits or phylogeny (as a proxy for functional traits) of native and alien species influence the outcome of invasion events. Darwin's naturalisation hypothesis and the adaption hypothesis both assume that an alien species with traits similar to those of native species will more easily invade a novel environment, which can be explained by the two species relying on the same biotic but also abiotic niche processes. Limiting similarity assumes the opposite: that trait differences enable an alien species to fill available niches; it is therefore similar to the empty niche hypothesis. Limiting similarity has mostly been tested for plants and has received substantial support (Jeschke and Erhardt 2018). These hypotheses cover a wide range of processes, which makes it difficult to test them empirically. The choice of appropriate spatial scale and the invasion stage are crucial when evaluating the validity of these hypotheses (de Bello et al. 2013).

2.7 Invasion Hypotheses

Biotic interactions are portrayed here as the building blocks of an ecological network. Conceptualising biotic interactions as trait-mediated density-dependent game plays that are flexible (interaction rewiring and switching) and adaptive (co-evolving trait) is the foundation on which we build in subsequent chapters to explore in detail diverse aspects of ecological networks facing biological invasions, as well as the associated invasion hypotheses.

Chapter 3 explores how biotic interactions can steer the formation and emergence of ecological networks and their associated network topology and architectures. Although co-evolution can explain the change, and perhaps more importantly the emergence, of biotic inter-actions, it cannot explain the rapid fluctuation and turnover of biotic interactions observed in many natural communities (Petanidou et al. 2008). Janzen (1985) proposed the concept of ecological fitting to explain the establishment of novel interactions between species which had shared little evolutionary history. According to ecological fitting, no drastic evolutionary change is needed to allow invading species to fit into local assemblages in their invaded ranges. Instead, readjustment from resident species through ecological fitting could provide sufficient opportunity for the establishment of novel biotic interactions. Interaction rewiring and switching are crucial steps of ecological fitting. Many successful invasive alien plant species have promiscuous interactions and experience host-jumps in their novel range (Richardson et al. 2000); such cases support the notion of novel interactions through ecological fitting. Invasion performance can be further connected to specific net-work topology and structures. For instance, the specialist-generalist hypothesis (Callaway et al. 2004) relies on a combination of co-evolution and ecological fitting to explain biological invasions: the enemies present in the introduced range must be specialists, and therefore less likely to affect novel species with which they have not co-evolved; in contrast, the mutualists should be generalists, to benefit the alien species (Sax et al. 2007).

In Chapter 4 we examine how invasion performance and invasibility are related to the loss of network stability or, more precisely, network instability.

Several invasion hypotheses relating to the invasibility of recipient ecosystems have been proposed to identify these aggravators of ecosystem instability (Hui et al. 2016). In an ecological network, numerous pairwise and higher-order interactions between resident and introduced species are in play, simultaneously and instantaneously, potentially forming complex reinforcing and balancing feedbacks that drive the adaptive cycles (panarchy) of complex system evolution (Gunderson and Holling 2002). Biotic interactions that foster reinforcing feedbacks, through for instance ecosystem engineering and intransitive interactions (Gaertner et al. 2014; Gallien et al. 2018), therefore profoundly affect the instability and invasibility of recipient ecosystems. Consequently, the recipient ecosystem is often susceptible to the invasion of alien species with specific functional and demographic traits (Hui et al. 2016). Related to this, the invasional meltdown hypothesis, which posits that the presence of some alien species can favour invasion by other species (Simberloff and Von Holle 1999), is well supported in the literature, especially for plants (Braga et al. 2018; Kuebbing 2020). Invasional meltdown can be caused by various processes, including direct biotic interactions (e.g., pollination, host-parasite interactions, herbivory, etc.), the effects of which are amplified or intensified by changes in the abiotic environment, such as increased disturbance or altered fire regimes.

Chapter 5 explores the community dynamics of an ecological network that result from invasion-induced instability. Most invasion hypotheses assume static native environments, i.e., that ecosystems have a set of constant characteristics that mediate invasibility. However, ecosystems are dynamic and their composition and environmental conditions fluctuate profoundly over time. It is therefore logical to conclude that biotic interactions between alien species or between alien and native species, and the resulting level of invasion success, vary accordingly. The opportunity windows hypothesis (Johnstone 1986) is an extension of the empty niche hypothesis that assumes that niche availability fluctuates, therefore offering different invasion opportunities at different points in time. Similarly, the dynamic equilibrium hypothesis (Huston 1979) assumes that abiotic and biotic conditions fluctuate, influencing the competition of local species, thereby potentially offering invasion opportunities to alien species. The importance of the dynamic aspect of ecosystems for biological invasions is apparent, for example, for herbivorous insects that rely on plants with annual cycles, which must rely on specific periods during which host plants are suitable, such as budburst (Ward and Masters 2007).

Chapter 6 builds on the premise that invaded ecosystems function as open dynamical systems (Hui and Richardson 2019). There, we explore the implication of ecological networks not being isolated, but embedded in much larger systems (e.g., a meta-community). Such systems experience a constant influx of alien species from the regional or external pool and outflux of extirpated species, driven by many intertwined factors (e.g., global environmental change, disturbance and biological invasions). For instance, the sampling hypothesis (Crawley et al. 1999) assumes that a large species pool will lead to more invasion events because there will be a greater chance that a species will have the characteristics to fill any available niche.

In Chapter 7 we deal with trait-mediated interactions in an eco-logical network and explore those specific alien traits that could experi-ence greater population growth (invasion performance) and cause greater impacts in the recipient communities in any specific ecological networks with trait-mediated interactions. Some invasion models have highlighted factors that can impose evolutionary barriers and create empty niche opportunities – largely through biotic interactions that can drastically change the invasion fitness landscape and give rise to unfilled pockets of positive fitness in trait space. For instance, an introduced species with a similar trait to that of a resident species might only suffer from weak biotic resistance – because the presence of the similar trait would ensure the introduced species would easily flourish. In contrast, an introduced species with its trait sitting between the traits of two resident species faces an uncertain outcome: either there is an empty niche to allow invasion or no niche is available to allow invasion (Hui et al. 2016). The imposed evolutionary barriers prohibit trait radiation in the resident species of a community; consequently, this would constrain the distribution of functional traits in the community and potentially create empty niches (note, disturbance can also create temporally available empty niches). Such empty niches set up by evolutionary barriers also highlight the priority effect: the traits and sequence/history of invasion can greatly affect how the invasibility of a community unfolds. Alien species could encounter empty realised niches with positive invasion fitness that have not been exploited (from disturbance) or are incapable of being exploited (from evolutionary barriers) by resident species; this could greatly increase the chances of invasion success (Elton 1958; MacArthur 1970).

References

Abrams PA (1990) The effects of adaptive behavior on the type-2 functional response. *Ecology* 71, 877–885.

Abrams PA, et al. (1993) Evolutionarily unstable fitness maxima and stable fitness minima of continuous traits. *Evolutionary Ecology* 7, 465–487.

Adler LS, et al. (2006) Leaf herbivory and nutrients increase nectar alkaloids. *Ecology Letters* 9, 960–967.

Aizen MA, et al. (2012) Specialization and rarity predict nonrandom loss of interactions from mutualist networks. *Science* 335, 1486–1489.

Aizen MA, Vázquez DP (2006) Flowering phenologies of hummingbird plants from the temperate forest of southern South America: Is there evidence of competitive displacement? *Ecography* 29, 357–366.

Aktas ME, et al. (2019) Persistence homology of networks: methods and applications. *Applied Network Science* 4, 61.

Allen WJ (2020) Indirect biotic interactions of plant invasions with native plants and animals. In Traveset A, Richardson DM (eds.) *Plant Invasions: The Role of Biotic Interactions*, pp. 308–323. Wallingford: CAB International.

Allesina S, Levine JM (2011) A competitive network theory of species diversity. *Proceedings of the National Academy of Sciences USA* 108, 5638–5642.

Allesina S, Tang S (2012) Stability criteria for complex ecosystems. *Nature* 483, 205–208.

Arditi R, Ginzburg LR (2012) *How Species Interact: Altering the Standard View on Trophic Ecology*. Oxford: Oxford University Press.

Axelrod R (1984) *The Evolution of Cooperation*. New York: Basic Books.

Axelrod R, Hamilton WD (1981) The evolution of cooperation. *Science* 211, 1390–1396.

Bairey E, Kelsic E, Kishony R (2016) High-order species interactions shape ecosystem diversity. *Nature Communications* 7, 12285.

Basilio AM, et al. (2006) A year-long plant-pollinator network. *Austral Ecology* 31, 975–983.

Basten U, et al. (2011) Trait anxiety modulates the neural efficiency of inhibitory control. *Journal of Cognitive Neuroscience* 23, 3132–3145.

Becerra JX, Venable DL (1999) Macroevolution of insect–plant associations: The relevance of host biogeography to host affiliation. *Proceedings of the National Academy of Sciences USA* 96, 12626–12631.

Bello FD, et al. (2013) Hierarchical effects of environmental filters on the functional structure of plant communities: A case study in the French Alps. *Ecography* 36, 393–402.

Bergelson J, et al. (2001) Models and data on plant-enemy coevolution. *Annual Review of Genetics* 35, 469–499.

Birnbaum C, et al. (2016) Nitrogen-fixing bacterial communities in invasive legume nodules and associated soils are similar across introduced and native range populations in Australia. *Journal of Biogeography* 43, 1631–1644.

Blackburn TM, et al. (2011) A proposed unified framework for biological invasions. *Trends in Ecology & Evolution* 26, 333–339.

Blossey B, Notzold R (1995) Evolution of increased competitive ability in invasive nonindigenous plants: A hypothesis. *Journal of Ecology* 83, 887–889.

Blumenthal DM (2006) Interactions between resource availability and enemy release in plant invasion. *Ecology Letters* 9, 887–895.

Bolnick DI, et al. (2011) Why intraspecific trait variation matters in community ecology. *Trends in Ecology & Evolution* 26, 183–192.

Bonawitz E, et al. (2014) Win-stay, lose-sample: A simple sequential algorithm for approximating Bayesian inference. *Cognitive Psychology* 74, 35–65.

Bosc C, et al. (2018) Interactions among predators and plant specificity protect herbivores from top predators. *Ecology* 99, 1602–1609.

Braga RR, et al. (2018) Structuring evidence for invasional meltdown: Broad support but with biases and gaps. *Biological Invasions* 20, 923–936.

Broekman MJE, et al. (2019) Signs of stabilisation and stable coexistence. *Ecology Letters* 22, 1957–1975.

Callaway RM, Ridenour WM (2004) Novel weapons: Invasive success and the evolution of increased competitive ability. *Frontiers in Ecology and the Environment* 2, 436–443.

Callaway R, et al. (2004) Soil biota and exotic plant invasion. *Nature* 427, 731–733.

Callaway RM, et al. (2008) Novel weapons: Invasive plant suppresses fungal mutualists in America but not in its native Europe. *Ecology* 89, 1043–1055.

Catford JA, Jansson R, Nilsson C (2009) Reducing redundancy in invasion ecology by integrating hypotheses into a single theoretical framework. *Diversity and Distributions* 15, 22–40.

Champagnat N, et al. (2001) The canonical equation of adaptive dynamics: A mathematical view. *Selection* 2, 73–83.

Chesson P (2000) Mechanisms of maintenance of species diversity. *Annual Review of Ecology and Systematics* 31, 343–366.

Christiansen F (1991) On conditions for evolutionary stability for a continuously varying character. *The American Naturalist* 138, 37–50.

Cohen Y (2009) Evolutionary distributions. *Evolutionary Ecology Research* 11, 611–635.

Colautti RI, et al. (2004) Is invasion success explained by the enemy release hypothesis? *Ecology Letters* 7, 721–733.

Crawley MJ, et al. (1999) Invasion-resistance in experimental grassland communities: species richness or species identity? *Ecology Letters* 2, 140–148.

Cressman R, Tao Y (2014) The replicator equation and other game dynamics. *Proceedings of the National Academy of Sciences USA* 111, 10810–10817.

Darwin CR (1859) *On the Origin of Species by Means of Natural Selection, or the Preservation of Favoured Races in the Struggle for Life.* London: John Murray.

Darwin CR (1862) *On the Various Contrivances by Which British And Foreign Orchids Are Fertilised by Insects, and on the Good Effects of Intercrossing.* London: John Murray.

Darwin CR (1868) *Variation of Plants and Animals under Domestication.* London: John Murray.

Dercole F, Rinaldi S (2008) *Analysis of Evolutionary Processes: The Adaptive Dynamics Approach and Its Applications.* Princeton: Princeton University Press.

Dieckmann U, Law R (1996) The dynamical theory of coevolution: A derivation from stochastic ecological processes. *Journal of Mathematical Biology* 34, 579–612.

Doebeli M, Dieckmann U (2000) Evolutionary branching and sympatric speciation caused by different types of ecological interactions. *The American Naturalist* 156, S77–S101.

Duncan R, Williams P (2002) Darwin's naturalization hypothesis challenged. *Nature* 417, 608–609.

Egas M, et al. (2005) Evolution of specialization and ecological character displacement of herbivores along a gradient of plant quality. *Evolution* 59, 507–520.

Egerton F (2007) Understanding food chains and food webs, 1700–1970. *Bulletin of the Ecological Society of America* 88, 50–69.

Ehrlich PR, Raven PH (1964) Butterflies and plants: a study in coevolution. *Evolution* 18, 586–608.

Elton C (1958) *The Ecology of Invasions by Animals and Plants*. London: Methuen.

Enders M, Jeschke JM (2018) A network of invasion hypotheses. In Jeschke JM, Heger T (eds.) *Invasion Biology: Hypotheses and Evidence*, pp. 49–59. Wallingford: CAB International.

Enders M, et al. (2018) Drawing a map of invasion biology based on a network of hypotheses. *Ecosphere* 9, e02146.

Enders M, et al. (2020) A conceptual map of invasion biology: Integrating hypotheses into a consensus network. *Global Ecology and Biogeography* 29, 978–991.

Eppinga MB, et al. (2006) Accumulation of local pathogens: A new hypothesis to explain exotic plant invasions. *Oikos* 114, 168–176.

Eshel I (1983) Evolutionary and continuous stability. *Journal of Theoretical Biology* 103, 99–111.

Eubanks MD, Denno RF (2000) Health food versus fast food: The effects of prey quality and mobility on prey selection by a generalist predator and indirect interactions among prey species. *Ecological Entomology* 25, 140–146.

Fantinou AA, et al. (2009) Preference and consumption of *Macrolophus pygmaeus* preying on mixed instar assemblages of *Myzus persicae*. *Biological Control* 51, 76–78.

Felker-Quinn E, et al. (2013) Meta-analysis reveals evolution in invasive plant species but little support for evolution of increased competitive ability (EICA). *Ecology and Evolution* 3, 739–751.

Fortheringham AS, Wong DWS (1991) The modifiable areal unit problem in multivariate statistical analysis. *Environment and Planning A* 23, 1025–1044.

Fortuna MA, Bascompte J (2006) Habitat loss and the structure of plant–animal mutualistic networks. *Ecology Letters* 9, 281–286.

Fossette S, et al. (2012) Does prey size matter? Novel observations of feeding in the leatherback turtle (*Dermochelys coriacea*) allow a test of predator–prey size relationships. *Biology Letters* 8, 351–354.

Gaertner M, et al. (2014) Invasive plants as drivers of regime shifts: Identifying high-priority invaders that alter feedback relationships. *Diversity and Distributions* 20, 733–744.

Gallien L, et al. (2017) The effects of intransitive competition on coexistence. *Ecology Letters* 20, 791–800.

Gallien L, et al. (2018) Emergence of weak-intransitive competition through adaptive diversification and eco-evolutionary feedbacks. *Journal of Ecology* 106, 877–889.

Gause GF (1932) Experimental studies on the struggle for existence: I. Mixed population of two species of yeast. *Journal of Experimental Biology* 9, 389–402.

Geritz SAH, et al. (1997) Dynamics of adaptation and evolutionary branching. *Physical Review Letters* 78, 2024–2027.

Geritz SAH, et al. (1998) Evolutionarily singular strategies and the adaptive growth and branching of the evolutionary tree. *Evolutionary Ecology* 12, 35–57.

Gigerenzer G, Gaissmaier W (2011) Heuristic decision making. *Annual Review of Psychology* 62, 451–482.

Gintis H (2000) *Game Theory Evolving*. Princeton: Princeton University Press.

Gintis H (2009) *The Bounds of Reason: Game Theory and the Unification of the Behavioural Sciences*. Princeton: Princeton University Press.

Gioria M, et al. (2018) Timing is everything: Does early and late germination favor invasions by herbaceous alien plants? *Journal of Plant Ecology* 11, 4–16.

Golubski AJ, et al. (2016) Ecological networks over the edge: Hypergraph trait-Mediated indirect interaction (TMII) structure. *Trends in Ecology & Evolution* 31, 344–354.

Gomulkiewicz R, Ridenhour BJ (2010) When is correlation coevolution? *The American Naturalist* 175, 525–537.

Grainger TN, et al. (2019) Applying modern coexistence theory to priority effects. *Proceedings of the National Academy of Sciences USA* 116, 6028–6210.

Gregory TR (2009) Understanding natural selection: Essential concepts and common misconceptions. *Evolution: Education and Outreach* 2, 156–175.

Grilli J, et al. (2017) Higher-order interactions stabilize dynamics in competitive network models. *Nature* 548, 210–213.

Gunderson LH, Holling CS (eds.) (2002) *Panarchy: Understanding Transformations in Human and Natural Systems*. Washington, DC: Island Press.

Hamilton WD (1964) The genetical evolution of social behaviour. *Journal of Theoretical Biology* 7, 1–52.

Heleno RH, et al. (2013) Seed dispersal networks in the Galapagos and the consequence of alien plant invasions. *Proceedings of the Royal Society B: Biological Sciences* 280, 20122112.

Herrnstein RJ, Vaughan W Jr (1980) Melioration and behavioral allocation. In Staddon JER (ed.) *Limits to Action: The Allocation of Individual Behavior*, pp. 143–176. London: Academic Press.

Hofbauer J, Sigmund K (1998) *Evolutionary Games and Population Dynamics*. Cambridge: Cambridge University Press.

Holling CS (1959) Some characteristics of simple types of predation and parasitism. *Canadian Entomologist* 91, 385–398.

Holt RD (2009) Bringing the Hutchinsonian niche into the 21st century: Ecological and evolutionary perspectives. *Proceedings of the National Academy of Sciences USA* 106, 19659–19665.

Hui C, McGeoch MA (2014) Zeta diversity as a concept and metric that unifies incidence-based biodiversity patterns. *The American Naturalist* 184, 684–694.

Hui C, Richardson DM (2017) *Invasion Dynamics*. Oxford: Oxford University Press.

Hui C, Richardson DM (2019) Network invasion as an open dynamical system: Response to Rossberg and Barabás. *Trends in Ecology & Evolution* 34, 386–387.

Hui C, et al. (2015) Adaptive diversification in coevolutionary systems. In Pontarotti P (ed.), *Evolutionary Biology: Biodiversification from Genotype to Phenotype*, pp. 167–186. Cham: Springer.

Hui C, et al. (2016) Defining invasiveness and invasibility in ecological networks. *Biological Invasions* 18, 971–983.

Hui C, et al. (2018a) Modelling coevolution in ecological networks with adaptive dynamics. *Mathematical Methods in the Applied Sciences* 41, 8407–8422.

Hui C, et al. (2018b) *Ecological and Evolutionary Modelling*. Cham: Springer.

Hui C, et al. (2020) The role of biotic interactions in invasion ecology: Theories and hypotheses. In Traveset A, Richardson DM (eds.) *Plant Invasions: The Role of Biotic Interactions*, pp. 26–44. Wallingford: CAB International.

Huston M (1979) A general hypothesis of species diversity. *The American Naturalist* 113, 81–101.

Hutchinson GE (1959) Homage to Santa Rosalia or why are there so many kinds of animals? *The American Naturalist* 93, 145–159.

Imhof LA, et al. (2005) Evolutionary cycles of cooperation and defection. *Proceedings of the National Academy of Sciences USA* 102, 10797–10800.

Janzen DH (1985) On ecological fitting. *Oikos* 45, 308–310.

Jaworski CC, et al. (2013) Preference and prey switching in a generalist predator attacking local and invasive alien pests. *PLoS ONE* 8, e82231.

Jeschke JM, Erhard F (2018) Darwin's naturalization and limiting similarity hypotheses. In Jeschke JM, Heger T (eds.) *Invasion Biology: Hypothesis and Evidence*, pp. 140–146. Wallingford: CAB International.

Jeschke JM, et al. (2002) Predator functional responses: Discriminating between handling and digesting prey. *Ecological Monographs* 72, 95–112.

Johnstone IM (1986) Plant invasion windows: A time-based classification of invasion potential. *Biological Reviews* 61, 369–394.

Jones LE, et al. (2009) Rapid contemporary evolution and clonal food web dynamics. *Philosophical Transactions of the Royal Society B: Biological Sciences* 364, 1579–1591.

Jordano P, et al. (2003) Invariant properties in coevolutionary networks of plant–animal interactions. *Ecology Letters* 6, 69–81.

Kaiser-Bunbury CN, et al. (2010) The robustness of pollination networks to the loss of species and interactions: A quantitative approach incorporating pollinator behaviour. *Ecology Letters* 13, 442–452.

Keane RM, Crawley MJ (2002) Exotic plant invasions and the enemy release hypothesis. *Trends in Ecology & Evolution* 17, 164–170.

Koch H, et al. (2014) Why rapid, adaptive evolution matters for community dynamics. *Frontiers in Ecology and Evolution* 2, 17.

Kolling N, et al. (2012) Neural mechanisms of foraging. *Science* 336, 95–98.

Kraines DP, Kraines VY (2000) Natural selection of memory-one strategies for the iterated Prisoner's Dilemma. *Journal of Theoretical Biology* 203, 335–355.

Kuebbing SE (2020) How direct and indirect non-native interactions can promote plant invasions, lead to invasional meltdown and inform management decisions. In Traveset A, Richardson DM (eds.) *Plant Invasions: The Role of Biotic Interactions*, pp. 153–176. Wallingford: CAB International.

Latombe G, et al. (2018) Drivers of species turnover vary with species commonness for native and alien plants with different residence times. *Ecology* 99, 2763–2775.

Le Roux JJ, et al. (2017) Co-introduction vs ecological fitting as pathways to the establishment of effective mutualisms during biological invasions. *New Phytologist* 215, 1354–1360.

Lehtonen J (2018) The Price equation, gradient dynamics, and continuous trait game theory. *The American Naturalist* 191, 146–153.

Lengyel S, et al. (2010) Convergent evolution of seed dispersal by ants, and phylogeny and biogeography in flowering plants: A global survey. *Perspectives in Plant Ecology Evolution and Systematics* 12, 43–55.

Letten AD, Stouffer DB (2019) The mechanistic basis for higher-order interactions and non-additivity in competitive communities. *Ecology Letters* 22, 423–436.

Levine JM, et al. (2017) Beyond pairwise mechanisms of species coexistence in complex communities. *Nature* 546, 56–64.

Lucas-Barbosa D (2016) Integrating studies on plant–pollinator and plant–herbivore Interactions. *Trends in Plant Science* 21, 125–133.

MacArthur R (1970) Species packing and competitive equilibrium for many species. *Theoretical Population Biology* 1, 1–11.

MacArthur R, Levins R (1967) The limiting similarity, convergence, and divergence of coexisting species. *The American Naturalist* 101, 377–385.

May RM (1972) Will a large complex system be stable? *Nature* 238, 413–414.

Mayfield M, Stouffer D (2017) Higher-order interactions capture unexplained complexity in diverse communities. *Nature Ecology & Evolution* 1, 0062.

McGill BJ, et al. (2006) Rebuilding community ecology from functional traits. *Trends in Ecology & Evolution* 21, 178–185.

Metz JAJ, et al. (1992) How should we define 'fitness' for general ecological scenarios? *Trends in Ecology & Evolution* 7, 198–202.

Minoarivelo HO, et al. (2014) Detecting phylogenetic signal in mutualistic interaction networks using a Markov process model. *Oikos* 123, 1250–1260.

Mitchell CE, et al. (2006) Biotic interactions and plant invasions. *Ecology Letters* 9, 726–740.

Mohammed MMA, et al. (2018) Frugivory and seed dispersal: Extended bi-stable persistence and reduced clustering of plants. *Ecological Modelling* 380, 31–39.

Moran N, Baumann P (1994) Phylogenetics of cytoplasmically inherited microorganisms of arthropods. *Trends in Ecology & Evolution* 9, 15–20.

Morgan CL (1888) Natural selection and elimination. *Nature* 38, 370.

Morozov A, Petrovskii S (2013) Feeding on multiple sources: Towards a universal parameterization of the functional response of a generalist predator allowing for switching. *PLoS ONE* 8, e74586.

Mouritsen KN, et al. (2011) Food web including metazoan parasites for an intertidal ecosystem in New Zealand. *Ecology* 92, 2006.

Murdoch WW (1969) Switching in general predators: Experiments on predator specificity and stability of prey populations. *Ecological Monographs* 39, 335–354.

Nnakenyi CA, et al. (2019) Fine-tuning the nested structure of pollination networks by adaptive interaction switching, biogeography and sampling effect in the Galápagos Islands. *Oikos* 128, 1413–1423.

Novak M, et al. (2016) Characterizing species interactions to understand press perturbations: What is the community matrix? *Annual Review of Ecology, Evolution, and Systematics* 47, 409–432.

Nowak M, Sigmund K (1990) The evolution of stochastic strategies in the Prisoner's Dilemma. *Acta Applicandae Mathematicae* 20, 247–265.

Nowak M, Sigmund, KA (1993) A strategy of win-stay, lose-shift that outperforms tit-for-tat in the Prisoner's Dilemma game. *Nature* 364, 56–58.

Nowak MA (2006) Five rules for the evolution of cooperation. *Science* 314, 156–1563.

Nuismer SL, et al. (2010) When is correlation coevolution? *The American Naturalist* 175, 525–537.

Nuwagaba S, et al. (2015) A hybrid behavioural rule of adaptation and drift explains the emergent architecture of antagonistic networks. *Proceedings of the Royal Society B: Biological Sciences* 282, 20150320.

Olesen JM, et al. (2008) Temporal dynamics in a pollination network. *Ecology* 89, 1573–1582.

Openshaw S (1984) *Concepts and Techniques in Modern Geography, Number 38: The Modifiable Areal Unit Problem.* Norwich: Geo Books.

Parker GA, Begon M (1986) Optimal egg size and clutch size: Effects of environment and maternal phenotype. *The American Naturalist* 128, 573–592.

Pauw A, et al. (2009) Flies and flowers in Darwin's race. *Evolution* 63, 268–279.

Pearson SF, Rohwer S (2000) Asymmetries in male aggression across an avian hybrid zone. *Behavioral Ecology* 11, 93–101.

Petanidou T, et al. (2008) Long-term observation of a pollination network: Fluctuation in species and interactions, relative invariance of network structure and implications for estimates of specialization. *Ecology Letters* 11, 564–575.

Poveda K, et al. (2005) Effects of decomposers and herbivores on plant performance and aboveground plant-insect interactions. *Oikos* 108, 503–510.

Queller DC (2017) Fundamental theorems of evolution. *The American Naturalist* 189, 345–353.

Ramanantoanina A, et al. (2014) Spatial assortment of mixed propagules explains the acceleration of range expansion. *PLoS ONE* 9, e103409.

Reinhart KO, Callaway RM (2006) Soil biota and invasive plants. *New Phytologist* 170, 445–457.

Rezende E, et al. (2007) Non-random coextinctions in phylogenetically structured mutualistic networks. *Nature* 448, 925–928.

Richardson DM, Pyšek P (2006) Plant invasions: Merging the concepts of species invasiveness and community invasibility. *Progress in Physical Geography* 30, 409–431.

Richardson DM, et al. (2000) Naturalization and invasion of alien plants: Concepts and definitions. *Diversity and Distributions* 6, 93–107.

Richardson LL, et al. (2016) Nectar chemistry mediates the behavior of parasitized bees: consequences for plant fitness. *Ecology* 97, 325–337.

Rieck B, et al. (2018) Clique community persistence: A topological visual analysis approach for complex networks. *IEEE Transactions on Visualization and Computer Graphics* 24, 822–831.

Rosati AG, Stevens JR (2009) Rational decisions: The adaptive nature of context-dependent choice. In Watanabe S, Blaisdell AP, Huber L, Young A (eds.) *Rational Animals, Irrational Humans*, pp. 101–117. Tokyo: Keio University Press.

Santamaría L, Rodríguez-Gironés MA (2007) Linkage rules for plant–pollinator networks: Trait complementarity or exploitation barriers? *PLoS Biology* 5, e31.

Sax DF, et al. (2007) Ecological and evolutionary insights from species invasions. *Trends in Ecology & Evolution* 22, 465–471.

Sher AA, Hyatt LA (1999) The disturbed resource-flux invasion matrix: A new framework for patterns of plant invasion. *Biological Invasions* 1, 107–114.

Simberloff D, Von Holle B (1999) Positive Interactions of nonindigenous species: Invasional meltdown? *Biological Invasions* 1, 21–32.

Smith CH (2012) Natural selection: A concept in need of some evolution? *Complexity* 17, 8–17.

Smith J, Price G (1973) The logic of animal conflict. *Nature* 246, 15–18.

Spencer H (1864) *The Principles of Biology*. Vol 1. London: Williams and Norgate.

Staddon JER (2010) *Adaptive Behavior and Learning*. Cambridge: Cambridge University Press.

Staddon JER, Hinson JM (1983) Optimization: A result or a mechanism? *Science* 221, 976–977.

Staniczenko PPA, et al. (2010) Structural dynamics and robustness of food webs. *Ecology Letters* 13, 891–899.

Stephens DW, Krebs JR (1986) *Foraging Theory*. Princeton: Princeton University Press.

Strauss SY, et al. (2006) Evolutionary responses of natives to introduced species: What do introductions tell us about natural communities? *Ecology Letters* 9, 357–374.

Suweis S, et al. (2013) Emergence of structural and dynamical properties of eco-logical mutualistic networks. *Nature* 500, 449–452.

Terry JCD, et al. (2017) Trophic interaction modifications: An empirical and theoretical framework. *Ecology Letters* 20, 1219–1230.

Thompson JN (2012) The role of coevolution. *Science* 335, 410–411.

Thompson JN (2013) *Relentless Evolution*. Chicago: The University of Chicago Press.

Toju H, Sota T (2006) Imbalance of predator and prey armament: Geographic clines in phenotypic interface and natural selection. *The American Naturalist* 167, 105–117.

Traveset A, Richardson DM (2014) Mutualistic interactions and biological invasions. *Annual Review of Ecology, Evolution, and Systematics* 45, 89–113.

Traveset A, Richardson DM (eds.) (2020) *Plant Invasions: The Role of Biotic Interactions*. Wallingford: CAB International.

van Baalen M, et al. (2001) Alternative food, switching predators, and the persistence of predator-prey systems. *The American Naturalist* 157, 512–524.

Vaughan W Jr, Herrnstein RJ (1987) Choosing among natural stimuli. *Journal of the Experimental Analysis of Behavior* 47, 5–16.

Wallace AR (1877) The colors of animals and plants. *The American Naturalist* 11, 641–662.

Wang Z, et al. (2014) Rewarding evolutionary fitness with links between popula-tions promotes cooperation. *Journal of Theoretical Biology* 349, 50–56.

Wandrag EM, Catford JA (2020) Competition between native and non-native plants. In Traveset A, Richardson DM (eds.) *Plant Invasions: The Role of Biotic Interactions*, pp. 281–307. Wallingford: CAB International.

Ward NL, Masters GJ (2007) Linking climate change and species invasion: An illustration using insect herbivores. *Global Change Biology* 13, 1605–1615.

Waser NM (2015) Competition for pollination and the evolution of flowering time. *The American Naturalist* 185, iii–v.

Waxman D, Gavrilets S (2005) 20 questions on adaptive dynamics. *Journal of Evolutionary Biology* 18, 1139–1154.

Weinberg S (1977) The forces of nature: Gauge field theories offer the prospect of a unified view of the four kinds of natural force-the gravitational and electromagnetic, and the weak and the strong. *American Scientist* 65. 171–176.

White EM, et al. (2006) Biotic indirect effects: A neglected concept in invasion biology. *Diversity and Distributions* 12, 443–455.

Whittall J, Hodges S (2007) Pollinator shifts drive increasingly long nectar spurs in columbine flowers. *Nature* 447, 706–709.

Wilsenach J, et al. (2017) Evolutionary fields can explain patterns of high-dimensional complexity in ecology. *Physical Review E* 95, 042401.

Wootton J (1994) The nature and consequences of indirect effects in ecological communities. *Annual Review of Ecology and Systematics* 25, 443–466.

Worthy DA, Maddox WT (2014) A comparison model of reinforcement-learning and win-stay-lose-shift decision-making processes: A tribute to W.K. Estes. *Journal of Mathematical Psychology* 59, 41–49.

Wright S (1932) The roles of mutation, inbreeding, crossbreeding, and selection in evolution. *Proceedings of the Sixth International Congress on Genetics* 1, 355–366.

Yang YH, Hui C (2021) How competitive intransitivity and niche overlap affect spatial coexistence. *Oikos* 130, 260–273.

Yoshida T, et al. (2003) Rapid evolution drives ecological dynamics in a predator-prey system. *Nature* 424, 303–306.

Zhang F, Hui C (2014) Recent experience-driven behaviour optimizes foraging. *Animal Behaviour* 88, 13–19.

Zhang F, et al. (2010) The evolution of cooperation on fragmented landscapes: The spatial Hamilton rule. *Evolutionary Ecology Research* 12, 23–33.

Zhang F, et al. (2011) An interaction switch predicts the nested architecture of mutualistic networks. *Ecology Letters* 14, 797–803.

Zhang F, et al. (2013) Adaptive divergence in Darwin's race: How coevolution can generate trait diversity in a pollination system. *Evolution* 67, 548–560.

Zhao ZH, et al. (2019) The failure of success: Cyclic recurrences of a globally invasive pest. *Ecological Applications* 29, e01991.

3 · *Network Assembly*

All free living systems are autopoietic, collectively autocatalytic, systems. If capable of heritable variation, such systems are undergoing natural selection and form evolving biospheres.

Stuart A. Kauffman (2019)

3.1 Succession and Assembly

To assess community assembly via natural colonisation and the potential ceiling of species richness in local communities, Wilson and Simberloff (1969) fumigated nine red mangrove (*Rhizophora mangle*) islands in Florida Bay, United States. This exemplifies the need in ecology to elucidate the concepts regarding community succession and assembly. New species arrive at a site predominantly via chance and dispersal, while resident species interact with each other via eco-evolutionary games (Chapter 2). Biotic interactions act as engineers to form ecological networks. Together with filters and forces from environmental and disturbance gradients, these ecological interaction networks define realised ecological niches and mediate community assembly rules and trajectories, thereby building an ecological house on the hill. With limited space and resource and the inevitable minimum sustainable size required for a viable population to survive stochasticity and disturbance, there must be an upper bound on the number and kinds of species that can be accommodated in a community, either via natural or human-mediated colonisation of both regional endemics and alien species. For this reason, questions pertaining to the ways in which an ecological community absorbs new arrivals have been on the agenda of community ecology since its inception. Despite progress on that front, making precise predictions about the trajectory of community assembly, the characteristics of the eventual resident species and the realised number of resident species in a local community remains a formidable challenge.

The view that a community assemblage is not at equilibrium but is in transition can be traced back to early discussions on forest succession by Comte de Buffon and on vegetation types in biogeography by Alexander von Humboldt and Jean-Baptiste Lamarck. Deliberation on the theme intensified in 1916 when Frederic Clements proposed his concept of secondary succession driven by disturbance. Clements (1916) argued that pioneer colonisers that established early in succession gradually give way to stronger competitors as a result of biotic interactions and organism–environment feedbacks within the ecological network. He posited that these processes can generate a self-organising, deterministic trajectory towards the community climax, akin to the birth and aging of a super-organism. Gleason (1927) and Tansley (1935) challenged Clements' view by arguing that species in a community behave individually with neither regard for such a climax nor community integrity. For example, Gleason (1927) wrote,

..., since it appears entirely possible that our deductions, however praiseworthy in their purpose, may sometimes have gone too far into a rigidity of concept and a formalism of expression and classification not fully justified by the facts in hand.

Tansley (1935) coined the term ecosystem to replace community and argued for the potential of multiple climaxes, while suggesting that a community can settle at one of many possible alternative stable states with each traversing a distinct path and thus experiencing a unique scenery of community assembly (Egler 1954). This means that the history and timing of species arrival at a community could impose priority effects on succession, potentially altering its final trajectory and destination (Chase 2003; Fukami 2015). In other words, the path of community assembly is driven both by the legacy of past dwellers and the gaming of current residents, including the impacts caused by recently arrived aliens. Together the roles of these drivers are manifested by the dynamic and adaptive fitness landscape (Chapter 2).

Following the seminal contributions of Robert MacArthur (1930–1972) on niche and limiting similarity, Jared Diamond (1975) introduced the term and concept of assembly rules based on his work on the birds of New Guinea and satellite islands to define a set of empirical patterns for depicting feasible community compositions. Diamond, however, went a step further by proposing the existence of permissible and forbidden species pairs (as assembly rules) due to competition that can form a checkerboard pattern of co-occurrence. These rules are nevertheless descriptions of co-occurrence patterns in an observed species-by-site matrix. They do not necessarily reflect the assembly processes that could have driven and produced such patterns; this gap between patterns and processes leaves space for debate (Figure 3.1). In particular, Connor and Simberloff (1979) criticised such one-to-one pattern-to-process inference and argued that chance alone, as generated from a null model of reshuffling species between sites, can produce similar checkerboard patterns. They thus questioned Diamond's assembly rules that competition

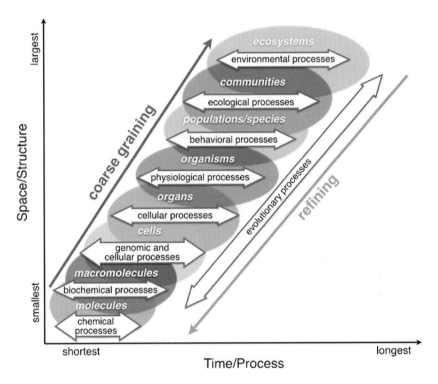

Figure 3.1 Hierarchical organisation of spatial structures (patterns) and temporal processes at work in an ecological community. The pattern-process is not 1-to-1 (a line) but n-to-m (a belt-shaped hierarchical ladder). The A hierarchical view of an ecosystem's organisational scale is envisioned as being composed of constituent communities, which in turn consist of species assemblages, populations of organisms and so on all the way down to the constituent molecules of cells. The organisation depicted here is neither unique nor complete but includes a number of typical focal levels at which biological systems are described and modelled. The axes indicate a notion of increasing time and spatial scales at increasingly higher levels of organisation. From Getz et al. (2018), reprinted with permission.

has prevailed over other assembly processes in structuring community assemblages (Gotelli and Graves 1996). The details of this debate are pivotal in the history of ecology, as was highlighted in a recent meta-analysis (Götzenberger et al. 2012),

... non-random co-occurrence of plant species is not a widespread phenomenon. However, whether this finding reflects the individualistic nature of plant communities or is caused by methodological shortcomings associated with the studies considered cannot be discerned from the available metadata.

Community ecology has subsequently sought clarity on the origin, evolution, maintenance and dynamics of biodiversity in varying and heterogeneous environments (Vellend 2016; Leibold and Chase 2017). Much of this work has focussed on developing and testing theories of species coexistence (Gause 1934; MacArthur and Levins 1967; Tilman 1982; Chesson 2000; Hubbell 2001). Different assembly processes have been emphasised by various schools of ecologists, leading to heated debates, such as those between advocates of Hubbell's (2001) neutral theory and supporters of the well-established niche theory (Tilman 1982; Clark 2012; Rosindell et al. 2012). Modern coexistence theory unites niche and neutral theories and emphasises the compound effect of niche overlap and fitness equivalence (Chesson 2000; see Section 2.3). Consideration of spatial scale and embedded environmental heterogeneity adds additional, but essential, complexity to the neutral-niche debate (Leibold and Chase 2017; Chase et al. 2018). Recent developments in community ecology have emphasised the possibility that the interplay of multiple processes in mediating the complexity of local context can simultaneously give rise to the full set of community patterns (e.g., Tilman 2004; Gravel et al. 2006; Latombe et al. 2015, 2021; Vellend 2016).

Unlike community ecology, invasion biology has, as discussed in Chapter 1, focussed largely on explaining why some introduced species are more successful than others in new environments (Hui and Richardson 2017). Key objectives in the field are to provide robust predictions of the invasiveness and dynamics of alien species and the susceptibility (invasibility) of recipient ecosystems to invasion (Lonsdale 1999; Richardson and Pyšek 2006; Hui and Richardson 2017). The field thus draws heavily on concepts that are crucial for understanding community assembly, but adds complexities invoked by human-mediated introductions and other modifications to ecosystems. Biological invasions not only impose direct impacts on the resident biota of the recipient ecosystems (e.g., Gaertner et al. 2009) but also establish often non-linear and complex feedback loops that potentially trigger ecosystem regime shifts (Gaertner et al. 2014). Consequently, management efforts have been focused on preventing and mitigating negative impacts from biological invasions (Hulme 2009; Epanchin-Niell and Hastings 2010). This implies a different disciplinary tradition from community ecology. Probing dialogue between the two fields is underway within the scope of global changes to address altered pathways, structures and functions facing both novel ecosystems and the transforming global biosphere. To this end, biological invasions provide natural experiments or biogeographical assays with great potential to answer fundamental questions in ecology (Richardson et al. 2004; Sax et al. 2005; Cadotte et al. 2006; Rouget et al. 2015). To utilize the full potential of the natural experiment provided by invasions (and us humans), we need to move tentatively away from considering only the traditional setting of closed systems with coexisting species that is the primary domain of community ecology. We need to tread, gingerly at first, beyond the framework of locally open but regionally closed metacommunities (Leibold and Chase 2017; King and Howeth 2019), to grasp the view of open systems with co-evolving components and high rates of species and propagule exchange through permeable system boundaries (Hui and Richardson 2019).

As mentioned in Chapter 1, invasion ecology has traditionally sought syntheses with population ecology, mirroring invasion success by the failures and extinctions of other resident species (Colautti et al. 2017). As soon as an alien species enters a novel environment, it starts interacting with resident species to become part of an invaded ecological network (Hui and Richardson 2017, 2019). Invaded communities are neither super-organisms nor ensembles of individualistic species; they are an entangled web with silver lines of constantly evolving and fitting biotic interactions. It is widely accepted that alien species now make up a significant portion of assemblages of many ecological communities (McGrannachan and McGeoch 2019; Vizentin-Bugoni et al. 2019; Fricke and Svenning 2020). Although restoration to achieve pre-invasion conditions is a worthwhile aim in some situations, the escalating scale of invasions worldwide means that attempts to remove all alien species from all ecosystems, especially those in or near human-dominated habitats, are futile and counterproductive (Hobbs et al. 2014). For this reason, invasion ecology is currently exploring the roles of functional traits and biotic interactions between enemies or mutualists in an attempt to better predict an alien species' performance (e.g., Traveset and Richardson

2020). Community ecology has also started to embrace the intricate effects of diverse types of biotic interactions besides traditionally only competition (e.g., Vellend 2016). These shifts in research agendas are paving the way for a closer marriage of the two fields to jointly address challenges facing transformed ecosystems and their altered structures and functions in the Anthropocene era (Shea and Chesson 2002; MacDougall et al. 2009; Pearson et al. 2018; Latombe et al. 2021).

With the contextual uncertainty and interaction complexity in ecological communities, diverse models and theories have been proposed, with each arguably receiving some support, crowding the knowledge landscape and blurring the knowledge architecture (Latombe et al. 2021). The ethical and societal relevance, especially in invasion science, has further fuelled continuous debates and conflicts (e.g., Pauchard et al. 2018; Schlaepfer 2018). It is worth noting that a parallel history occurred during the Modern Synthesis, championed by R. A. Fisher, Sewall Wright, Theodosius Dobzhansky, Ernst Mayr, George Gaylord Simpson and G. Ledyard Stebbins, where diverse views and evidence of evolution and Mendelian genetics were integrated into a cohesive field. Such syntheses and integration are also happening within both community ecology and invasion science and between the two fields. Inspired by the four forces at work in evolution (selection, drift, gene flow and mutation), Vellend (2016) synthesised a multitude of models and theories pertaining to community assembly using analogously similar high-level processes: selection, drift, speciation and dispersal. A plethora of hypotheses pertaining to invasion performance have also been found clustering around three key components: propagule pressure, traits of alien species and characteristics of abiotic environment (Catford et al. 2009). Together, six key processes can be highlighted for a meaningful synthesis that drives the assembly and disassembly of an open ecological network facing biological invasions (Figure 3.2; Latombe et al. 2021): dispersal, drift, abiotic interactions, within-guild interactions, cross-guild interactions and evolution (genetic changes).

Expert classification of existing models in community ecology and invasion science based on these six assembly processes has opened up some issues pertinent to each field. In particular, the biggest module in the bi-adjacency matrix (Figure 3.2), which describes models of community ecology and invasion science implementing shared processes, contains the necessary combinations of all types of interactions, with the current invasion hypotheses and models relying heavily on the role of cross-guild interactions (Latombe et al. 2021). Such reliance on

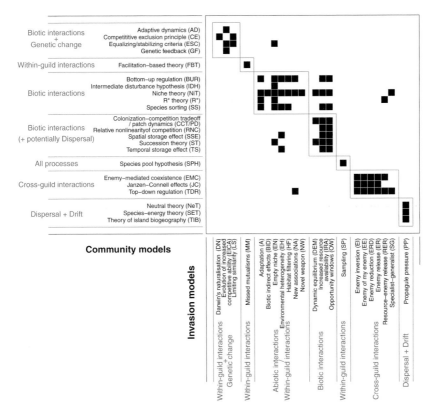

Figure 3.2 Relationship between community (rows) and invasion (columns) models presented as a bipartite network. The modules were identified using the Dormann–Strauss algorithm based on the model-by-process matrix from classicisation by experts. The main processes characterising each module are indicated in red; "/" represents at least one process in the models, whereas "+" indicates that the processes are combined in the models. From Latombe et al. (2021), under CC-BY Licence.

cross-guild interactions allows us to arrange invasion models according to a hierarchy of hypotheses (Jeschke and Heger 2018) that exposes the subtlety of cross-guild interactions in diverse scenarios. Related invasion hypotheses include enemy inversion and enemy reduction (Colautti et al. 2004), the enemy of my enemy (Eppinga et al. 2006) and enemy release (Keane and Crawley 2002). However, the subtlety also causes ambiguities in hypothesis testing and ferments debates (Farji-Brener and Amador-Vargas 2014; Latombe et al. 2019). Moreover, although the joint roles of cross-guild interactions and other processes have been well elucidated in invasion hypotheses, the interplay between other processes is often ignored when seeking to predict invasion performance. Although these other processes, especially genetic changes, propagule pressure and drift, are invoked in some invasion hypotheses (Figure 3.2), they clearly require further unpacking given the complexity of each. The smaller number of multi-process models in invasion ecology is consistent with the current integration of case-specific explanations of biological invasions under the banner of 'invasion syndromes' (Kueffer et al. 2013; Perkins and Nowak 2013; Novoa et al. 2020). In contrast, although community ecology has only started to appreciate the role of genetic changes, key processes are nonetheless evenly covered in the field.

Community assembly rules have been classified into evolutionary and ecological rules (HilleRisLambers et al. 2012). Evolutionary assembly rules refer to historical processes, including evolution, that contribute to the formation of regional species pools; ecological assembly rules include dispersal, biotic interactions and abiotic environmental factors (Götzenberger et al. 2012; HilleRisLambers et al. 2012). This divides assembly processes into those responsible for the arrival/emergence of a species and those that mediate the performance in a community after its arrival. This predominant view in community ecology was further elaborated in invasion science to differentiate three stages (introduction, naturalisation/establishment, invasion/spread) and the processes set up between these stages as ecological and evolutionary filters and barriers (Richardson et al. 2000a; Blackburn et al. 2011; Richardson and Pyšek 2012). To untangle the puzzle of how an assemblage passes through these interweaved filters, we can sort the degree of non-random patterns in invaded communities along three spectrums: temporal, spatial and ecological dispersion.

First, temporal dispersion acknowledges the non-equilibrial, dynamic nature of assembly and disassembly. While the impacts of biological invasions on the abundance and diversity of the resident community are often clearly evident (Gaertner et al. 2009; Vilà et al. 2011), the temporal assembly and disassembly can play out in an invaded community in many different ways. On the one hand, the impacts and stages of biological invasions often depend on the residence time of these alien species in the recipient ecosystems. Rapid evolutionary changes can occur during time of residency of an alien species and have been shown to be associated with invasion success in many cases (Whitney and Gabler 2008). As postulated in Hanski's (1982) core-satellite hypothesis, a community often comprises both inhabiting members (core) and visitors (satellite), with each following different rates and rules of assembly.

While post-introduction alien species progress along the invasion continuum, successful ones are also moving from the satellite end to become the core of the invaded community. In accordance with the core-satellite hypothesis, the role of stochasticity can be augmented due to the sequence of events playing out (Cuddington and Hastings 2016; Legault et al. 2019), while its signal can be weakened towards later stages through the selective barriers and filters along the invasion continuum (Hui et al. 2013). On the other hand, the progress of invasion stages and associated impacts are often not immediately observed during community assembly but lag behind the actual events, resulting in an extinction debt (Tilman et al. 1994; Kuussaari et al. 2009) and a mirror-image invasion debt (Rouget et al. 2016). In addition, sampling effects can also cause the discovery rate of reporting alien incursions to be underestimated and thus to lag behind the actual introduction rate (Costello and Solow 2003; Coutts et al. 2018). This means that the assemblage of an invaded ecological network is in temporal dispersion and transition. To capture temporal dispersion, species richness is often not a sensitive measure of biodiversity shift from environmental changes; instead, temporal turnover should be preferred to capture the change of species dominance and community identity over time (Figure 3.3; Dornelas et al. 2014; Hillebrand et al. 2018). Temporal dispersion in an invaded community is related to the instability and dynamics of the embedded ecological network; we will expand on these points in Chapters 4 and 5, respectively.

Second, spatial dispersion emerges from the co-occurrence and co-distribution of resident species (Diamond 1975; Bell 2001), with the spatial association or dissociation between species potentially reflected by the level of aggregation and compositional similarity (e.g., Hui et al. 2006, 2010; Hui and McGeoch 2014). On the one hand, the size and shape of species occupancy/range size are indicative of a species' performance in a community (Gaston 2003) and therefore a strong predictor of invasion success (Hayes and Barry 2008; Hui et al. 2011, 2014). On the other, to reveal associated assembly processes responsible for within- and between-site variations in network structures, the biodiversity of a community can be partitioned, typically, into alpha and pairwise beta diversity components. Moreover, as pairwise beta diversity cannot capture the full spectrum of spatial dispersion in ecological communities because ofits incapability of differentiating assembly processes of the turnover of common species versus that of rare ones (McGeoch et al. 2019), the complete biodiversity partitioning can be achieved using zeta

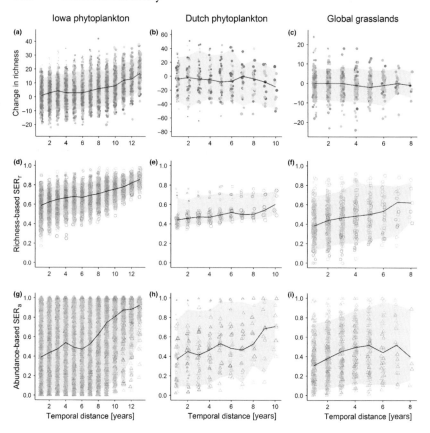

Figure 3.3 Change in species richness (a–c), richness-based species exchange ratio (SERr) (d–f), and abundance-based species exchange ratio (SERa) (g–i), with increasing temporal distance between years, based on annual mean presence and abundance. Different colours represent different sites within each of the datasets. Grey shading represent 5–95% quantiles, the dark grey lines represents the median. From Hillebrand et al. (2018), under CC-BY Licence.

diversity of different orders (Hui and McGeoch 2014). As the number of sites of interest increases, the shared compositional similarity (zeta diversity) declines. Such declines of zeta diversity typically follow either a negative exponential form or a power law in real communities (Figure 3.4 Top), reflecting stochastic versus non-stochastic assembly processes, respectively, at work. The decay of compositional similarity, thus, not only follows the pattern of distance decay of similarity (Soininen et al. 2007) but also declines with respect to involved sites and the level of species commonness and rarity (along the order of zeta

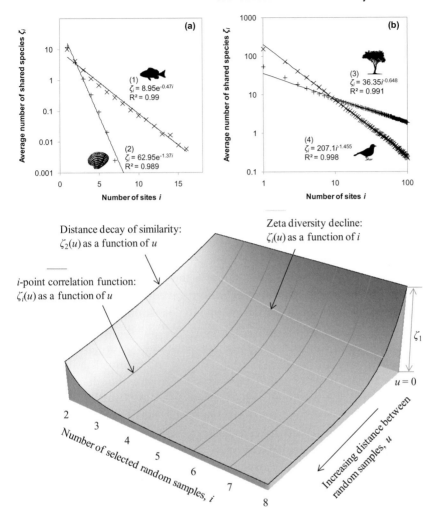

Figure 3.4 Spatial dispersion from community assembly captured by zeta diversity. Top: a, b, the two dominant forms of zeta decline (the relationship between zeta and sample number): The negative exponential (a) and the power law (b). b, Freshwater fish in the Greenbrier River, West Virginia, with 35 species occurring in 30 sites (1; Hocutt et al. 1978); 41 marine fouling organisms on 12 tile plates (2; Sutherland and Karlson 1977). b, 20 × 20-m quadrate samples of 307 tree species in the 50-ha plot on Barro Colorado Island (3; Condit 1998; Hubbell et al. 1999, 2005); quarter-degree cells of 761 bird species in southern Africa (4; Harrison et al. 1997). An exponential decline represents a stochastic assembly, while a power law decline represents non-stochastic assembly and the role of niche differentiation. Bottom: zeta diversity as a function of the number of random samples (green lines of zeta diversity declines) and the average distance between random samples (distance decay of zeta diversity; red lines). From Hui and McGeoch (2014), reprinted with permission.

diversity; Figure 3.4 Bottom). Different orders of zeta diversity (including alpha and beta components as special cases) can allow us to discern assembly processes responsible for the isolation by distance that reflects the distance decay of similarity (e.g., from dispersal limitation) and the isolation by resistance that reflects the Moran effect from the spatial autocorrelation of environmental and disturbance gradients (Latombe et al. 2018, 2019; Hansen et al. 2020). Considering the feedback between dispersal (a stochastic force) and biotic interactions (a non-stochastic force), together with the role of spatial correlations of stochasticity in population size and growth in boosting invasion performance (Cuddington and Hastings 2016; Hui et al. 2017), will likely reveal unexpected trajectories for both the invaders and the structure of the recipient community (Latombe et al. 2015). Such spatial dispersion during community assembly and disassembly from biological invasions are explored in more detail in Chapter 6.

Finally, ecological dispersion indicates the functional similarity or dissimilarity between resident species, where 'divergence' and 'convergence' are typically considered the two opposite directions of functional trait dispersion (Stubbs and Wilson 2004; Kraft et al. 2008). Trait convergence reflects shared ecological filters that hierarchically or sequentially select resident species (Cornwell et al. 2006; Ackerly and Cornwell 2007), similar to the concept of eco-evolutionary barriers along the introduction–naturalisation–invasion continuum (Richardson et al. 2000a; Blackburn et al. 2011). Trait divergence, on the other hand, indicates the evidence of limiting similarity (Watkins and Wilson 2003) and often suggests that resident species are arranged by their traits along opposite directions in the niche spectrum (MacArthur and Levins 1967; Stubbs and Wilson 2004). With a sufficiently wide range of niche spectra, multimodal clusters of species can be packed into a community, with species within a cluster showing emergent neutrality (Figure 3.5; Scheffer and van Nes 2006; D'Andrea et al. 2020). Alien traits and trait dispersion in recipient ecological community can profoundly affect interaction strength (Chapter 2), thereby steering community assembly; they are therefore crucial in determining the invasiveness of alien species and the invasibility of the recipient ecological network (Hui et al. 2016). We unpack this aspect of non-random ecological dispersion here and in more detail in Chapter 7. Taken together, we can dissect the issues related to the pattern formation in an invaded ecological network into its temporal dynamics, spatial structures and functional architectures. To infer assembly processes from these patterns and to eventually derive predictive models, we now turn to

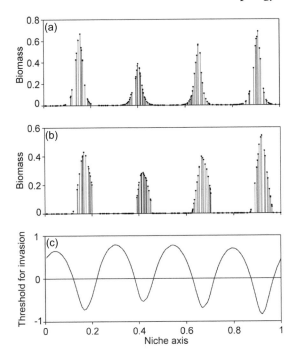

Figure 3.5 Self-organized lumpy patterns in the abundance of competing species along a niche axis. (a) A transient state after a simulation run of 1,000 generation times. (b) A stable pattern of species abundance reached after 5,000 generation times in the presence of mild density-dependent losses. (c) The competitive threshold for invasion of a new species expressed as percentage deviation of its carrying capacity relative to that of the resident species is lowest in the species lumps, showing that these represent relative windows of opportunity for invasion, and attractors in the fitness landscape. From Scheffer and van Nes (2006), reprinted with permission.

network metrics and emergence models, and to studies that examine network patterns before and after invasions.

3.2 Network Topology and Architecture

Network science emerged from graph theory, which was initiated in 1736 when Leonhard Euler described the well-known mathematical problem, The Seven Bridges of Königsberg. There has been sporadic progress on this front over the ensuing years. A few milestones deserve mention here. In 1959 Paul Erdős and Alfréd Rényi developed a model to explain the formation of random graphs. In the 1930s Jacob Moreno developed the sociogram, the social structure of school kids, thereby paving the way for the rise of social network analysis. At the turn of the millennium, three works on complex networks heralded the emergence of network science (Figure 3.6; Molontay and Nagy 2019): (1) Watts and Strogatz (1998) developed small-world networks, popularly known as six degrees of separation (Milgram 1967), through interaction rewiring. (2) Barabási and Albert (1999) developed scale-free networks, with power law node-degree distributions, through preferential attachment during the addition of new vertices. (3) Girvan and Newman (2002) developed a method to identify community structures in a network, which led to the flourishing of algorithms to detect network modules and compute the level of modularity. These three works had been cited 45,041, 38,590 and 15,292 times, respectively, by 18 February 2021 according to Google Scholar, indicating their immense influence. The number of studies on complex networks and their topologic metrics has grown rapidly over the past two decades, while the related fields of informatics and data science have

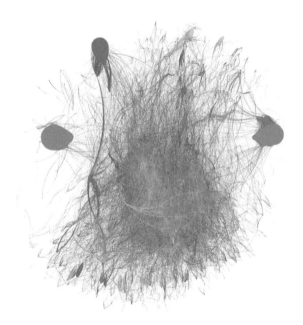

Figure 3.6 The largest connected component of the co-authorship network of 52,406 scientists coloured by communities who cited the three works: Watts and Strogatz (1998); Barabási and Albert (1999); Girvan and Newman (2002). The network was based on the publication dataset from the Web of Science retrieved on May 16, 2019. From Molontay and Nagy (2019), reprinted with permission.

mushroomed. Readers particularly interested in these developments should consult the excellent reviews on metrics of complex networks and tips for analysing network data (e.g., Albert and Barabási 2002; Newman 2003; Boccaletti et al. 2006; Miele et al. 2019).

Summerhayes and Elton (1923) documented the first ecological network of a trophic food web from data collected during their expedition to Spitsbergen. The term *food chain* was used and all food chains of a community were named a *food cycle*, rather than the term *food web* that is currently used. Other ecological networks such as bipartite networks comprising two functional guilds (between hosts and parasites; between plants and their mutualists) only began to be studied recently with the rise of molecular ecology and as ecological data have become easier to access. Both the topology and stability of ecological networks are crucial to the proper functioning of its embedded ecosystems. An ecological network, with both its nodes and edges being adaptive, represents an important type of complex adaptive network, and the analysis of ecological interaction network has therefore become an integral part of wider network science. We only introduce a few basic metrics pertinent to the analysis and our discussions on the structure and function of invaded ecological networks. The utility and interpretation of these network metrics are tailored for ecological relevance (Hui et al. 2018; Delmas et al. 2019; Dale and Fortin 2021).

Following the elaboration on the nature of biotic interactions in Chapter 2, we are now ready to explore how these building blocks form an ecological network and how the structure and architectural style of emerged ecological networks can be described by a set of network topological metrics. A network is a graphical representation of interactions between components of a system (Figure 3.7). Let there be a set of nodes or vertices in the focal system and L edges or links that connect specific pairs of nodes. In an ecological network, nodes typically represent resident species, while edges represent the interaction strength between linked species. For small networks consisting of relatively low numbers of nodes and edges, a simple graphical representation of the network with points and arrows/lines can clearly illustrate the network structure (Figure 3.7). However, for large networks comprising tens of thousands of nodes (e.g., Figure 3.6), such visual interpretation is uninformative. Underlying network visualisation lies an interaction matrix $\langle a_{ij} \rangle$, known as the adjacency matrix, with its entry a_{ij} representing the interaction strength from node j to node i (see Chapter 2). A network is unweighted when interaction strengths are represented by a binary

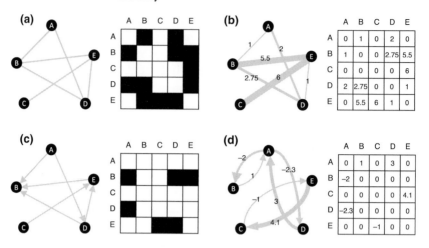

Figure 3.7 Categorisation of ecological networks, according to link directionality and weight. Black (white) entries in matrices **a, c** represent presence (absence) of interaction. (**a**) Unweighted undirected; (**b**) weighted undirected; (**c**) unweighted directed; (**d**) weighted directed. Note that links point to the affected species. For example, species A in d is positively affected by species B and D, while negatively affecting species B and D. From Landi et al. (2018), under CC-BY Licence.

variable (0 or 1) describing the absence or presence of an interaction; now, the adjacency matrix $\langle a_{ij} \rangle$ becomes an incidence matrix, also known as a qualitative matrix (Figure 3.7a). A network is weighted when interaction strengths are measured as real numbers, with the adjacency matrix a quantitative one (Figure 3.7b). A network is undirected if species i and j affect each other equally. A network is directed if species i affects species j at a different magnitude from the effect of species j on i (Figure 3.7c, d). A network is bipartite when the nodes can be classified into two distinct non-overlapping groups (guilds) and the interactions (edges) exist only between groups; otherwise, the network is unipartite or multipartite. For instance, a trophic food web is normally described as a directed unipartite network (e.g., Figure 2.1), while pollination networks involving a group of insect pollinators and a group of flowering plants are typically considered as forming a bipartite mutualistic network (note, competition between plants or between pollinators is normally omitted in such cases, although not necessarily so).

The adjacency matrix of interaction strength allows us to define network size, complexity, topology and architecture, as well as the stability of an ecological network. Here we focus on three network structural features, and leave the stability measures to Chapter 4.

The size of an ecological network typically refers to the richness of its resident species (S). However, using species richness as network size implicitly treats each species the same when counting them. A more generic measure could assign weights to each species (e.g., relative biomass or abundance; let p_i be the weight for species i in the network) and use the Hill numbers as a generic measure of network size,

$$H_\alpha = \left(\sum_{i=1}^{S} p_i^\alpha \right)^{1/(1-\alpha)}. \tag{3.1}$$

For $\alpha = 0$, the Hill number is simply the species richness, $H_0 = S$. When α approaches 1, the Hill number converges to the exponential of the Shannon-Wiener entropy index ($H_1 = \exp(H_{SW})$). For $\alpha = 2$, the Hill number becomes the reciprocal of the Simpson index. Therefore, the Hill numbers can be termed the effective sizes of an ecological network. If the weight is chosen to be relative abundance, when $\alpha \to +\infty$ the Hill numbers converge to the reciprocal of dominance, $H_{+\infty} = 1/\max(p_i)$; when $\alpha \to -\infty$ the Hill numbers converge to the number of rarest species needed to fill the community $H_{-\infty} = 1/\min(p_i)$. Heavy weighted species are rare in an ecological network, which means that H_α declines with increasing α. In addition, nodes in an ecological network do not preclude the inclusion of non–species nodes, such as social–environmental factors, as long as the quantity and the response of such non–species nodes to quantitative changes of other nodes can be estimated. For instance, the movement and social network of anglers were integrated with ecological networks in Nebraska reservoirs to assess the spread and impact of the invasive freshwater snail *Bellamya chinensis* (Haak et al. 2017).

Network complexity represents the global and local topology of a network and measures the level of interconnectedness and centrality of nodes. The global topology includes typically connectivity, defined as the total number of edges (i.e., L) in a network, and its normalised form, connectance (C), defined as the proportion of realised edges among all possible edges. For instance, for a unipartite network $C = L/(S(S-1))$, L is the number of interactions, whereas for a bipartite network $C = L/(S_C \times S_R)$, with S_C and S_R representing the numbers of species in the two guilds. The level of generalisation of a network can be measured by linkage density (LD); this is typically used to explore whether the network is dominated by generalists or specialists. For a unipartite network, it is simply the average number of links connecting each species, $LD = L/S$. For quantitative networks, we could use weighted linkage density (LD_w) to measure the level of generation (Ulanowicz 1997),

$$LD_w = \frac{1}{2}\left(\sum_{k=1}^{S} \frac{a_{k\cdot}}{\bar{a}} 2^{H_{O,k}} + \sum_{k=1}^{S} \frac{a_{\cdot k}}{\bar{a}} 2^{H_{I,k}}\right), \tag{3.2}$$

where $a_{k\cdot}$ and $a_{\cdot k}$ represent the column and row sums, respectively, of the adjacency matrix for species k; $\bar{a}$ is the matrix total. Parameter $H_{O,k} = -\sum_{i=1}^{S}(a_{ik}/a_{\cdot k})\ln(a_{ik}/a_{\cdot k})$ represents the Shannon diversity index for outflow interactions from species k, and $H_{I,k} = -\sum_{j=1}^{S}(a_{kj}/a_{k\cdot})\ln(a_{kj}/a_{k\cdot})$ is the Shannon diversity of inflow interactions to species k. Weighted connectance (C_w) is the quantitative version of connectance $C_w = LD_w/S$. Clustering coefficient or transitivity (CC) measures the extent to which neighbours of a node are also connected. The local clustering coefficient of a node i with k number of neighbours is first defined as $CC_i = L_i/T$, where L_i is the number of links connecting neighbours of i, and T the number of possible interactions among neighbours of i. If the network is undirected, the number of possible links between k nodes is $T = C_k^2 = k(k-1)/2$; i.e., for the binomial coefficient of k choose 2. The network clustering coefficient is then computed as the average CC_i over all nodes,

$$CC = \frac{1}{N}\sum_{i=1}^{N} \frac{2L_i}{k(k-1)}. \tag{3.3}$$

Average path length measures the efficiency of information transfer in a network. The shortest path length (D) between two nodes is the minimum number of edges connecting the two nodes. The average path length of a network is the average of the shortest path lengths for all pairs of nodes in the network. Most real-world complex networks, such as the World Wide Web, social networks and metabolic networks, possess the small-world feature, connecting any two nodes in the network via a short path, typically fewer than six steps (i.e., the six degrees of separation; Milgram 1967). In a small-world network, the shortest path length grows proportionally to the logarithm of the number of nodes ($D \sim \ln S$; Watts and Strogatz 1998).

The local topology captures the role and topologic position of a node within a network and is typically quantified by its centrality and peripherality. The simplest is degree centrality, or node degree. It accounts for the total number of links connected to a focal node. In a directed network, we can also differentiate the in-degree from the out-degree of a node, representing the counts of links reaching and leaving the node, respectively. Node strength is the quantitative extension of node degree for weighted networks. It is measured as the sum of interaction strengths of all links connected to a node. Complex networks also often exhibit a scale-free node-degree distribution (Barabási and Albert 1999), $prob(k) \sim k^{-\gamma}$, which is particularly common for large values of k (well-connected nodes), with the exponent γ ranging between 2 and 4. In contrast, the node-degree distribution of a random network follows a binomial distribution (Erdős and Rényi 1959), converging to an exponential degree distribution for large networks, $prob(k) \sim \exp(-\lambda k)$. Most food webs display exponential degree distributions, with those of high connectance even showing a uniform distribution (Camacho et al. 2002; Dunne et al. 2002). Recent large-scale comparisons of nearly 1,000 real-world complex networks suggest that the scale-free structure is actually empirically rare, while most complex networks have a log-normal node-degree distribution, $prob(k) \sim \exp\left(-(\log(k) - \mu)^2/(2\sigma^2)\right)/k$, or follow a truncated power law, $prob(k) \sim k^{-\gamma} \exp(-\lambda k)$ (Broido and Clauset 2019). For instance, the degree distribution of mutualistic networks mostly follows the truncated power law (Jordano et al. 2003). As node-degree distributions are marginal distributions of the network adjacency matrix, it is worth noting that both truncated power law and lognormal have come on top when fitting the marginal distributions of empirical incidence and abundance species-by-site matrices of ecological communities (McGill et al. 2007; Hui 2012). Besides the many existing ecological and emergence models on the genesis of such marginal distributions that will be discussed later in the chapter, it is worth noting that classification and cataloguing of natural systems in such a matrix/tabular fashion could inherently result in such highly skewed distributions, a phenomenon which has been explored in great detail since the 1950s (Aitchison and Brown 1957; Brown and Sanders 1981).

Closeness centrality measures the capacity of a node to affect all other nodes in a network. The closeness C_i of a node i in a network of S nodes is measured as the reciprocal of the average shortest path between node i and all other nodes,

$$C_i = \frac{S - 1}{\sum\limits_{j=1, j \neq i}^{S} d_{ij}}, \tag{3.3}$$

where d_{ij} denotes the shortest path between node i and j, with the length of a path being measured as a number of steps of consecutive edges. Betweeness centrality measures the number of times a node acts as a bridge between two other nodes. Betweeness centrality of node i is the fraction of the shortest paths between all pairs of nodes in the network which pass through node i,

$$B_i = 2 \frac{\sum_{j \neq i \neq k}^{S} g_{jk} g_{jk}(i)}{(S-1)(S-2)}, \qquad (3.4)$$

where S is the total number of nodes in the network, g_{jk} is the number of shortest paths between node j and node k, and $g_{jk}(i)$ is the number of shortest paths between j and k that pass through node i.

Network architecture represents clustering and hierarchical structures and is typically measured as modularity and nestedness. Modularity or compartmentalisation depicts the extent to which interactions are organised into subsets of nodes, or modules, where nodes interact more frequently with nodes of the same module than with nodes from other modules. The property of modularity is a common feature of food webs (Moore and Hunt 1988) and pollination networks with high species richness (Olesen et al. 2007). Although a number of metrics have been developed to quantify the level of compartmentalisation in a network, the index of modularity (Newman and Girvan 2004) has become the most widely accepted. This measure assumes that nodes in the same module have more links between them than expected for a random network. The modularity index is given by,

$$Q = \sum_{i=1}^{m} \left(\frac{l_i}{L} - \left(\frac{d_i}{2L} \right)^2 \right), \qquad (3.5)$$

where l_i denotes the number of edges in module i; d_i the sum of the degrees of the nodes in module i; and L the total number of edges in the network. However, the limitations of the modularity index are still being debated (Rosvall and Bergstrom 2007; Landi and Piccardi 2014). Calculating the level of modularity for a network requires us first to divide nodes into a specific module partition and then compute the modularity index (Q) for the particular partition. An optimisation algorithm is normally used for finding the partition that maximises the level of modularity. A number of optimisation algorithms exist for computing network modularity. One of the most commonly used is the simulated annealing algorithm (Guimerà and Amaral 2005). It is a stochastic optimisation technique that aims to minimise a given cost function. In the case of modularity, the cost function is $C_t = -Q$. Starting from each node in one module, the algorithm then tries to shift module membership by adding or removing one member node to or from a module, and chooses the one that minimises the cost function from a few trials at each step. Note that a number of partitions are tried per step to avoid getting stuck in a local minimum on the cost surface. The final solution is the partition showing a maximum modularity Q_{max}. Among other algorithms, the genetic algorithm and the extremal optimisation algorithm are often used (Schelling and Hui 2015).

Nestedness describes the hierarchal structure interactions in which specialists can only interact with those species with which generalists interact. In a nested network, both generalists and specialists tend to interact with generalists, whereas specialist-to-specialist interactions are extremely rare. A high level of nestedness is a common feature of mutualistic networks (Bascompte et al. 2003). Several metrics have been developed to quantify nestedness. Among the most commonly used is the *temperature* metric (Atmar and Patterson 1993), although there has been a recent shift to using the NODF metric (Nestedness metric based on Overlap and Decreasing Fill: Almeida–Neto et al. 2008). The NODF metric takes two components into account: the Decreasing Fill of the adjacency matrix (DF) and the Paired Overlap of its elements (PO). For an m-by-n interaction matrix of a bipartite network, the nestedness metric is first computed for each pair of rows and columns. Assuming that a pair of rows r_1 and r_2 (respectively, a pair of columns c_1 and c_2) and that r_1 is located above r_2 (respectively, c_1 to the left of c_2), the NODF metric penalises the situation where r_2 is fuller than r_1 (respectively c_2 is fuller than c_1). Hence, the decreasing fill metric (DF) of the pair of rows (respectively, columns) depends primarily on their marginal totals (MT),

$$DF_{r_1 r_2} = \begin{cases} 100 & (MT_{r_2} > MT_{r_1}) \\ 0 & (MT_{r_2} \leq MT_{r_1}) \end{cases}. \tag{3.6}$$

For non-penalised pairs (i.e., those with $MT_{r_2} > MT_{r_1}$), the Paired Overlap (PO) metric is computed as the percentage of filled cells in r_2 that are also filled in r_1. The NODF value for the pair of rows is then computed as $NODF_{r_1 r_2} = DF_{r_1 r_2} PO_{r_1 r_2}$; respectively, $NODF_{c_1 c_2} = DF_{c_1 c_2} PO_{c_1 c_2}$).

The NODF of the entire matrix of m rows and n columns is then given by the average NODF values of all possible pairs of columns and rows,

$$NODF = \frac{\sum\limits_{xy} NODF_{xy}}{\frac{m(m-1)}{2} + \frac{n(n-1)}{2}},$$ (3.7)

where xy corresponds to each pair of columns or rows. A weighted NODF, known as WNODF, can be used for estimating the level of nestedness of weighted networks (Almeida-Neto and Ulrich 2011). To calculate the WNODF, the number of non-empty cells is considered, rather than the marginal totals when computing the DF component for each pair of rows and columns, while the computed PO component is the fraction of non-empty cells in r_2 (respectively, c_2) that have values less than the corresponding values in r_1 (respectively, c_1). NODF and WNODF range from 0 to 100, and increase with the level of nestedness.

There are, however, a few issues that need to be flagged when using these network metrics. First, it is important to emphasize that as metrics of all existing network structures are essentially based on the same interaction matrix, they are not independent from each other but are intertwined (e.g., Figure 3.8; Vermaat et al. 2009; Nuwagaba et al. 2017). For instance,

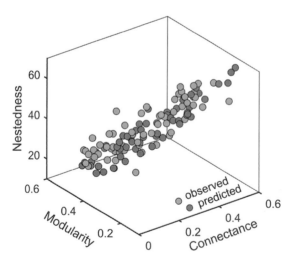

Figure 3.8 Observed and predicted relationships among connectance, modularity and nestedness for 60 bipartite antagonistic networks. For each network, simulations were run until equilibrium was reached based on a simple Lotka–Volterra model with Holling's type II functional response and implementing adaptive interaction switching. From Nuwagaba et al. (2017), under CC-BY Licence. *A black and white version of this figure will appear in some formats. For the colour version, please refer to the plate section.*

connectance has been found to affect the correlation between nestedness and modularity (Fortuna et al. 2010), with the correlation shifting from positive to negative in mutualistic networks with the increase of connectance, with the sign flipping at around connectance $= 0.2 \sim 0.3$. The node degree distribution also shifts from predominately following a truncated power law to become exponential in antagonistic networks, with the increase of connectance from below 0.2 to above 0.3 (Nuwagaba and Hui 2015). Second, different proxies of interaction strength could yield different conclusions and interpretations in terms of the role and function of network structures (Lopezaraiza-Mikel et al. 2007); reconciling this requires one to recognise that proxies are only fitness components and they only contribute one part in the fitness equation. Finally, as discussed in Chapter 2, interaction strengths are scale dependent, as are the topology and architecture of networks. As area increases, the numbers of species and interactions increase in a nonlinear fashion, thereby affecting measured network structures (e.g., Figure 3.9; Galiana et al. 2018). With the increase of sampling duration, we expect to observe more species and infrequent interactions, increasing observed connectance and thus also change our views on observed network structures (e.g., Figure 3.10; CaraDonna and Waser 2020). Together, network structures can be portrayed as their size, complexity and architecture, with these measures interrelated and scale dependent. Conclusions regarding a particular non-random pattern need to be tested with proper null models and statistical tests to tease apart the confounding factors of inter- and scale-dependence (Dale and Fortin 2021).

3.3 Three Types of Network

3.3.1 Overview

Effects of biological invasions on the structure and function of ecological networks have only recently begun to receive attention. A search of the ISI Web of Science on 4 September 2020 for ecological networks with three types of interactions, $+|+$ (mutualistic), $+|-$ (antagonistic) and $-|-$ (competitive) yielded, respectively, 3347, 7297 and 2124 publications, of which 8.2 per cent, 7.3 per cent and 11.4 per cent also explored issues related to biological invasion (Table 3.1). Although we used inclusive and flexible search phrases and thus did not capture the precise contour of the knowledge landscape, the results are nevertheless informative. Moreover, as most scientific publications report significant results (close to 90 per cent; Mlinarić et al. 2017), a high count or

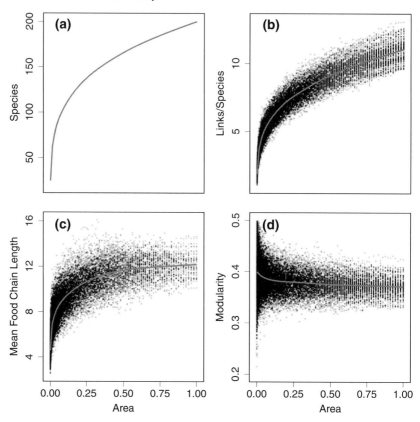

Figure 3.9 Network-area relationships as demonstrated by a trophic sampling model. Area values close to 0 correspond to local communities and values close to 1 correspond to regional communities. a–d, Relationship of the number of species (a), the number of links per species (b), Mean Food Chain Length (MFCL) (c) and modularity with area (d). Notice that the SAR shown in a is given by $S = kA^z$, where A is area, $k = 10$ is a constant and $z = 0.27$ is the scaling factor of the number of species (S) with area. Lines represent a generalized additive model fitted to data points. From Galiana et al. (2018), reprinted with permission.

percentage is indicative of the support to specific network structures and the impact of biological invasions on these structures in natural and simulated ecological networks. As we acknowledge that this is only a rapid and crude way to skim through the vast literature, it is important to realise that a significant feature that is frequently reported does not necessarily mean that this feature is stronger in such ecological networks; a weaker feature can also be significantly different from the null. As such, after the overview in this subsection, we dive deeper to consider a few

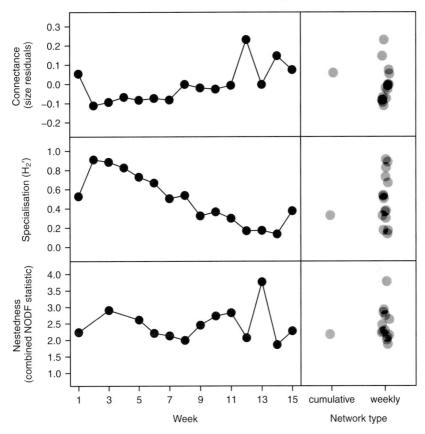

Figure 3.10 Temporal variation in the structure of plant–pollinator networks from week to week across the 2014 growing season (left panels) and comparison of cumulative, season-long network metrics with weekly metrics (right panels). From CaraDonna and Waser (2020), reprinted with permission.

specific papers in the hope of providing a more reliable picture of the structure of ecological networks and the impact of biological invasions on these structures for three types of ecological networks.

The top five structures reported in studies on mutualistic networks, from our literature search, are, from high to low: nestedness, species richness, connectivity, modularity and species (node) degree (Table 3.1). This indicates that nestedness has been reported as a dominant feature of mutualistic networks. Of those studies on invaded mutualistic networks, instead, the top five features are species richness, node degree, nestedness, species strength and interaction strength. The highest increase in reported

Table 3.1 *Network structure counts and percentages in literature, based on the ISI Web of Science. (12,768 publications were identified using inclusive search phrases with limited overlaps; 4 September 2020). Columns of N are counts of studies for mutualistic (+|+), antagonistic (+|−) and competitive (−|−) networks, while columns of M are filtered counts for biological invasion-related studies from the searched results; the last row reports the total counts. Columns for M/N report percentages of ecological network studies and filtered counts, while the last three columns report changes in percentage studies*

Network metrics	N			M			M/N			M/Total(M) − N/Total(N)		
	+|+	+|−	−|−	+|+	+|−	−|−	+|+	+|−	−|−	+|+	+|−	−|−
Interaction strength	201	505	197	26	37	21	12.94%	7.33%	10.66%	3.52%	0.02%	−0.63%
Jacobian element	3	5	3	0	0	0	0.00%	0.00%	0.00%	−0.09%	−0.07%	−0.14%
Species richness	278	515	151	37	53	26	13.31%	10.29%	17.22%	5.25%	2.89%	3.59%
Connectance	137	163	50	19	13	7	13.87%	7.98%	14.00%	2.87%	0.21%	0.53%
Connectivity	250	192	62	20	13	4	8.00%	6.77%	6.45%	−0.14%	−0.19%	−1.27%
Centrality	63	58	21	7	0	0	11.11%	0.00%	0.00%	0.68%	−0.79%	−0.99%
Species degree	207	391	143	31	25	17	14.98%	6.39%	11.89%	5.17%	−0.67%	0.26%
Linkage density	8	42	10	0	8	1	0.00%	19.05%	10.00%	−0.24%	0.93%	−0.06%
Species strength	192	449	186	28	38	19	14.58%	8.46%	10.22%	4.52%	0.98%	−0.94%
Species dependence	78	168	76	9	10	8	11.54%	5.95%	10.53%	0.97%	−0.43%	−0.29%
Modularity	210	124	61	15	9	3	7.14%	7.26%	4.92%	−0.78%	−0.01%	−1.64%
Nestedness	296	114	71	28	8	7	9.46%	7.02%	9.86%	1.41%	−0.06%	−0.46%
Interaction similarity	54	80	50	10	5	9	18.52%	6.25%	18.00%	2.05%	−0.16%	1.35%
Network similarity	82	72	55	13	5	8	15.85%	6.94%	14.55%	2.31%	−0.05%	0.70%
Total (Average)	3347	7297	2124	273	533	243	10.81%	7.12%	9.88%	1.96%	0.18%	0.00%

features due to invasion are species richness and node degree (+5.2 per cent), while the highest decline is modularity (−0.8 per cent). Interaction strength, species strength, connectance and network similarity have clearly been reported more frequently in invasion-related studies of mutualistic networks. This suggests that biological invasions could enhance species richness and connectance and reduce the modularity of local mutualistic networks.

There are more studies on antagonistic networks (more than double the number of studies of mutualistic networks) largely due to the disciplinary tradition. Of the top five features studied for antagonistic networks, species richness, interaction strength, species strength and species degree featured prominently (each >5 per cent) while connectivity also featured importantly (2.6 per cent). Antagonistic networks under biological invasions reported largely similar features. The feature reported more often in invaded relative to generic antagonistic networks is species richness (+2.9 per cent), suggesting that the effect of biological invasions on the species richness of recipient ecosystems has been the most notable impact, while the changes in reporting of other features due to biological invasions are minute (<1 per cent each). The most markedly underreported feature in antagonistic networks experiencing biological invasions is centrality (−0.8 per cent). This tentatively suggests that network size and interaction strength are of particular concern in antagonistic networks, while node degree and connectivity are also key features for describing antagonistic networks. The impacts of biological invasions in such networks are largely measured by the changes in species richness, while invasiveness could be related to centrality or the lack of centrality in antagonistic networks.

Competitive interactions are normally not considered as part of ecological networks. However, as clarified in Chapter 2, there is no reason to ignore this obvious interaction type which has been the main focus in community ecology. Perhaps due to this tradition, there are fewer publications that consider competitive interactions as part of an ecological network than those that consider the other two interactions. Again, features discussed in competitive networks are centred on interaction strength, species strength, species richness, node degree and species dependence, which are clearly necessary when exploring species coexistence and biodiversity maintenance in ecological communities. Under biological invasions, exactly the same five features have been prioritised, together with network similarity (related to compositional turnover). In ecological communities facing biological invasions, more studies have explored the changes in species richness (+3.6 per cent). Interestingly, there are a number of features that are reported less frequently in invaded communities than in generic communities: modularity, connectivity and centrality.

Taking our crude literature review further, we note some emerging features (Figure 3.11): First, nestedness, modularity and connectivity have been particularly well studied in mutualistic networks, while richness, interaction strength and node degree have been the focus for antagonistic networks including food webs. In contrast, interaction strength and species strength are the key topics of studies on competitive communities. Second, with biological invasions, linkage density has been nearly exclusively explored for antagonistic networks, while centrality is mainly discussed in invaded mutualistic networks. Invasion science has focused more on the invasions in mutualistic networks and competitive communities, and less on antagonistic networks. Limited reviews are available on ecological networks under biological invasions. For instance, Frost et al.

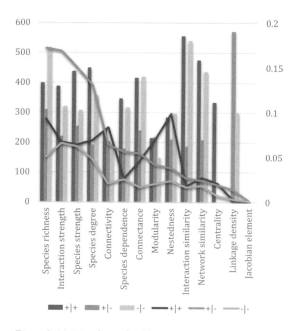

Figure 3.11 Number of publications using specific metrics of network topology and architecture in generic literature of ecological networks (lines, left axis) and the percentage of these also exploring biological invasion–related issues (bars, right axis). Data from Web of Science (accessed on 4 September 2020).

(2019) reviewed 52 ecological networks experiencing biological invasions but with competitive interactions $(-\,|\,-)$ excluded. Only three features – connectance, nestedness and interaction evenness – have each more than 5 networks (35 networks in total with about two-thirds showing no changes in network features). It can hardly be considered as a robust consensus, but it does appear that invasions increase connectance and interaction symmetry evenness in antagonistic networks, and nestedness and interaction unevenness in mutualistic networks. There are two possible reasons for the inconclusiveness and low predictability from these studies: small numbers and contextual dependence (Beckage et al. 2011). As the number of studies is rather low, and since there were inconsistent conclusions between many studies, we will need to rely on the context to draw a conclusion. We dissect and relate this contextual dependence to network stability and function in Chapter 4, but let us first look at some emerging threads. In the following section we examine specific studies and attempt to weave a more congruent picture on the structures of ecological networks and the role of biological invasions. Again, it is not conclusive but selective; our chosen narrative could be biased. In the following subsections, based on updates of several recent reviews (Hui and Richardson 2017; Landi et al. 2018; Hui et al. 2020), we first go through the typical patterns for each type of ecological network and then discuss the impacts of biological invasions on their typical network structures.

3.3.2 Competitive Communities

As mentioned in Section 3.1, multispecies interactions were first studied for systems with competition between species only (e.g., MacArthur 1972), although the term 'network' was not explicitly used. Competitive interactions of such a 'flat' community are structurally simpler than other types of ecological networks due to the single type of biotic interactions and the single functional guild of concern. The simplicity has allowed our understanding of the structure and function of ecological communities to be greatly refined, more so than for other types of ecological networks. The most important contributions from community ecology that have relevance to network science come from the rich literature on the relationship between community complexity and stability (May 1972; Alesina and Tang 2012; see review by Landi et al. 2018). Numerous experimental (e.g., Lehman and Tilman 2000) and theoretical (e.g., Cottingham et al. 2001) studies have been conducted to elucidate the relationships between species richness and community stability. There is also a vast literature on identifying drivers behind the assembly of different biodiversity components. Stability is the main theme of Chapter 4; here, we touch briefly on a few issues related to community/network structures.

Ecological communities have been traditionally recorded as a species-by-site data table. The marginal totals in this table (row sums: occupancy and abundances; column sums: site-level richness and diversity) have been the main focus, whereas much less attention has been given to the interaction structure. Diamond (1975) provided a novel approach for analysing and interpreting such species-by-site matrices. He highlighted that the co-occurrence patterns in such matrices can be used to infer the presence and role of competition. This notion stimulated the development of matrix structural metrics (e.g., the checkboard score; Stone and Roberts 1990) and approaches for testing null models (Connor and Simberloff 1983; Gotelli 2000). These developments effectively paved the way for network analyses that relied almost entirely on matrix visualisation and analyses. Moreover, a species-by-site matrix, $M_{S \times n}$, documenting S number of species in a community sampled at n sites can be simply transformed to produce proxies of interaction strength. For instance, the adjacency matrix of a co-occurrence network can be defined as the product of $M_{S \times n}$ and its transpose, $A_{S \times S} = M_{S \times n} M_{n \times S}^{T}$, and can then be visualised and analysed following standard network analyses (see Chapter 6). The key relevance of studies on competitive communities to network science is probably the proposed concept of nestedness, which was initially designed to explore nested niche-overlap among species in species-by-site matrices (Atmar and Patterson 1993). Insights from work in this area have been drawn into network ecology to provide one of the most important measures of network architecture (Bascompte and Jordano 2007; Almeida-Neto and Ulrich 2011).

Analyses of nearly 300 species-by-site matrices of real ecological communities suggest a strong negative correlation between species richness and connectance (defined as the number of presences divided by matrix size) (Figure 3.12 left; Hui 2012). Regarding the occupancy rank curve (related to node-degree distribution), when compared between a variety of model forms, the truncated power law was found to provide the best fit for 44 per cent of communities, followed by concave exponential (35 per cent) and lognormal (11 per cent) (Hui 2012). The level of nestedness, when measured using both Atmar and Patterson's (1995) temperature metric and Guimarães and Guimarães's (2006) implementation

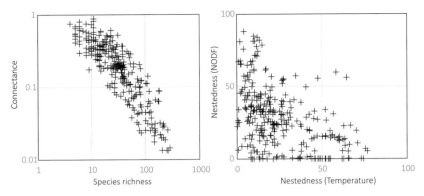

Figure 3.12 Relationships between species richness and connectance (left) and between two measures of nestedness (right) for 291 competitive communities. Each community is described by its species-by-site matrix, $M_{S \times n}$, with connectance defined as the total number of presences divided by the matrix size ($S \times n$). Nestedness is measured as Temperature from nestcalc.exe (Atmar and Patterson 1995) and as NODF from Aninhado 3 (Guimarães and Guimarães 2006). For data used to produce this figure, see Appendix S1 in Hui (2012).

of NODF, spans the entire range from 0 to100 but highlights the lack of communities with high temperature and high NODF (Figure 3.12 right). The reason for the lack of real ecological communities to fill up the entire spectrum of network structures, in terms of connectance and nestedness, is explored in Chapter 4.

Competition between resident and alien species is a reason for biotic resistance to invasion (Parker et al. 2020). Negative effects of competition on alien species are expected to intensify with increasing richness of resident species (Levine et al. 2004; Fargione and Tilman 2005). However, the consequence of competitive interactions is confounded by spatial scales and associated eco-environmental heterogeneity which can alter the anticipated negative correlation between native and alien richness to eventually result in a positive correlation at large spatial scales (Figure 3.13; Levine 2000; Stohlgren et al. 2003; Davies et al. 2005; Richardson et al. 2005; Tomasetto et al. 2019). Certainly, the exact relationship can be system- and scale-dependent (Jeschke 2014). Moreover, with fewer natural enemies in the invaded range, alien species can become more competitive (Blossey and Nötzgold 1995). In contrast, native plants could be intolerant of allelopathic compounds produced by invasive plants and become competitively weaker. In this regard, the translocation of co-evolved species that are allelopathically

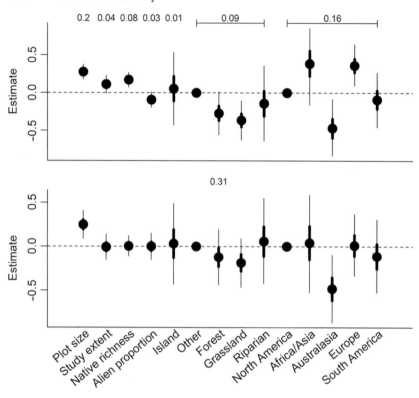

Figure 3.13 Parameter estimates for covariates (standardised by subtracting the mean and dividing by the standard deviation) when included alone (top panel) and together in multiple regression (bottom panel). Thick lines show the 50% credible intervals and thin lines the 95% credible intervals around the means (filled circles) of the posterior distributions. The marginal R^2 values for each covariate when included alone are shown at the top of the first panel, and the marginal R^2 for all covariates in the multiple regression model is shown at the top of the bottom panel. From Tomasetto et al. (2019), reprinted with permission.

tolerant to each other can collectively outcompete resident species in the novel communities, leading to invasional meltdown (e.g., Callaway and Aschehoug 2000; Bais et al. 2003; Callaway and Ridenour 2004; Vivanco et al. 2004).

An ecological community could contain empty niches that can be potentially exploited by alien species with fitting traits and niche requirements (Elton 1958), while biotic interactions are key processes to prune a species' realised niche. The realised niche, estimated based on data from a species' native range, is often much smaller due to the presence of many

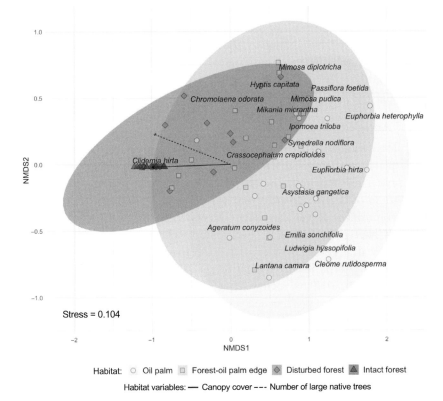

Figure 3.14 Non-metric multidimensional scaling (NMDS) of species composition based on Bray–Curtis index for oil palm, forest edge, disturbed forest and intact forest. Each symbol represents one transect and species names are in italics. For visualisation, random x-axis jitter has been added to the intact forest points, which were composed of a single species and so had no variation in NMDS scores. From Waddell et al. (2020), under CC-BY Licence.

co-evolved competitors. Many species can expand their realised niches in invaded ranges due to release from key enemies (Hierro et al. 2005; Traveset et al. 2015). As the extent of empty niches in a community depends on the level of saturation in species packing (a feature of evolutionary instability; Hui et al. 2016; see Chapter 4), biological invasions can modify the realised niche of each species through altering biotic interactions in ecological communities, capable of triggering assemblage-level succession or community disassembly (Ramsey-Newton et al. 2017). When anchored with single species vectors in the ordination space (Figure 3.14), such assembly trajectories and occupied niches are clearly visible (Waddell et al. 2020), with the peripheries of the ellipses in the ordination space pinpointing potential empty niches that may be temporally available. Notably, the trait vector along the successional direction highlights the transition of traits suitable at different stages of invasion.

3.3.3 Antagonistic Networks

While investigating the relationship between butterflies and their food/host plants in their seminal study, Ehrlich and Raven (1964) referred to reciprocal selective responses between ecologically linked organisms as the origin of organic diversity and biotic interactions (Figure 3.15). Such co-evolutionary forces can drive Darwin's race between the involved traits (Chapter 2) and lead to specific network structures over phylogeographic scales (Ferrer-Paris et al. 2013). However, bipartite antagonistic networks are typically of a small to medium size (Fortuna et al. 2010; Krasnov et al. 2012). A review of 61 such networks revealed a network size (the richness of resident species) ranging from 18 to 130 and number of interactions ranging from 33 to 736, while connectance ranged from 0.0987 to 0.556 (Nuwagaba and Hui 2015). Only 8 per cent of the networks had a scale-free structure (i.e., a power law node-degree distribution); 15.3 per cent follow a truncated power law; 31.15 per cent exponential; 12.02 per cent negative binomial and 33.3 per cent uniform. Connectance and functional guilds, as well as the method to rank the goodness of fit, also greatly affect the shape of node-degree distributions (Nuwagaba and Hui 2015). For these networks, modularity is high (mean 0.327, range 0.140–0.547). However, when compared with null models, only 11 out of 61 are significantly modular, 26 not significant and 24 significantly less modular. Nestedness is also high (NODF, mean = 42.97; range 13.07–74.31) with 55 and 49 significantly nested when compared to fixed and probabilistic null models. The three-way relationship among connectance, modularity and nestedness is strong (Figure 3.8). No consensus has yet to be reached in the literature regarding the architecture of antagonistic networks.

Food webs are a special type of antagonistic network; they can be huge and are the default model systems of complex networks. Figure 2.1 visualised a food web of 1205 trophic links and 123 free-living species. For food webs of this size it is still possible to differentiate nodes and links. For even bigger food webs (e.g., Figure 3.16) we need to aggregate nodes and links even if just for visualisation. The need for network

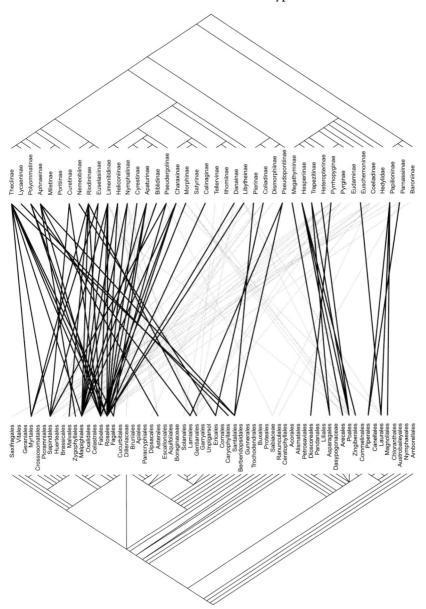

Figure 3.15 Congruence among plant (right) and butterfly (left) phylogenies. Lines between the phylogenies indicate associations based on the interaction matrix of important links. Black lines represent congruent links (p < 0.05) according to the ParaFitLink2 test. From Ferrer-Paris et al. (2013), under CC-BY Licence.

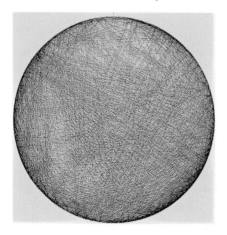

Figure 3.16 An aggregated food web of the Greater Antillean coral reef, with 265 trophic guilds (from 750 species) and 4656 interactions (34,465 species-to-species interactions). From proopnarine.wordpress.com (24/02/2010), under CC-BY Licence.

metrics for capturing topological features becomes obvious when dealing with large networks. Early work suggested a linear relationship between the number of interactions and the number of species (thus a decline of connectance with network size; Cohen and Briand 1984; Cohen and Newman 1985). However, this view was contested by Martinez (1992) who proposed a constant connectance hypothesis in food webs. With the proliferation of data quality and quantity, this constant connectance hypothesis has been criticised (Havens 1992; Dunne et al. 2002; Banašek-Richter et al. 2009). Nevertheless, there seems to be a reasonable degree of consensus that real food webs normally have low levels of connectance (~0.1; Havens 1992; Martinez 1992) – much lower than many real-world complex networks (Dunne et al. 2002). Node-degree distributions of food webs can be diverse, but mostly follow exponential form (Camacho et al. 2002; Dunne et al. 2002). Food webs with high connectance can exhibit a uniform node-degree distribution, whereas a power law or truncated power law provides a better fit for food webs with low connectance (Dunne et al. 2002; Montoya and Solé 2002). The distribution of interaction strength is highly skewed in food webs; there are many weak links but few strong ones (Paine 1992; Berlow 1999; Berlow et al. 2004; Wootton and Emmerson 2005). Due to the direct trophic interactions, species on higher trophic levels are often more generalised than those on lower trophic levels; this forms a nested diet structure (Williams and Martinez 2000; Neutel et al. 2002; Cattin et al. 2004; see Section 3.3.4). However, it has long been acknowledged that food webs typically contain modules of tightly coupled trophic groups (Moore and Hunt 1988), although these two concepts – modules and trophic groups – could mean different things (Gauzens et al. 2015). The level of nestedness in food webs is typically lower than in mutualistic networks (Bascompte et al. 2003). This is probably because food webs typically comprise nested subwebs, whereas the difference in nestedness between food webs and mutualistic networks is rather insignificant (Figure 3.17; Kondoh et al. 2010). Overall, all structures of antagonistic networks are highly correlated, while species richness and connectance can be used to explain much of the variations in most other network structures (Vermaat et al. 2009; Nuwagaba et al. 2015).

The loss of interactions with natural enemies in the novel range of an alien species allows the alien to achieve a higher level of fitness and greater abundance than in its native range; this is known as the enemy release hypothesis (Keane and Crawley 2002; Colautti et al. 2014). Many plant species can buffer damage from natural enemies through compensatory growth and reproduction, which complicates the assessment of the real effect of enemy release (Müller-Schärer et al. 2004). Also, trade-offs

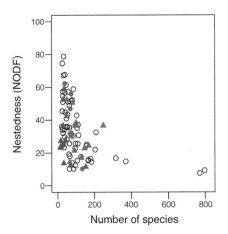

Figure 3.17 Species richness and nestedness (NODF) of 31 representative food webs (triangles and filled circles) and 59 mutualistic networks (open circles). Filled circles are for guild webs with significant (light grey) and non-significant (dark grey) nestedness, respectively. Triangles represent complex webs with significant (light grey) and non-significant (dark grey) nestedness, respectively. From Kondoh et al. (2010), reprinted with permission.

between resource acquisition and defence imply that fast-growing plant species adapted to high resource availability have a weaker defence against enemies and thus benefit more from enemy release in the novel range (Blumenthal et al. 2009). As generalist enemies are widespread and abundant and are likely to appear in both native and novel ranges, the enemies that alien species escape in novel ranges are often specialists (Andow and Imura 1994; Hinz and Schwarzlaender 2004; Knevel et al. 2004; Torchin and Mitchell 2004; Van der Putten et al. 2005). Consequently, alien species are primarily released from specialist enemies, and thus only reduce their production of defence compounds that specifically target specialist enemies but not generalist enemies (Müller-Schärer et al. 2004; Joshi and Vrieling 2005; Stastny et al. 2005). Although the notion of lower pathogen and parasite loads and fewer herbivores on alien plants has received much support in the literature (e.g., Torchin and Mitchell 2004), there are exceptions, especially within local networks (Agrawal et al. 2005; Van der Putten et al. 2005).

The evolutionary consequences of enemy release could be an increase in resource allocation to growth and reproduction at the expense of allocation to defence. This has been termed the evolution of increased competitive ability hypothesis (Blossey and Nötzold 1995, Maron et al. 2004, Bossdorf et al. 2005). However, such benefits resulting from the release could further subside with increasing residence time, once resident species begin responding to the newcomers as alternative targets. In particular, the biotic resistance of the receiving ecosystem to invasion starts when resident enemies begin targeting introduced species to form novel antagonistic interactions (Elton 1958; Mack 1996; Maron and Vilà 2001; Levine et al. 2004). Such biotic resistance relies on the adaptability and flexibility of the foraging/host-selection behaviour of resident enemies; it often emerges only after a delay during which resident enemies learn the potential benefits of exploiting the introduced species (Mitchell and Power 2003; Zhang and Hui 2014). This means that introduced species as potential hosts and resources are engaging in apparent competition with native host/resource species by sharing resident natural enemies (Connell 1990; van Ruijven et al. 2003; Mitchell and Power 2006).

As a result of the lack of shared co-evolutionary history with the alien species, those resident enemies in the novel range that start to target aliens could have a much stronger effect than those co-evolved enemies from the native range; this is known as the new associations hypothesis (Hokkanen and Pimentel 1989). This is because co-evolution often leads to better defence in hosts and reduced impact from natural enemies (e.g., stronger resistance and reduced virulence) (Jarosz and Davelos 1995; Parker and Gilbert 2004; Carroll et al. 2005; Parker and Hay 2005). Because generalist enemies are often promiscuous and thus likely to exploit potential new hosts or resources, alien species could evolve greater defences against the new associations with generalist enemies (Joshi and Vrieling 2005). Together, these hypotheses imply that introduced species often experience temporary enemy release, followed by greater biotic resistance from new associations of resident enemies. This suggests a temporal succession in the antagonistic interactions experienced by introduced species, although the evidence for such succession or enemy accumulation is not convincing (Andow and Imura 1994; Torchin and Mitchell 2004; Carpenter and Cappuccino 2005).

Intuitively, biological invasions can alter the structure of antagonistic networks. The evidence for this is, however, rather sparse. This is probably because of the lack of consensus regarding the topology and architectures of antagonistic networks, which can only be resolved with better metrics and data. Another reason could be that the impacts of biological invasions are inherently unpredictable and/or context-dependent (Chapter 4). Consequently, most studies have examined biological invasions into model food webs or antagonistic networks. It is unclear whether the results are contingent on model structures or reflect responses in real antagonistic networks. Despite the lack of clarity, it is worth picking some results from the modelling exercise (more details on network emergence models in the next section). For instance, in assembled model food webs (Baiser et al. 2010), invasion success can be explained by network connectance, fractions of native basal, native herbivore, native omnivore and native carnivore species, with connectance emerging as the best predictor of invasion success for basal, herbivore and omnivore invaders. Carnivore invasion success was best predicted by the fraction of herbivores in the food web. The effect of connectance was found to depend largely on the trophic level of the invader due to interactions that can occur only between species of adjacent trophic levels. If multiple consumers prey on an invader, the invasion has a higher probability of failing. If multiple prey species are available, the invasion has a higher probability of success. Similar studies on invading modelled ecological networks abound (Galiana et al. 2014; Lurgi et al. 2014). However, when exploring empirical food webs, higher connectance in networks was found to exert stronger biotic resistance and therefore associated with fewer invaders (Figure 3.18; Smith-Ramesh et al. 2017). However, these correlations do not necessarily infer a

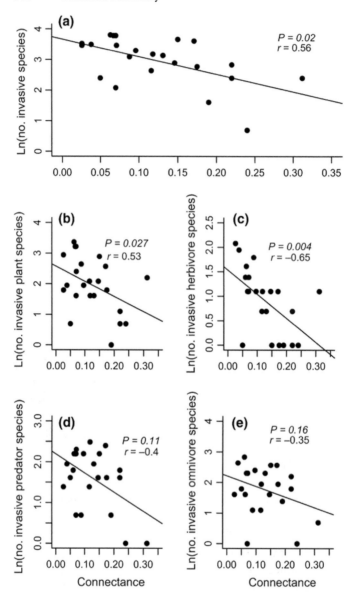

Figure 3.18 The log-transformed number of invaders listed in the Global Invasive Species Database (GISD) regressed against food web connectance for (a) all invaders, (b) plant invaders, (c) herbivore invaders, (d) predator invaders and (e) omnivore invaders. Higher-connectance food webs were paired with the closest possible habitat and region listed in the GISD. Connectance is negatively correlated with the number of potential invaders in a given habitat. Each point represents a well-characterized food web from the literature; 23 food webs and estimated connectance included in analyses, with habitat and regional associations matched to the Global Invasive Species Database (GISD). From Smith-Ramesh et al. (2017), reprinted with permission.

causal link between food web connectance and habitat invasibility. Reconciling the inconsistency between model predictions and empirical cases will probably follow the same path that has resolved the invasion paradox of 'the rich get richer'.

3.3.4 Mutualistic Networks

Investigations on the structure and associated functions of bipartite mutualistic networks heralded the arrival of network ecology, starting with Jordano's (1987) meta-analysis of network topologies of empirical pollination and seed-dispersal communities. This particular work reports a negative correlation between network size (species richness) and connectance, invariant linkage density, skewed distributions of relative interaction strength (dependence) and compartmentalisation (the presence of modules and the level of modularity). The network size–connectance relationship was further narrowed to a negative exponential form, while the range of connectance spanned from 0.1 to 0.2 (Olesen and Jordano 2002; Rezende et al. 2007). This publication, the first paper to invoke network analyses for exploring patterns of ecological communities, only mentioned the phrase 'mutualistic network' once (Jordano 1987),

The main result for the individual species is an *a priori* limitation of its possibilities of interaction to certain parts of the mutualistic network.

This paper achieved a milestone of similar stature in the literature as Diamond's (1975) paper that initiated the pattern-to-process inference in community ecology (Section 3.1). Both drew on system interaction matrices, species-by-site co-occurrence matrices in the case of Diamond, and species-by-species interaction matrix in the case of Jordano. Jordano's innovation quickly led to the trialling of other network topological metrics for unveiling the hidden non-random structures in mutualistic network data, such as the level of specialisation and generalisation (Waser et al. 1996; Memmott 1999; Vázquez and Aizen 2003). However, it also triggered an ongoing debate on the nature of interaction strength (Chapter 2); this is because shifting from one metric of interaction strength to another can yield different conclusions on how an ecological network structures and consequently functions (Figure 3.19; Lopezaraiza-Mikel et al. 2007; Novella-Fernandez et al. 2019).

As network data have accumulated, marginal distributions of the network adjacency matrices have become the focal point. A truncated

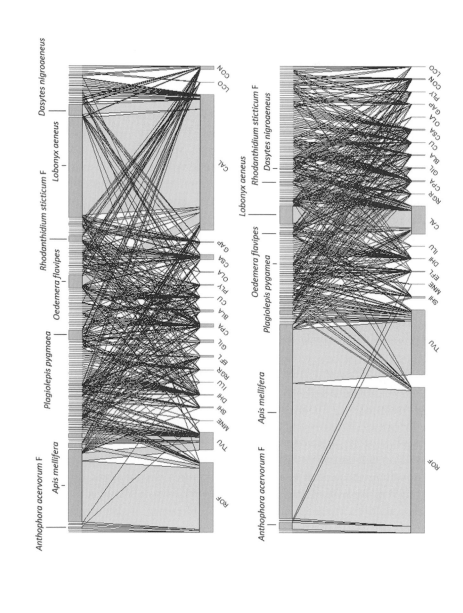

power-law or even a gamma distribution of node degrees have been found to be dominant in plant–pollinator and plant–frugivore networks (Jordano et al. 2003; Okuyama and Holland 2008), suggesting the presence of rare super generalists surrounded by many specialists. This provides strong support for the notion that the assembly in bipartite mutualistic networks is highly non-random and nested (Bascompte et al. 2003); it also indicates that the relative interaction strengths within and between functional guilds are highly skewed and asymmetric (Bascompte et al. 2006). The prevalence of nested structures in mutualistic networks has fostered a debate in the field. Mutualism begets nested networks while antagonism generates compartmentalised ones; the debate hinges on whether such a proposition is due to the contrasting roles of mutualistic versus antagonistic interactions in affecting the stability of ecological networks (Chapter 4; Thébault and Fontaine 2010; Fontaine et al. 2011). This proposition is further challenged by studies that reveal clear compartmentalised structures in many mutualistic networks, especially for those with large network sizes (Olesen et al. 2007; Mello et al. 2011; Dupont and Olesen 2012). This apparent contradiction can be resolved by the network-area relationship (Figure 3.9; Galiana et al. 2018). For instance, although the collective pollination network of the Galápagos Islands exhibits a strong compartmentalised structure (Traveset et al. 2013), closer examination suggests that each module corresponds largely to a highly nested local network (Nnakenyi et al. 2019).

Many species require mutualists to complete parts of their life cycle or to increase their reproductive output. When such species are moved to new areas, progression along the introduction-naturalisation-invasion continuum relies on the replacement of the lost services of co-evolved mutualists; this is known as the mutualist facilitation hypothesis (Richardson et al. 2000b). Introduced species often suffer from the loss of mutualistic partners, and the identity of potential new mutualists in the novel range can affect

---

Figure 3.19 A plant–pollinator network in a Mediterranean scrubland according to encounter-based (top) and visit-based (bottom) measures of interaction strength. Pollinator species in the top bar of a plot, plant species int the bottom bar. Width of horizontal bars denotes interaction frequency of each species. Grey bar width denotes interaction strength. TVU, *Thymus vulgaris*: EFL, *Euphorbia flavicoma*: MNE, *Muscari neglectum*: SHI, *Sideritis hirsute*: ROF, *Rosmarinus officinalis*: DHI, *Dorycnium hirsutum*: GAP, *Galium aparine*: BLA, *Biscutella laevigata*: CPA, *Centaurea paniculata*: ILU, *Iris lutescens*: OLA, *Orobanche latisquama*: RGR, *Ranunculus gramineus*: CAL, *Cistus albidus*: CSA, *Cistus salvifolius*: CLI, *Centaurea linifolia*: LCO, *Leuzea conifera*: PLY, *Phlomis lychnitis*: GIL, *Gladiolus illyricus*: CON, *Convolvulus althaeoides*. From Novella-Fernandez et al. (2019), under CC-BY Licence.

invasion success (Schemske and Horvitz 1984; Bever 2002, Klironomos 2003). Consequently, introduced species often experience high perform-ance when the richness of mutualists is high (van der Heijden et al. 1998, Jonsson et al. 2001), when specialised mutualists are widespread, occurring in both native and invaded ranges, or when specialised mutualists can be replaced by closely related similar taxa. Moreover, introduced plants often rely on both above-ground and below-ground mutualists that create syn-ergies through their interactions with the plants. For example, arbuscular mycorrhizas can promote an increase in floral display and/or the quantity and quality of nectar in some plant species, thereby directly increasing the flower visitation rates and effective pollination (Wolfe et al. 2005). Mutualism can further facilitate subsequent invasions through mediated interactions and the alteration of environments (Bruno et al. 2005) through, for example, increased soil nitrogen levels or fire frequency, a phenomenon termed invasional meltdown (Simberloff and Von Holle 1999; see Traveset and Richardson 2014 for other examples). For instance, many successful invasive legumes, e.g., *Acacia* species, form new mutualisms with bacteria found in the introduced environment (the host jumping hypothesis). However, symbionts can also be co-introduced with host plants, either intentionally as inoculants for species used in forestry or for other purposes, or accidentally by hitchhiking on introduced plant material and/or in soil (the co-introduction hypothesis; Figure 3.20) (Le Roux 2020). Australian *Acacia* species are nodulated predominantly by strains of *Bradyrhizobium*, many of which were probably co-introduced with the plants to new geographical regions (Birnbaum et al. 2016). The promiscuous nature of these and other woody legumes and the co-introductions of native sym-bionts are both important factors in explaining invasion success in this group (Rodríguez-Echeverría et al. 2011; Le Roux et al. 2017).

In many mutualistic networks, alien species have become dominant, moving from the periphery of the network, measured by node-degree and centrality, towards the central position. In such ecological networks with dominant alien species, nodes with high centrality often represent generalist invaders; such dominant invaders can serve as network hubs and connectors to link across modules and consequently reduce the level of modularity in the recipient ecosystem (Figure 3.21; Albrecht et al. 2014; Hui 2021). Biological invasions can, however, trigger adaptive niche and interaction changes (Valdovinos 2019), thereby gradually filling potential empty niches in the system, temporarily or permanently enhancing the richness of invaded mutualistic networks and subsequently increasing niche overlaps and thus the level of nestedness (Figure 3.22; Stouffer et al. 2014).

With increased residence time, such an invaded network reduces the relative role of stochastic processes (Hui et al. 2013) and becomes both highly nested and compartmentalised (Figure 3.23; Vizentin-Bugoni et al. 2019). Consequently, it is possible to use network structures to indicate the level of invasion in an ecosystem, and the topological position of an invader in the network to indicate the potential invasion stage it can reach.

Most of the literature on mutualistic networks deals with bipartite networks involving interactions only among above-ground mutualists. Advances in molecular approaches have paved the way for the development of plant-symbiont and below-ground soil microbial networks, thereby paving the way for a more comprehensive understanding of the role of mutualisms in all processes that mediate the performance of plants in an assemblage. For instance, the highest modularity of bacterial and archaeal mutualistic networks occurs in the topsoil, indicating a relatively higher system resistance to environmental change in topsoil than in communities in deeper soil layers (Bai et al. 2017). It is necessary to merge above- and below-ground ecological networks to elucidate the ways plants interact with enemies, symbionts and decomposers, testing the rich invasion hypotheses using cases of invasions in microbial communities (Figure 3.24; Ramirez et al. 2018). This also calls for consideration of networks of different types of biotic interactions – a multilayer network – to arrive at an adequate representation of a focal ecosystem. A multiplayer network, with the strength measures of different types of interactions standardised and expressed in the same unit, has been coined an equal-footing network (Figure 3.25; García-Callejas et al. 2018). This is consistent with the interaction strength measures defined in Chapter 2, where the interaction strength from species A to B can be measured as the sensitivity of species B's demographic performance (fitness) to the change of abundance in species A. In this way, the strength of different types of biotic interactions can be standardised, with the unit a focal species' demographic performance. In this regard, modelling an ecological network essentially requires four mappings: genotype to phenotype mapping ($G \rightarrow P$; e.g., Pigliucci 2010); phenotype to interaction [strength] mapping ($P \rightarrow I$; e.g., McGill et al. 2006); interaction strength to fitness mapping ($I \rightarrow F$); fitness to abundance (or biomass) mapping ($F \rightarrow A$). We have covered $P \rightarrow I$ and $I \rightarrow F$ mappings in Chapter 2 as interaction kernels ($P \rightarrow I$) and Lotka–Volterra type phenomenological models ($I \rightarrow F$), respectively. Notably, two consecutive mappings can be merged, e.g., $P \rightarrow I \rightarrow F$ as $P \rightarrow F$; see Eq. (2.14). In the next section we merge the last three mappings to synthesise network emergence models.

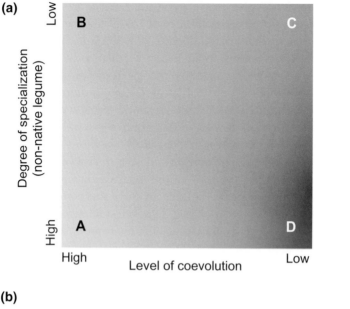

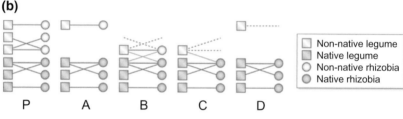

Figure 3.20 Conceptual framework for inferring establishment success and invasion performance of introduced legumes from their symbiotic interaction with rhizobia. (a) The performance (light grey, high; dark grey, low) is visualised on the plane of co-evolution (shared phylogenetic history between the invading legume and the rhizobia with which it interacts in the invasion range) and interaction specialization (ranging from interactions between specialist legumes and specialist rhizobia to those between generalist legumes and generalist rhizobia). Capital letters indicate four possible introduction scenarios. (b) Expected interaction network structures under each of these scenarios are indicated (A–D). Legumes are on the left of networks and rhizobia are on the right. P, pre-introduction networks (dashed lines, missing co-evolved mutualisms; light grey lines, potential new associations). Scenario A: legumes co-introduced with specialized and co-evolved rhizobia have a high level of performance. Co-introduced specialized species pairs form separate, unconnected motifs in the mutualistic network (network modularity). Thus, while they do not disrupt native interaction networks directly, they can rapidly alter soil nutrient cycling through ecosystem engineering, with native plants and soil rhizobia being replaced/excluded through altered abiotic conditions. Scenario B: legumes co-introduced with co-evolved but promiscuous rhizobia lack specialized mutualists

3.4 Network Emergence

3.4.1 Structural Emergence Models

Without fumigating entire ecosystems, network scientists rely on structural emergence models to make sense of observed patterns in complex networks. These models typically mimic certain key structuring processes that can explain non-random network topologies (e.g., Watts and Strogatz 1998; Barabási and Albert 1999). As studies of ecological networks started with food webs, two standard stochastic models have been proposed to explain the assembly and structural emergence of food webs (and other types of networks): the cascade model (Cohen et al. 1990) and the niche model (Williams and Martinez 2000). Considering only species richness (S) and connectance (C) as input parameters, the cascade model is based on two simple rules (Cohen et al. 1990):

Rule 1: Species are placed in a one-dimensional feeding hierarchy;
Rule 2: Species can only feed on those that are lower in the hierarchy than themselves.

The first rule captures the hierarchy along the major ordination axis of the adjacency matrix and often represents the body size spectrum. The second rule means that, after sorting species from low to high along this hierarchy in the adjacency matrix, the lower-left entries of the matrix become zeros (i.e. $a_{ij} = 0$ for $i \geq j$). Note that, in the absence of cannibalism in the food web, the diagonal entries should also be zeros. However, when considering cannibalism, we can modify the diagonal entries. The rest of the matrix entries can be randomly assigned to 1 with probability $2SC/(S-1)$ and 0 otherwise. The cascade model performed reasonably well when predicting food web topology, showing that the observed non-random

---

Figure 3.20 (*cont.*) and have a moderately high level of performance. Introduced legumes can establish symbioses with resident generalist rhizobia, thus affecting resident native legumes interacting with the same generalist rhizobia. Introduced promiscuous rhizobia could jump hosts in the new environment and negatively affect resident native legumes. Scenario C: promiscuous legumes can establish symbioses with existing (native or non-native) soil rhizobia in the new environment and have a moderately low level of performance. Introduced legumes are likely to establish symbioses with promiscuous resident rhizobia. Scenario D: specialized legumes lacking co-evolved rhizobia in the new environment have poor performance, which may lead to failed establishment. From Le Roux et al. (2017), reprinted with permission.

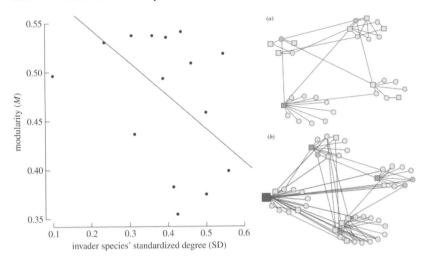

Figure 3.21 (Left) Relationship between species generalization of the principal invader plant, measured as standardised degree (SD), and modularity (M) of invaded plant–pollinator networks. (Right) Example of the modular structure of (a) an uninvaded plant–pollinator network and (b) a network invaded by an alien plant invader (*Carpobrotus affine acinaciformis*; large dark grey square). Interaction networks represent Mediterranean shrubland communities sampled at two locations at Cap de Creus, Spain. Plants are represented by squares, whereas pollinators by circles. Different levels of grey represent different topological species' roles (from light to dark grey): peripheral species, connector, module hub and network hub. From Albrecht et al. (2014), reprinted with permission.

structures could emerge from these simple stochastic rules. The niche model relaxes the second rule of the cascade model by allowing a species to feed only on those within its feeding range (the Goldilocks factor; Williams and Martinez 2000); this can be used to evaluate the chance of invasion for alien species with known niche positions in the feeding hierarchy, dietary widths and optimal diets (Figure 3.26; Romanuk et al. 2009).

The one-dimensional feeding hierarchy of the cascade and the niche models can be further expanded, as in the nested-hierarchy model that considers phylogenetic constraints, prohibiting trophic interactions between species with close-by niche positions (Cattin et al. 2004). The nested-hierarchy model can be implemented as follows:

Step 1: Assign each species a random niche value from 0 to 1;
Step 2: For species i, first assign a random prey j ($n_j < n_i$);
Step 3: For species i, assign an additional prey k ($n_k < n_i$) if prey k does not have consumers; if prey k has other consumers, then select a

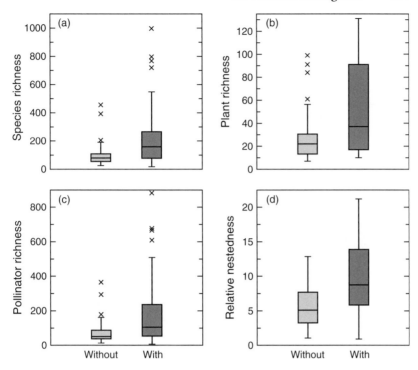

Figure 3.22 Properties of pollination networks with and without exotic plants. (a) Species richness, (b) plant richness, (c) pollinator richness and (d) relative nestedness were significantly different between the two types of communities. In all panels, the box covers the 25th–75th percentiles, the middle line marks the median and the maximum length of the whiskers is 1.5 times the interquartile range. Points outside this range show up as outliers. From Stouffer et al. (2014), reprinted with permission.

prey species at random from the diet of prey k's consumers that also consume j, and replace prey k with this selected prey; if the additional prey k has other consumers but none preys on j, choose k randomly from $n_k < n_i$; otherwise, from $n_k > n_i$;

Step 4: Repeat Step 3 until the number of prey objects of species i follows a Beta distribution that produces a network of connectance C.

However, besides the complexity, the improvement that the nested-hierarchy model brings to predicting the structure of natural food webs is not substantial. It appears that these structuring processes in the three structural emergence models follow random and constrained interaction topologies, which do not reflect the real phylogenetic structure and evolutionary history of an ecological network.

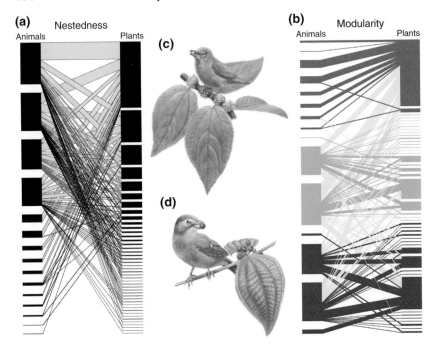

Figure 3.23 Structure of the island-wide seed dispersal network on Oʻahu and illustration of two emblematic interactions. (a and b) The novel network was nested [specialist species interacting with proper subsets of partners of the most generalist species (WNODF) = 48.67; 95% confidence interval (CI) = 34.24 to 46.66] (a) and modular [subsets of species interacting preferentially with each other, forming modules of highly connected species (QW) = 0.24; 95% CI = 0.07 to 0.09] (b). Species and links from distinct modules are depicted by different shades of grey, and grey links are interactions connecting modules. (c) Japanese white-eye feeding on *Pipturus albidus*, the most commonly consumed native plant. (d) A red-billed leiothrix feeding on *Clidemia hirta*, the most widely consumed and widespread introduced plant. Illustration credit: P. Lorenzo. From Vizentin-Bugoni et al. (2019), reprinted with permission.

As biotic interactions are often closely linked to the morphology and thus the underlying genotypes of involved species, they could potentially persist over evolutionary time scales, and species that emerge through co-evolution could inherit those interaction partners of their ancestors (Rezende et al. 2007). To incorporate real phylogenetic signals (Figure 3.27) into a network emergence model, a procedure using the framework of continuous-time Markov process has been proposed (Minoarivelo et al. 2014). For instance, for a bipartite network, the presence and absence of an interaction between species from the two functional guilds can be considered a binary trait (interacting or not). Following this approach makes the network emergence model essentially similar to the modelling of binary traits along the

Plots without legumes   Plots with legumes

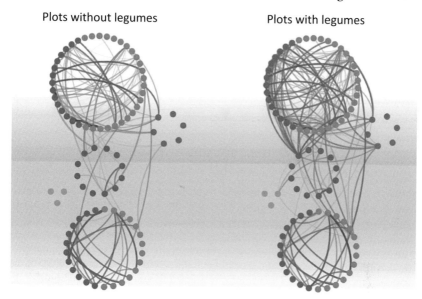

Figure 3.24 Network analyses for exploring function in above-belowground (AG-BG) interactions. Network visualization of plots with (n = 43) and without (n = 39) legumes. Plots with legumes have many more ground–AG-BG interactions when compared to plots with legumes. Plots without legumes had fewer connections (740 compared to 811), and perhaps more biologically interesting was that plots without legumes had many more negative connections than plots with legumes (without legumes, 254 negative out of 740; with legumes, 110 negative out of 811). The thickness and transparency of the lines designate the degree of correlation (darker, thicker lines have higher correlation values). From Ramirez et al. (2018), reprinted with permission.

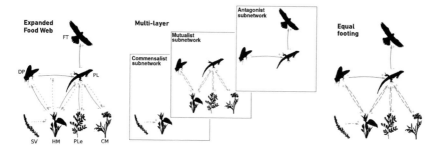

Figure 3.25 Three approaches for constructing and analysing networks with multiple interaction types. In the first panel, solid lines represent trophic interactions, dotted lines non-trophic ones. Note that frugivory and pollination have both trophic and non-trophic components. In the second and third panels, solid lines represent antagonistic interactions, dashed lines mutualistic ones and dotted-dashed lines commensalistic ones (competitive interactions). Data for building the network taken from the Aire Island community. From García-Callejas et al. (2018), reprinted with permission.

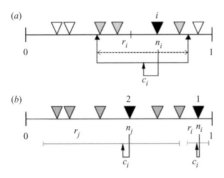

Figure 3.26 Diagram of the niche model and the invasion sequence. (a) S (trophic species richness) and C (connectance) are set at the desired values. Each of S species (here $S = 7$, shown by inverted triangles) is assigned a 'niche value' (n_i) drawn uniformly from the interval $(0,1)$. Species i consumes all species falling in a range (r_i) that is placed by uniformly drawing the centre of the range (c_i) from the interval $(r_i/2, n_i)$. Thus, in this diagram, species i consumes four species (grey and black triangles) including itself. The size of r_i is assigned by using a beta function to randomly draw values from the interval $(0, 1)$ whose expected value is $2C$ and then multiplying that value by n_i to obtain a web with C that matches the desired C. These rules stochastically assign each invader three fundamental niche values (n_i, r_i, c_i). These values determine the invader's fundamental niche and, in concert with the fundamental niches of species in the invaded web, determine the realized niche of the invader. Thus, for example, an invader i with a specific r_i and c_i has higher generality in an invaded web when relatively many species' n_j fit within i's feeding range than when invading a web with relatively few species' n_j fitting within i's r_i. (b) Example of attempted invasions by two different invaders into the same web. Invader 1 cannot invade because no species fall within its feeding range. Invader 2 can invade as it has prey (five grey triangles and it self) and therefore is allowed. From Romanuk et al. (2009), reprinted with permission.

combined phylogeny of the two guilds (Figure 3.27; Minoarivelo et al. 2014). In particular, the Markov process explores a set of two states: absence or presence of an interaction. The state of an interaction for a node on the joint phylogeny (at a particular time) is inherited from the state of its parental node. The inheritance of biotic interactions is imperfect and follows a transition probability of $\exp(q_{ij}t)$, where i and j represent the interaction states (i.e., presence or absence) and t is the branch length from the focal node to its parental node; q_{ij} represents the instantaneous rate of state transition,

$$\langle q_{ij} \rangle = \begin{array}{c} \\ \text{Absence} \\ \text{Presence} \end{array} \begin{array}{cc} \text{Absence} & \text{Presence} \\ \left[\begin{array}{cc} -\mu\pi_1 & \mu\pi_0 \\ \mu\pi_0 & -\mu\pi_1 \end{array} \right] \end{array}, \qquad (3.8)$$

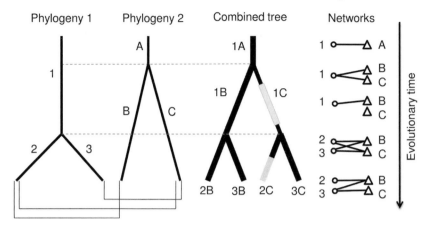

Figure 3.27 Example of the combination of two phylogenetic trees: every combination of branches of the two phylogenetic trees is represented by a corresponding branch in the combined tree. The branch in the combined tree is coloured grey when the interaction between the corresponding species pair is absent, and black when the interaction is present. The right panel shows resulting interaction networks. From Minoarivelo et al. (2014), reprinted with permission.

where μ is the interaction gain-loss rate; π_0 and π_1 are the equilibrium frequencies of absences and presences of biotic interactions, respectively. These three parameters can be inferred directly from observed phylogenies and the interaction matrix, while the model can explain 34 per cent variation in the topology and architecture of mutualistic networks (Minoarivelo et al. 2014). The key drawback of these three structural emergence models is that biotic interactions are pre-determined or inherited, and are neither consequential nor adaptive.

For the emergence of ecological networks with dynamic and adaptive interaction structures, we need to follow assembly-level network models which have a long history in community ecology (e.g., Drake 1990; Morton and Law 1997). Such models often assume infrequent colonisation of new species from a pre-determined regional pool while the resident species engage in pre-defined interactions. Some recent assembly-level models incorporate limited evolutionary processes (e.g., Drossel et al. 2001; McKane 2004) and adaptive response to disturbance (Kondoh 2003; Zhang et al. 2011; Suweis et al. 2013; Nuwagaba et al. 2015; Hui et al. 2015; Valdovinos 2018; Nnakenyi et al. 2019). In particular, the model proposed by Loeuille and Loreau (2005) can depict the emergence of complex food webs through ecological and evolutionary processes involving trait-mediated interactions. Targeting only functional traits that appear especially important to the population demography of focal species makes model parameterization more tractable. A few studies have attempted to develop modelling procedures and routines for exploring biological invasions in dynamic ecological networks (Figure 3.28; Lurgi et al. 2014). Even only characterising the directionality of interactions among species without estimates of interaction strength can be effective in predicting invasiveness and invasibility (Rossberg et al. 2010). Some would say that structural emergence models developed to generate ecological networks and their responses to biological invasions have become excessively complex. Although we need to acknowledge the structural and process complexity of both real and thus modelled ecosystems (Figure 3.1) the utility of the model is especially relevant (Figure 3.29) to design a model to serve its audience and purpose (Getz et al. 2018). McCullagh and Nelder (1983) remind us of George Box's aphorism,

Modelling in science remains, partly at least, an art. Some principles do exist, however, to guide the modeller. The first is that all models are wrong; some, though, are better than others and we can search for the better ones. At the same time we must recognize that eternal truth is not within our grasp.

As emerged properties of an ecological network, the invasiveness of introduced species and the invasibility of recipient ecological networks can be explored in two ways (Hui et al. 2016). First, many studies have followed a simple procedure of randomly assigning trait values and parameters for all initial species, running the model until equilibrium is reached, and then removing those species with population sizes below a certain threshold (Holland and DeAngelis 2010). At this stage the network is considered to be at its equilibrium. Once the recipient community has reached its equilibrium, we could consider the invasiveness of a potential introduced species as its invasion fitness (Chapter 2). Invasion fitness is a good proxy of invasiveness for an introduced species: if the trait of an introduced species lies within the positive intervals along the zero invasion fitness line (Figure 3.30a), the introduced species will experience positive invasion fitness and thus be able to establish and invade the resident community. If trait values land within the negative intervals, the species will experience negative invasion fitness and thus be repelled by the resident community (Figure 3.30a). Clearly, not all species can invade the resident network. For a given introduced species with a particular trait, if there is a resident species having an identical/similar trait (i.e., the trait of introduced species is close to any one solid dot [traits of resident species] in Figure 3.30a), the invasion fitness will then become close to zero. Because of the zero population growth, such species are less likely to establish simply due to demographic stochasticity (the case of neutral coexistence). Even if these species establish they will not become invasive but persist at low abundance until either eliminated via eco-logical drift or increasing opportunistically in response to disturbance. If the trait of an alien species is quite different from those of any resident species (i.e., sitting between two dots in Figure 3.30a), it is then likely to invade quickly (peaks in positive zones [darker shading]) or be quickly expelled from the network (valleys in negative zones [lighter shading]), with a 50/50 chance for successful invasion in a species-rich network due to the constraints on any dynamic systems given the continuity of the invasion fitness function (from the Fundamental Theorem of Algebra and the Central Limit Theorem). To this end, the invasibility of the recipient ecological network can be defined as the total width of the opportunity niche in the trait space (i.e., the summation of all the positive intervals). Network invasibility receives further attention in Chapter 7.

As an alternative to the static trait approach taken here we could also generate a model community as an adaptive network, where species within the network can co-evolve according to adaptive dynamics (Chapter 2), or where species with different traits can be continuously introduced into the local community from a large species pool (Gilpin and Hanski 1991; Hubbell 2001). This approach will potentially, but not always, lead to a saturated ecological network (Figure 3.28b). No alien species can invade a saturated network as the invasion fitness of any introduced species is equal to or less than zero. These two ways of generating community assemblages – sitting either at the equilibrium of ecological dynamics (Figure 3.30a) or the saturated assembly (Figure 3.30b) – offer two different scenarios for explor-ing the standard naturalisation hypothesis (Duncan and Williams 2002). Of course, even if the saturated assembly does exist, a community under constant bombardment of new arrivals is not likely to be either at ecological equilibrium or remain saturated, but somewhere between the two extremes (see Chapter 4). The next section looks at details of two specific types of ecological network models that consider adaptive processes and therefore can elucidate the eco-evolutionary responses of an ecological network to biological invasions.

Chapter 2 introduced how biotic interactions between two species can be established via co-evolution or interaction rewiring. Using both pro-cesses in standard dynamic ecological models allows us to capture the

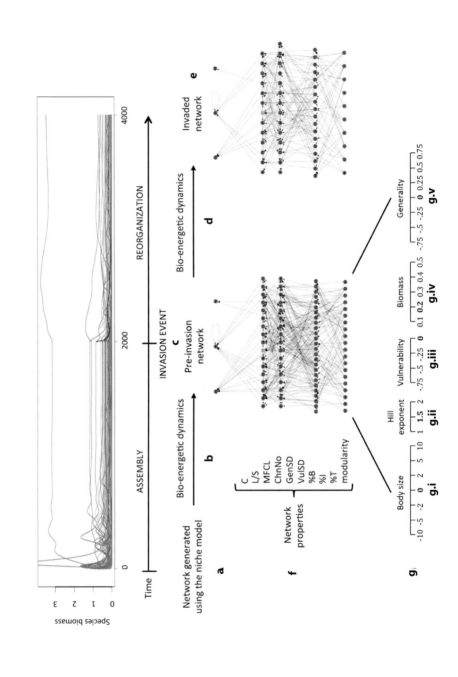

assembly of adaptive ecological communities, with the strength and partner of interactions changing adaptively to cope with the eco-evolutionary experiences in the network. Adaptive ecological networks allow the processes of co-evolution (Ehrlich and Raven 1964) and ecological fitting (Janzen 1985) to come into play. The next two sections expand on models that implement co-evolution and ecological fitting and can explain the emergence of adaptive networks and their response to biological invasions. This discussion draws largely on the concepts developed in Chapter 2. Given the particular interests of mutualistic networks and their responses to biological invasions, we provide further details when introducing adaptive models for mutualistic interactions. Guidelines are also offered for building models for other types of ecological networks.

3.4.2 Co-evolution and Invasion

In this section we introduce the equal-footing formulation of a multi-layer network under biological invasions. Let us consider the demographic performance of animal species i and its mutualistic plant partner species j, with their population sizes given by A_i and P_j and their traits x_i and y_j, respectively. Using the concepts introduced in Chapter 2, we could consider the demographic performance, the per-capita growth rate and the interaction strength expressed as interaction kernel functions of involved traits. To be more realistic, we can consider a nonlinear functional response (specifically here, Holling's (1959) type II functional response) to describe the saturation effect in consumption. This then leads to a commonly used trait-mediated mutualistic network model (e.g., Holland et al. 2006; Zhang et al. 2011; Hui et al. 2015, 2018; Minoarivelo and Hui 2016a, b),

---

Figure 3.28 Conceptual diagram of the *in silico* experiments performed. (a) The initial food web structure of the community is generated using the niche model. (b) Community dynamics through a transient period (from iteration 0 to 2,000) are simulated using the bio-energetic model (see text). Communities that are stable at time step 2,000 are subject to invasion (c). After invasion, the community evolves further 2,000 time steps (to 4,000) (d) until it reaches a new state (e). Before the introduction of the invasive species, a series of network properties are measured (f). These properties are again measured in (e) for comparison against their original values in (c) (before invasion). This process is repeated with several values for five different species traits (g.i–v) in order to test for the influence of variability within these traits on invasion success. (See description of these experiments in the text, including the meaning of each trait value). From Lurgi et al. (2014), under CC-BY Licence.

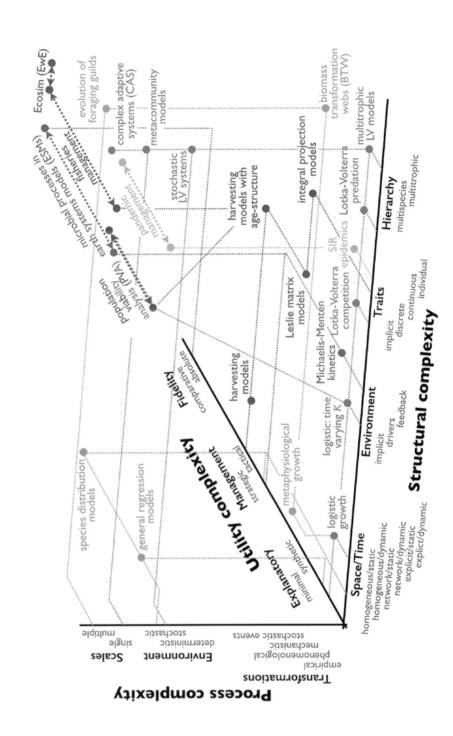

$$\frac{dA_i}{A_i dt} = f_A(x_i) = r_A - \frac{r_A \sum_k \alpha(x_i, x_k) A_k}{K_A(x_i)} + \frac{\sum_j b_{A_i P_j} w_{A_i P_j} P_j}{1 + h \sum_j w_{A_i P_j} P_j}$$

$$\frac{dP_j}{P_j dt} = f_P(y_i) = r_P - \frac{r_P \sum_k \alpha(y_j, y_k) P_k}{K_P(y_j)} + \frac{\sum_j b_{P_j A_i} w_{P_j A_i} A_i}{1 + h \sum_i w_{P_j A_i} A_i}$$

$$(3.9)$$

Looking at the top equation describing the dynamics of animal species i, we see that the per-capita population growth rate, $(1/A_i) dA_i/dt$, is considered equivalent to the trait-mediated demographic fitness, $f_A(x_i)$, in constant with the notion in Chapter 2.

The explicit form of the demographic fitness has three parts. The first is the intrinsic growth rate of animal species i, $r_A(x_i)$, and is considered to be density independent. If the trait of concern (here, related to mutualistic interactions) is unrelated to the intrinsic rate (e.g., the proboscis length of a pollinator does not necessarily correlate with ovulation), we could ignore the trait dependence; that is, $r_A(x_i) = r_A$. The second part represents trait-mediated within-guild interactions (e.g., resource competition between pollinators for host plants). The collective within-guild interactions are felt by animal species i after being scaled by its carrying capacity K_A and intrinsic rate of growth r_A, following the standard Lotka–Volterra form (see Eq. (2.14)). We set the carrying capacity to be trait dependent and follow the standard Gaussian shape interaction kernel, $K_A(x_i) = k_A N(x_A^{max}, \sigma_A, x_i)$, with k_A, x_A^{max} and σ_A representing the maximum carrying capacity, the ideal trait for resource consumption and the resource breadth. Let the intra-guild interaction kernel follow a standard Gaussian form (see Eq. (2.12) and Figure 2.9), $\alpha(x_1, x_2) = \exp(-(x_1 - x_2)^2/(2\sigma_C^2))$, representing stronger competition between species with similar traits, with σ_C the width of competition kernel.

←

Figure 3.29 Model typology. A way of categorising models is to locate them in a structure-/process-/utility-complexity space. Although measures do not exist to ordinate this space, segments of it can include a notion of increasing complexity with respect to the named primary categories along each axis: descriptions within each segment are indicated by the smaller text underlying each category name, staggered left to right to indicated increasing levels of complexity. Shades denote related modelling threads. Placement of models in this space is approximate, to facilitate labelling. Dots connected by arrows imply the labels above pertain to models that span the indicated range. From Getz et al. (2018), reprinted with permission.

(a)

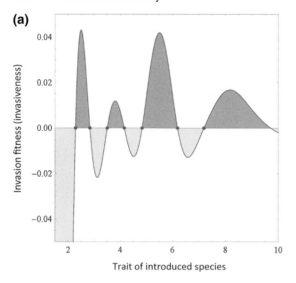

Trait of introduced species

(b)

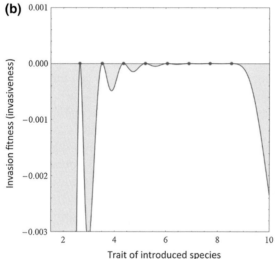

Trait of introduced species

Figure 3.30 Invasion fitness of an introduced species as a function of its trait value relative to the trait values of the resident species in the ecological network. Solid dots indicate the trait values of native resident species. **a**, A randomly generated network at its ecological equilibrium. Only introduced species with traits lying in the positive-fitness intervals can establish and invade the recipient ecological network; introduced species with traits in the negative-fitness intervals will be repelled by the network (i.e., wiped out from the resident species/ecosystem). Invasibility is thus defined as the total width of the positive-fitness intervals along the zero-fitness line. **b**, A saturated assembly is defined as the recipient network with zero invasibility. From Hui et al. (2016), reprinted with permission.

The third part represents cross-guild interactions in the form of a multispecies functional response, with the signs possibly tailored for other types of ecological networks (Hui et al. 2018). The mutualistic benefit, $b_{A_iP_j}$, represents trait-mediated assortative interactions that mutualistic partners with matched traits bring to each other in high fitness gain, also assumed to follow a Gaussian interaction kernel, $b_{A_iP_j} = b_{AP}(x_i, y_j) = c \cdot \exp(-(x_i - y_j)^2 / (2\sigma_m^2))$, with c maximum benefit and σ_m tolerance to trait difference between mutualistic partners. After a random encounter, the probability of actual interaction ($w_{A_iP_j}$) reflects interaction preference emerging from adaptive foraging behaviour based on the distribution of benefits from different mutualistic partners and their abundances (Egas et al. 2005),

$$w_{A_iP_j} = \frac{b_{A_iP_j}^{\beta} \sum_k A_k}{\sum_k \left(A_k b_{A_kP_j}\right)^{\beta}}, \tag{3.10}$$

where β is a parameter that determines whether the interaction is optimal ($\beta \gg 1$), suboptimal ($\beta = 1$) or neutral ($\beta = 0$), with the summation term $\sum_k A_k$ in the numerator for normalisation. As mutualistic benefit $b_{A_iP_j}$ and interaction preference $w_{A_iP_j}$ are both trait-mediated and rewarding more to assortative interactions, they can be combined by reinterpreting their product as realised benefit, with $b_{A_iP_j}$ replaced by a scale factor c and $w_{A_iP_j} = c \cdot \exp(-(x_i - y_j)^2 / (2\sigma_m^2))$ (Nuismer et al. 2010; Minoarivelo and Hui 2018). Parameter h represents the handling time and can be further assumed to be species specific. Interpretations of the plant equation in Eq. (3.9) can be mirrored.

As introduced in Chapter 2, the dynamics of trait evolution can be captured by the canonical equation of adaptive dynamics. However, we first need to calculate the invasion fitness of a rare mutant invading this system, $f_A(x_i')$ and $f_P(y_j')$, defined as the per capita growth rate when the rare mutants ($A_i' = 0$ and $P_j' = 0$) are invading the system at its asymptotically stable equilibrium ($\tilde{A}_i$ and $\tilde{P}_j$), with x_i' and y_j' representing mutant traits. This then allows us to calculate the selection gradient that drives directional selection, $g_{A_i}(x_i) = \partial f_A(x_i')/\partial x_i'\big|_{x_i'=x_i}$ and $g_{P_j}(y_j) = \partial f_P(y_j')/\partial y_j'\big|_{y_j'=y_j}$, which then drives the trait evolution,

$$\begin{aligned} \dot{x}_i &= m_A \tilde{A}_i g_{A_i} \\ \dot{y}_j &= m_P \tilde{P}_j g_{P_j} \end{aligned}, \tag{3.11}$$

where m_A and m_P are parameters proportional to the rate and variation of the mutation (in practice set to a small number to separate fast ecological events from slow evolutionary events). When directional trait evolution stops, other conditions need to be examined for disruption selection and potential evolutionary branching (Chapter 2). The previous text is solely explanatory and the model should be integrated for actual running, with only a few parameters as input mainly to control the shapes of interaction kernel functions, while the model outputs are typically population and trait dynamics.

From an initial pair of interacting animal and plant species, the system can undergo sequential diversification and give rise to an evolved ecological network (Figure 2.13). The interaction networks can be depicted as a quantitative interaction matrix with its elements (q_{ij}) defined as the non-linear functional response for both animal species i and plant species j (Berlow et al. 2004),

$$q_{ij} = \frac{1}{2} \left(\frac{A_i b_{A_i P_j} w_{A_i P_j} P_j}{1 + h w_{A_i P_j} P_j} + \frac{P_j b_{P_j A_i} w_{P_j A_i} A_i}{1 + h w_{P_j A_i} A_i} \right). \tag{3.12}$$

The model can produce mutualistic networks with structures and topologies comparable to real networks (Minoarivelo and Hui 2016a). First, quantitative connectance (= quantitative linkage density divided by the number of species in the network; Tylianakis et al. 2007) of emerged networks spans over 0.26 ± 0.14 [mean $\pm$ standard deviation], nestedness (WNODF; Almeida-Neto and Ulrich 2011) over 0.22 ± 0.25, and modularity (QuanBimo; Dormann and Strauß 2014) over 0.28 ± 0.22. The values of these three network features depend in an intricate way on the interplay of resource breadth (σ_A), tolerance to cross-guild trait difference (σ_m) and competition kernel breadth (σ_C); see Figure 3.31. Second, the level of specialization (H_2'; Blüthgen et al. 2006) increases only slightly with intra-guild resource breadth but declines notably with the tolerance to cross-guild trait difference. High resource accessibility makes cross-guild mutualistic interactions redundant, thus diminishing the potential for highly structured networks to emerge, while low tolerance to cross-guild trait difference (i.e., highly assortative interaction) facilitates reciprocal specialisation and can break down nested structures. Only a moderate level of specialisation can foster a high level of nestedness, which can be achieved through a moderate level of resource accessibility and tolerance to trait difference. Finally, network structures are intricately interlinked (Figure 3.32). After removing some strongly correlated structures, network size (total species richness, $N = n + m$), trait dispersion of plants (FDis;

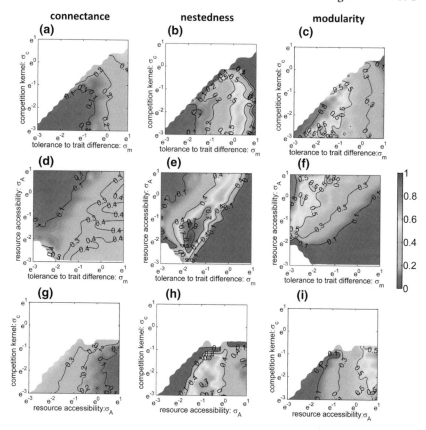

Figure 3.31 Quantitative connectance (first column), weighted nestedness (second column) and quantitative modularity (third column) of emerged bipartite mutualistic networks as a function of model parameters. In each plot we varied two parameters and kept the third parameter fixed. The white area in each plot represents an unfeasible parameter zone where the system becomes monomorphic with no network metrics calculated. From Minoarivelo and Hui (2016a), reprinted with permission. *A black and white version of this figure will appear in some formats. For the colour version, please refer to the plate section.*

Laliberté and Legendre 2010) and cross-guild trait complementarity ($Cp = -\ln D$, where $D = \left(\sum_i \sum_j D_{ij}\right)/(n \cdot m)$ is the average trait difference between interacting pairs, weighted by the normalised interaction strength, $D_{ij} = |x_i - y_j| \cdot \bar{q}_{ij}$, modified by Minoarivelo and Hui (2016a) from Guimarães et al. (2011)), can be used to explain connectance, nestedness and modularity of evolved networks (Table 3.2). Connectance and modularity are well explained especially by trait complementarity, but not nestedness.

Table 3.2 *Network architectures, including connectance, nestedness and modularity, explained by network size, trait complementary and trait dispersion in evolved bipartite mutualistic networks. In brackets: regression coefficient and the decrement in R^2 after removing the variable from the full model linear regression. From Minoarivelo and Hui (2016a)*

	R^2	Network size	Complementarity	Trait dispersion
Connectance	0.83	(−0.032,0.06)	(−0.116,0.64)	(−0.001[ns],0.01)
Nestedness	0.11	(0.035, 0.02)	(0.066,0.07)	(−0.059,0.06)
Modularity	0.72	(−0.0003,0.01)	(0.073,0.4)	(0.667,0.09)

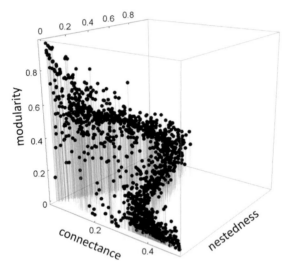

Figure 3.32 Relationships between connectance, nestedness and modularity for simulated networks. Each dot represents an evolved bipartite mutualistic network. From Minoarivelo and Hui (2016a), reprinted with permission.

Further manipulation of such evolved mutualistic networks can be done to assess the performance and impact of introduced species. There are different ways of introducing an alien species into such an evolved bipartite mutualistic network. Minoarivelo and Hui (2016b) simulated the propagule pressure equivalent to 5 per cent, 10 per cent and 25 per cent of the average population size of the recipient community. These alien propagules were introduced in five different ways,

- All individuals of the alien species were introduced only once before letting population dynamics change;
- Individuals of the alien species were divided into two groups of equal size. The first group was introduced at the initial time step, the second group after five time steps.
- Individuals of the alien species were introduced at three consecutive times separated by an interval of five time steps. The number of individuals introduced increased each time, representing 20 per cent, 30 per cent and 50 per cent of the total propagule pressure.
- Individuals were introduced three times but with declining numbers each time (50 per cent, 30 per cent and 20 per cent).
- Individuals were introduced in five consecutive times with equal densities (20 per cent each time), with introductions separated by five time steps.

Both the number of introductions and the propagule pressure at each introduction event are important determinants of invasion success (Figure 3.31A). Even if the dependence of invasion success on the number of introductions showed contingent patterns on the generalisation level of an invader, a general pattern still acknowledges the importance of multiple introductions, especially with decreasing propagule pressure, consistent with previous studies (Jeschke and Strayer 2006; Simberloff 2009). Indeed, a high number of introductions could help in lessening potential effects of environmental stochasticity (Simberloff 2009) or rescuing the establishment of each introduction as in the phenomenon of invasional meltdown (Traveset and Richardson 2014). This is probably achieved by the indirect positive effect of mutualism. In particular, once some individuals of the invader establish in the system, they proliferate the population densities of their mutualistic partners and subsequently facilitate the establishment of new arrivals from future introductions, potentially forming a positive feedback between aliens and natives in mutualistic networks (Memmott and Waser 2002; Bartomeus et al. 2008; Traveset and Richardson 2014; Minoarivelo and Hui 2016b). Moreover, the additional effect of decreasing propagule pressure in multiple introductions suggests that such proliferation from earlier introductions is diminishing or saturating with the number of established individuals.

The roles of the trait and the level of mutualism generalization of the invader, relative to the resident species in recipient communities, have also been explored (Minoarivelo and Hui 2016b). First, animal species were introduced with different trait values, ranging evenly from the lowest to the highest trait value of the natives. For simplicity, the trait value of the invader was reported as the relative trait value and scaled between 0 (lowest trait value) to 1 (highest trait value), relative to the traits of resident species. Second, the level of mutualism generalisation was measured as the tolerance of the invader to trait difference (i.e., σ_m) for feasible mutualistic interactions. A high tolerance to trait difference (large σ_m) suggests that mutualistic benefits can be assured by interacting with mutualistic partners with a wide range of traits, making the focal species a generalist. Different levels of generalization of the invader were considered relative to the generalisation level of the native community. Two measurements of invasion success were considered: invasiveness of the alien species, and the impact it has on the native community. Invasiveness (INVn) was defined as the relative growth rate of the invader: $\text{INVn} = \ln \left(A_{\text{final}}^{\text{alien}} / A_{\text{initial}}^{\text{alien}} \right)$, where $A_{\text{final}}^{\text{alien}}$ is the population density of the invader measured at a specified time after invasion, and $A_{\text{initial}}^{\text{alien}}$ is the total number of propagules introduced. The impact of the invasion (IMP) was measured as the absolute magnitude of change in the relative growth rate of the native species: $\text{IMP} = \left| \ln \left(A_{\text{final}}^{\text{resident}} / A_{\text{initial}}^{\text{resident}} \right) \right|$, where $A_{\text{final}}^{\text{resident}}$ is the total population size of resident species after invasion at the same specified time, and $A_{\text{initial}}^{\text{resident}}$ is the total population size before invasion.

Simulating invasions into evolved bipartite ecological networks helped us to clarify some important points (Figure 3.33). It allowed us to explore the determinants of invasion success for different combinations of invader characteristics (trait and level of generalization) and characteristics of the recipient community. Minoarivelo and Hui (2016b) found that the effect of invader characteristics on its invasion success is not unidirectional but intertwined. Furthermore, alien species with traits dissimilar to those of the natives can be the most invasive ones, supporting findings of other studies (Aizen et al. 2008; Campbell et al. 2015). The importance of high interaction generalization to invasiveness as observed by others (Aizen et al. 2008; Bartomeus et al. 2008; Vila et al. 2009; Albrecht et al. 2014) was only discovered when the traits of the invader were dissimilar to the average resident traits. The most invasive species (most abundant or widespread) is not always the one that has the biggest impact, highlighting the need to differentiate highly invasive species from those with big impact when prioritizing management. Besides trait distinctiveness, a high level of interaction generalisation is also a strong predictor of big impacts (Aizen et al. 2008; Albrecht et al. 2014), often through the cascading effect of interactions that are strongly associated with generalists. Different from

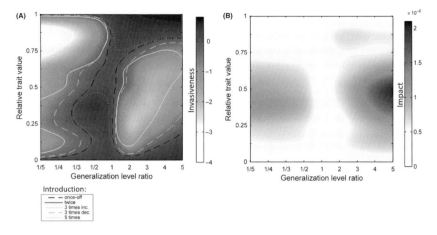

Figure 3.33 (a) Invasiveness and (b) impact of the invader as a function of its relative trait value and generalization level ratio, relative to those of the native community. Invasiveness and impact values represent the average over 100 medium-sized networks. Lines represent the zero level of invasiveness under different introduction modes. From Minoarivelo and Hui (2016b), under CC-BY Licence.

Campbell et al. (2015) but consistent with Morales and Traveset (2009), the evolved mutualistic networks felt the least impact from invaders with traits atypical of the native community. Interestingly, the overall impact from invasion observed in the evolved mutualistic network is detrimental rather than proliferating, probably because the detrimental effect from intensified competition has overridden the proliferation effect from added mutualism (Minoarivelo and Hui (2016b).

The impact of biological invasions on native population densities is small in mutualistic networks; it thus has a negligible effect on network archi-tecture. Such limited impact has been previously documented (Padron et al. 2009; Vila et al. 2009) and can be caused by the peripheral role of the invader in the network. In particular, Albrecht et al. (2014) found that the overall number of modules in an empirical pollination network was not altered by invasion, but only that modules were more connected by the super-generalist invaders. The trait value and node degree (level of inter-action generalization) of an invader determines its invasiveness and impact in the recipient network. Again, this can be explained by the balance of two forces (Minoarivelo and Hui 2016b): detrimental effects of competi-tion and beneficial effects arising from mutualism. While a high level of interaction generalisation often implies large benefits from mutualism, a trait atypical of resident species allows escape from competition.

Consequently, a generalist invader also possessing traits atypical of resident species is the most successful invader. In contrast, to have the highest impact on the recipient network, the invader's trait should be similar to those of an average resident species so that competition can be intensified. For an invader to have a big impact, it should either be an extreme generalist, allowing it to monopolize mutualistic benefits from most resident species, or be an extreme specialist, thereby depriving rival resident species of benefits from targeted mutualists. We elaborate the relationship between network structure and stability, and trait-mediated network invasibility in Chapters 4 and 7.

3.4.3 Ecological Fitting

Even for intricate and interlinked network structures, adaptive network models implementing co-evolution can still capture the many features of network topology, in particular connectance and modularity. However, their power to predict the level of nestedness is rather unsatisfactory (Table 3.2). Seeking a reasonable explanation of the high level of nestedness has been a key quest regarding structural emergence in mutualistic networks (Bascompte and Jordano 2014). Ecological fitting, the formation of biotic interactions through compatibility of traits after rapid trial-and-error matching, has been proposed as an alternative mechanism to co-evolution for establishing novel biotic interactions (Janzen 1985; Raimundo et al. 2018). Faced with a rapidly shifting eco-environmental context, species can undergo adaptive switching and readjust their interaction partners for short-term fitness gain, regardless of whether they share any joint evolutionary history (Agosta and Klemens 2008). While this can be achieved by adaptively changing a species' life-history strategies, a more straightforward explanation is that species simply enter or leave specific positions in a local community (Figure 3.34; e.g., Mora et al. 2020). Because ecological fitting is also a sorting process, whereby only 'fits' can persist, its products resemble those that are expected in co-evolution (Agosta 2006). Previous studies have highlighted the importance of species rewiring in promoting species coexistence and persistence in species-rich or nutrient-poor communities (Murdoch 1969; Kaiser-Bunbury et al. 2010; Staniczenko et al. 2010; Valdovinos et al. 2010; Ramos-Jiliberto et al. 2012; Suweis et al. 2013; but see Gilljam et al. 2015). In particular, a species can switch its partners based on the quality and quantity of available resources, as well as on the cost to acquire such resources (Whittall and Hodges 2007; Valdovinos et al. 2010).

a Species per position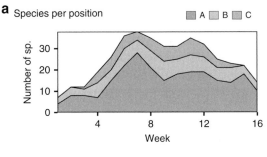

b Position properties

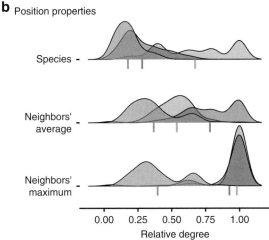

Figure 3.34 Distinguishing properties of the distinct groups of pollinator positions. **a**, Number of pollinator species in each group of positions over time. Each level of grey represents a different group of positions. **b**, Different properties summarising the species forming each group. The top panel shows the relative degree distribution of species in each group. The middle panel shows the neighbours' average relative degree for the species in each group. The bottom panel depicts the relative degree distribution of the most connected neighbour of every species in each group. The coloured segments depicted under the distributions characterise the mean of each distribution. From Mora et al. (2020), under CC-BY Licence.

Ecological fitting requires species to switch their partners in response to changes of context such as the temporal presence/absence of previous and new mutualists from disturbance and biological invasion (Valdovinos et al. 2010). This adaptive nature of selecting and switching interaction partners (i.e., tuning target resources) in an ecological network is essential for the survival of consumers that are competing for available resources (van Baalen et al. 2001; Kimbrell and Holt 2005;

Staniczenko et al. 2010). Mutualism is essentially a case of reciprocal exploitation; consequently, both species rewiring and ecological fitting can be framed under the theory of optimal and adaptive foraging, where species can adapt their diet based on current profitability, encounter rate and past experience (Stephens and Krebs 1986; Fossette et al. 2012; Zhang and Hui 2014). Consumer species often possess an adaptive diet that is constantly modifying through recent past experiences (Zhang and Hui 2014), which can lead to the heuristic behavioural strategy of Win-Stay, Lose-Shift (Nowak and Sigmund 1993) or habitat selection rule of Good-Stay, Bad-Disperse (Hui et al. 2012). Ecological fitting via interaction rewiring has received strong support in the field of adaptive foraging, where a species optimises its diet for maximum rate of energy intake (Fossette et al. 2012), and in the field of interaction fidelity and promiscuity (Sanchez 2015). This provides a behavioural strategy for adaptive interaction switching, which has been implemented in a number of network-emergence models for explaining network structures in mutualistic communities (Kondoh 2003, Staniczenko et al. 2010, Zhang et al. 2011, Suweis et al. 2013, Mougi and Kondoh 2016). With alien species facing evident ecological novelty in their new homes (Heger et al. 2019), ecological fitting can be a crucial prerequisite for establishment and invasion success.

Considering ecological fitting in structural emergence models could give rise to realistic network architectures. Different forms of interaction switching are available to achieve ecological fitting (Figure 2.8): survival through selecting and switching partners (Staniczenko et al. 2017); decision-making based on profitability and encounter rate (Zhang and Hui 2014); Wallace's (1889) elimination of the unfit; and Darwin's survival of the fittest (Kondoh 2003; Suweis et al. 2013). Several models have been proposed to integrate adaptive interaction switching into network emergence. There is, however, much scope for improvement and refinement in this area, especially regarding the predictive power of such models and the need to incorporate all essential forces of evolution (Saavedra et al. 2009, Suweis et al. 2013). Adaptive interaction switching implemented in most of these models only emphasizes adaptation from one or two particular processes, specifically by using the classical optimality methodology in evolutionary ecology (Suweis et al. 2013), or optimization-based analytical treatment for adaptive behaviour (Kondoh 2003; Valdovinos et al. 2010). To capture the essence of network evolution, the roles of both optimization and random drift need to be appreciated, as together they represent a combination of both adaptive and non-adaptive forces that potentially counter-balance each other in structuring network architecture. A hybrid model that emphasizes both adaptation and drift offers a more complete picture of network evolution through adaptive interaction switching (Zhang et al. 2011; Nuwagaba et al. 2015). Consumers in the model are allowed not only to selectively eliminate the unfit resources from their diets based on the benefits and abundance of these resources (i.e., an adaptive process) but also to try new resources randomly (i.e., random drift). This hybrid behavioural rule depicts adaptation in line with Alfred Russell Wallace's (1889) view of natural selection – via the elimination of the unfit and random drift as the innovation in foraging behaviours – separating it from other rules of adaptive interaction switching (Figure 2.8). This hybrid behavioural rule not only emphasises the adaptive process by which the consumer gradually improves its resource utilization efficiency by retaining highly beneficial and abundant resources and eliminating less beneficial and rare ones, but also allows for behavioural innovation whereby new resources can be exploited by consumers via the random drift of interactions.

Let us look into details of a bipartite mutualistic network model that implements ecological fitting via adaptive interaction switching. The model was used to explore the structural emergence in pollination networks across the Galápagos Islands (Traveset et al. 2013; Nnakenyi et al. 2019). The Galápagos flora consists of more than 1400 vascular plants, of which 59 per cent are aliens, 14 per cent are endemic and 27 per cent are native (Jaramillo and Guézou 2013). Of the pollination networks of ten Galápagos Islands, four were found to be significantly nested. The model implements the mutualistic relationship between plant and animal species using Holling's (1959) type II functional response and adaptive interaction switching of species interaction (Kondoh 2003; Zhang et al. 2011). Specifically, the pollination network was considered a bipartite network of plants (*P*) and animals (*A*) *for each island separately*. In a local network, the population dynamics for *m* number of plants and *n* number of animals can be described by the following (Okuyama and Holland 2008; Bastolla et al. 2009; Nnakenyi et al. 2019),

$$
\frac{dP_i}{dt} = r_i^{(P)} P_i - \alpha_i^{(P)} P_i^2 + \frac{\sum_{j=1}^{n} a_{ij} \beta_{ij}^{(P)} A_j P_i}{1 + h \sum_{k=1}^{n} a_{ik} A_k}
$$

$$
\frac{dA_j}{dt} = r_j^{(A)} A_j - \alpha_j^{(A)} A_j^2 + \frac{\sum_{i=1}^{m} a_{ij} \beta_{ij}^{(A)} P_i A_j}{1 + h \sum_{k=1}^{m} a_{jk} P_k}
$$

$$(3.10)$$

where P_i and A_j are population densities of plant species i and animal species j. Superscript (P) and (A) represent plant and animal species, respectively. The model has three similar parts on the right-hand side as Eq. (3.9), although the intra-guild interactions were removed for simplicity. In the functional response that describes the fitness gain from the mutualistic interactions, parameter a_{ij} is the element on row i and column j of the binary interaction matrix $M_{m \times n}$, and indicates whether plant i interacts with animal j ($a_{ij} = 1$) or not ($a_{ij} = 0$). Parameters $\beta_{ij}^{(P)}$ and $\beta_{ij}^{(A)}$ describe the per-capita benefits obtained per unit of time by plant i from interacting with animal j and by animal j from pollinating plant i, respectively. The parameter h stands for the handling time, representing the proportion of time a species spends on searching and handling resources (i.e., not all time was used for consuming resources).

To implement adaptive interaction switching, where species can adaptively rewire their interactions based on *the elimination of the unfit* algorithm (Zhang et al. 2011), we can follow a two-step procedure. Step 1 eliminates the least contributor to a species fitness, after evaluating the contribution from all its interaction partners. Afterwards, step 2 randomly rewires to a new species after dropping the least contributing partner species. For this model we could implement switching using the following pseudocode (Nnakenyi et al. 2019):

Step 1: At each time step when numerically solving the previous equations, a species is first selected at random. For example, say animal j was selected, we then evaluate the relative benefit contribution received from interacting with a plant species by comparing $a_{ij}\beta_{ij}^{(A)}P_i$ for $i = 1, \ldots, m$, and identify the plant species that interacts with animal j (those $a_{ij} = 1$) but contributes the least (say, plant k) to animal j's benefit/fitness gain (thus the one with the minimum nonzero relative contribution). Thereafter, animal j drops its interaction with plant k by setting $a_{kj} = 0$.

Step 2: We then choose a non–interacting plant species at random, say plant l (those currently with zero relative contribution, $a_{lj} = 0$); we then switch the interaction between animal j and plant k to between animal j and plant l by setting $a_{kj} = 0$ and $a_{lj} = 1$.

The initial interaction matrix a_{ij} can be randomly assigned as a binary 0/1 matrix of dimension $m \times n$ with c proportion of elements being 1 and the rest 0, where m represents the number of plant species, n the number of animal species and c the binary connectance observed. Other model parameters and initial population densities can also be randomly assigned from uniform and normal distributions, to ensure the persistence of all observed species. For each pollination network in an island, the model takes three numbers (m, n and c) as the input and the encountering matrix $\langle a_{ij} P_i A_j \rangle_{m \times n}$ as the output, from which we can calculate all necessary network metrics. For pollination networks on ten Galápagos Islands, this model, representing the process of ecological fitting via adaptive inter-action switching, together with island age, island area, isolation (biogeo-graphic factors) and sampling effort, explained 92 per cent of the variation in observed level of nestedness (Figure 3.35). The process of adaptive interaction switching is the greatest independent contributor (37.27 per cent), followed by sampling effort (27.76 per cent) and island area (22.11 per cent), while island isolation and age explained each less than 10 per cent of the variation in observed nestedness.

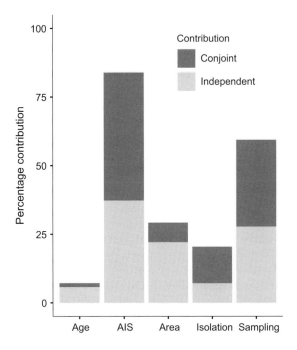

Figure 3.35 Nestedness and variance hierarchical partitioning showing the percentage of the independent and conjoint contribution of each of the explanatory variables: adaptive interaction switching model, age, area, isolation and sampling effort. From Nnakenyi et al. (2019), reprinted with permission.

When the performance of the generic adaptive interaction switching model was compared with two varieties (Zhang et al. 2011): benefit- and demographic-neutral, with the former representing equal benefit (β_{ij}) across species received from mutualistic interactions while the latter equal intrinsic growth rate (r_i) and density dependence (α_i) across species, interesting insights emerged (Figure 3.36). Besides good convergence and resemblance to observed networks, comparisons of the two varieties

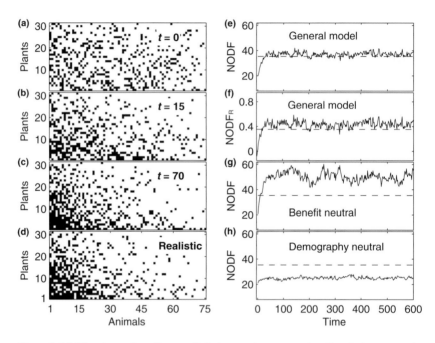

Figure 3.36 The dynamics of mutualistic interaction networks. Simulations started from a random interaction matrix, which has the same numbers of plant and animal species, as well as the same number of interactions, as the real mutualistic network of Elberling and Olesen (P11 in Mutualistic Network Database; Arctic tundra). Panels (a), (b) and (c) are snapshots of interaction matrices at the beginning and times 15 and 70, respectively. (Note that each time is m + n time steps.). Panel (d) illustrates the interaction matrix of Elberling and Olesens, real network. Panels (e) and (f) give the dynamical behaviour of nestedness (NODF) and relative nestedness (NODFR with the null model Ce provided by ANINHADO 3.0) over time for the general model. Panels (g) and (h) give the dynamics of nestedness for models with only benefit-neutral and demography-neutral interactions, respectively. Dashed lines indicate NODF or NODFR of the real network. See additional dynamics of absolute and relative nestedness using other null models in ANINHADO 3.0 in Section S3 of the supporting online materials. From Zhang et al. (2011), reprinted with permission.

suggest that equal benefit (i.e., lack of benefit heterogeneity among species) could foster an even higher level of nestedness, while equal demography (i.e., lack of demographic variations among species) hampers network nestedness. Real networks are compromised by benefit neutrality (or interaction neutrality) that enhances nestedness and demographic neutrality that suppresses nestedness. Consequently, we could expect that biological invasions could increase the level of nestedness in recipient mutualistic networks if the alien species has different demographic rates from the average resident species but offers an average interaction benefit. Along the same line, an alien species can reduce network nestedness by possessing average demography but offering a different level of benefit to resident species.

A dynamic ecological network model implementing adaptive interaction switching to emulate ecological fitting in real communities has performed very well in explaining the full suite of network structures when testing using real network data. It can explain 73 per cent of variation in nestedness for 48 pollination networks and 33 seed-dispersal networks (Figure 3.37; Zhang et al. 2011), 89.7–91 per cent of variation in modularity and 58–64 per cent of variation in nestedness for 28 plant–herbivore networks and host–parasite networks (Nuwagaba et al. 2015). This dynamic ecological network model can produce not only network architectures resembling real networks but also other network topologies, such as node-degree distributions (Figure 3.38). This has led to a bold claim: this adaptive interaction switching model almost perfectly explains fundamental network architecture as only 10 per cent of variance is unaccounted for – which could be due to many other stochastic factors or sampling artefacts.

3.4.4 Open Network Emergence

Together with structural emergence models and adaptive networks models (via co-evolution and ecological fitting), we can have a reasonable understanding of how ecological networks assemble and collectively and adaptively respond to incoming alien species and how an alien species performs if we know the alien species and recipient ecological networks well. Assembled ecological networks, via co-evolution, ecological fitting and biological invasions, and allowing local extinction and extirpation, provide us with an ideal model system and a set of models to understand open adaptive networks. A framework of open network emergence can be proposed (Figure 3.39). This framework highlights

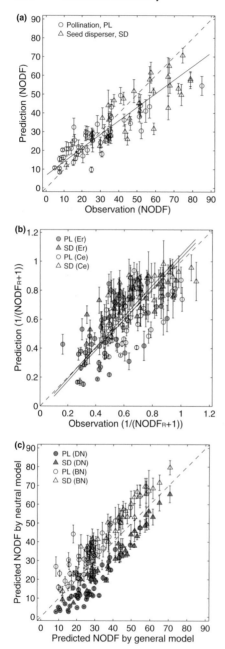

Figure 3.37 The predicted nestedness of 48 pollination (PL) and 33 seed-dispersal (SD) networks. (a) NODF predicted by the general interaction switch model vs. observations

three pathways of network structural emergence: co-evolution, eco-logical fitting and open flux (invasion and extinction). Key drivers behind the emergence of network architecture can now be elucidated. First, evolutionary history, community assembly processes, biological invasions and environmental characteristics (e.g., productivity and het-erogeneity) could largely determine the number of species (i.e., network size) and the composition/turnover of species and traits (i.e., model parameters) that a community can accommodate. Second, once the bounds of the arena are set (network size and connectance), co-evolution and ecological fitting can together fine-tune the network, representing the two important forces of community assembly – adaptation and drift. Adaptive interaction switching (ecological fitting) and the context of co-evolutionary history, local community composition and environmental features could together explain the bulk of network architectures. This conceptual framework of network emergence also highlights priorities for future research.

With the interplay of key processes in an ecological network, intricate and interlinked network structures and topologies emerge through co-evolution and ecological fitting. Each network structure represents one of many facets of how the community assembles and functions. Biological invasions have greatly sped up the flow and exchange of species within the community. Consensus has not been reached, though we are trying to suggest a way forward. Our aim in this chapter was to connect essential assembly processes to explain complex and diverse interweaved network patterns. As a prelude to Chapter 4, allow us to use some metaphoric language here. To make sense of this rapidly burgeoning research field with metrics, data, models and theories jousting for attention in the quest to explain pattern formation and emergence and *prophesy* the future, we are reminded of the words of the character Joseph Knecht in Hermann Hesse's (1943) novel, who asks,

←

Figure 3.37 (cont.) of real networks. (b) NODFR [transformed to 1 / (NODFR + 1)] predicted by the general model vs. observations of real networks. (c) Predictions from the general model vs. predictions from the model with benefit-neutral (BN) and with demography-neutral (DN) interactions. Predictions are the averages of 300 samplings each m + n time steps after 200 · (m + n) time steps in the simulations. Error bar indicates the standard deviation. The solid lines represent regression results using the reduced major axis and the dashed lines indicate perfect agreement between predictions and observations. From Zhang et al. (2011), reprinted with permission.

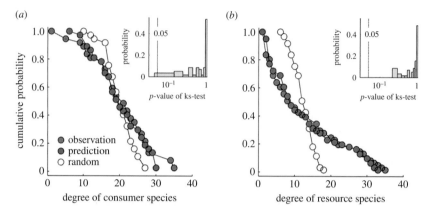

Figure 3.38 Comparison of the predicted and observed node–degree distributions of consumer species (a) and resource species (b) for the network from Townsend et al. (1998). 'Random' is the degree distribution of the initial random network at the beginning of the simulation. Inset shows the distribution of the p–value of the Kolmogorov–Smirnov (ks) test for 61 real bipartite antagonistic networks. From Nuwagaba et al. (2015), reprinted with permission.

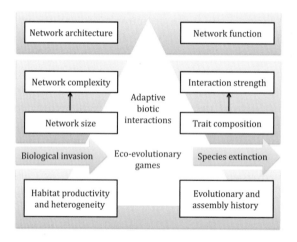

Figure 3.39 Open adaptive network emergence. Network architecture and function emerge from co-evolution and ecological fitting of assembled species through biological invasions and species extinction, within environmental context and evolutionary constraints.

If only there were a dogma to believe in. Everything is contradictory, everything is tangential; there are no certainties anywhere. Everything can be interpreted one way and then again interpreted in the opposite sense. The whole of world history can be explained as development and progress and can also be seen as nothing but decadence and meaninglessness. Isn't there any truth? Is there no real and valid doctrine?

We will tentatively address this question in Chapter 4. It only makes sense when we further relate these patterns to the functioning of an ecological network, in particular, to another multifaceted concept – network stability. By merging the two together – network structure and stability – we see important themes emerge that the organisation of natural systems – in the form of a flexible and agile ecological network – is operating at the margin of stability-instability. This is different from Tansley's (1935) conjecture of multiple climaxes and the jumping basins of attraction, from one regime to another. Instead, we will suggest an open-ended surfing at the boundary of chaos; although the short-term direction is predictable, the future alternative regime is undetermined and unformed. At this boundary with feasibility and stability, the system is more constrained than optimised by evolution. It is therefore important to differentiate a constrained system from an optimal one. An optimised system exhibits exponential patterns, while a constrained system sitting at the edge of collapse exhibits scale-free patterns (power law and alike such as truncated power law and lognormal structures). Adaptive biotic interactions via evolution, co-evolution and fitting provide means for the system to move towards the boundary, which affords a means of non-adaptive system-level selection.

Is the ESS better or ideal to its constituents? Not necessarily. It is rather a situation of the best of a sub-optimal set of options – an unsatisfying democracy, a forever-deflating balloon, a surfer riding never-ending waves. All network patterns emerge at this boundary, while the emergent patterns are merely the dance of survival.

References

Ackerly DD, Cornwell WK (2007) A trait-based approach to community assembly: Partitioning of species trait values into within- and among-community components. *Ecology Letters* 10, 135–145.

Agosta SJ (2006) On ecological fitting, plant–insect associations, herbivore host shifts, and host plant selection. *Oikos* 114, 556–565.

Agosta SJ, Klemens JA (2008) Ecological fitting by phenotypically flexible genotypes: implications for species associations, community assembly and evolution. *Ecology Letters* 11, 1123–1134.

Agrawal AA, et al. (2005) Enemy release? An experiment with congeneric plant pairs and diverse above- and belowground enemies. *Ecology* 86, 2979–2989.

Aitchison J, Brown JAC (1957) *The Lognormal Distribution with Special Reference to Its Uses in Economics*. Cambridge: Cambridge University Press.

Aizen MA, et al. (2008) Invasive mutualists erode native pollination webs. *PLoS Biology* 6, e31.

Albert R, Barabási A (2002) Statistical mechanics of complex networks. *Reviews of Modern Physics* 74, 47–97.

Albrecht M (2014) Consequences of plant invasions on compartmentalization and species' roles in plant–pollinator networks. *Proceedings of the Royal Society B: Biological Sciences* 281, 20140773.

Alcántara JM, Rey PJ (2012) Linking topological structure and dynamics in ecological networks. *The American Naturalist* 180, 186–199.

Allesina S, Tang S (2012) Stability criteria for complex ecosystems. *Nature* 483, 205–208.

Almeida-Neto M, et al. (2008) A consistent metric for nestedness analysis in ecological systems: reconciling concept and measurement. *Oikos* 117, 1227–1239.

Almeida-Neto M, Ulrich W (2011) A straightforward computational approach for measuring nestedness using quantitative matrices. *Environmental Modelling & Software* 26, 173–178.

Andow DA, Imura O (1994) Specialization of phytophagous arthropod communities on introduced plants. *Ecology* 75, 296–300.

Atmar W, Patterson BD (1993) The measure of order and disorder in the distribution of species in fragmented habitat. *Oecologia* 96, 373–382.

Atmar W, Patterson BD (1995) *The Nestedness Temperature Calculator: A Visual Basic Program, Including 294 Presence Absence Matrices AICS Res.* Chicago: NM and the Field Museum.

Bai R, et al. (2017) Microbial community and functional structure significantly varied among distinct types of paddy soils but responded differently along gradients of soil depth layers. *Frontiers in Microbiology* 8, 1–16.

Bais HP, et al. (2003) Allelopathy and exotic plant invasion: from molecules and genes to species interactions. *Science* 301, 1377–1380.

Baiser B, et al. (2010) Connectance determines invasion success via trophic interactions in model food webs. *Oikos* 119, 1970–1976.

Banašek-Richter C, et al. (2009) Complexity in quantitative food webs. *Ecology* 90, 1470–1477.

Barabási A, Albert R (1999) Emergence of scaling in random networks. *Science* 286, 509—512.

Bartomeus I, et al. (2008) Contrasting effects of invasive plants in plant–pollinator networks. *Oecologia* 155, 761–770.

Bascompte J, Jordano P (2007) Plant-animal mutualistic networks: the architecture of biodiversity. *Annual Review of Ecology, Evolution, and Systematics* 38, 567–593.

Bascompte J, Jordano P (2014) *Mutualistic Networks.* Princeton: Princeton University Press.

Bascompte J, et al. (2003) The nested assembly of plant–animal mutualistic networks. *Proceedings of the National Academy of Sciences USA* 100, 9383–9387.

Bascompte J, et al. (2006) Asymmetric coevolutionary networks facilitate biodiversity maintenance. *Science* 312, 431–433.

Bastolla U, et al. (2009) The architecture of mutualistic networks minimizes competition and increases biodiversity. *Nature* 458, 1018–1020.

Beckage B, et al. (2011) The limits to prediction in ecological systems. *Ecosphere* 2, a125.

Bell G (2001) Neutral macroecology. *Science* 293, 2413–2418.

Bellay S, et al. (2011). A host-endoparasite network of Neotropical marine fish: Are there organizational patterns? *Parasitology* 138, 1945.

Berlow E (1999) Strong effects of weak interactions in ecological communities. *Nature* 398, 330–334.

Berlow E, et al. (2004) Interaction strengths in food webs: Issues and opportunities. *Journal of Animal Ecology* 73, 585–598.

Bever JD (2002) Negative feedback within a mutualism: Host–specific growth of mycorrhizal fungi reduces plant benefit. *Proceedings of the Royal Society B: Biological Sciences* 269, 2595–2601.

Birnbaum C, et al. (2016) Nitrogen-fixing bacterial communities in invasive legume nodules and associated soils are similar across introduced and native range populations in Australia. *Journal of Biogeography* 43, 1631–1644.

Blackburn TM, et al. (2011) A proposed unified framework for biological invasions. *Trends in Ecology & Evolution* 26, 333–339.

Blossey B, Notzold R (1995) Evolution of increased competitive ability in invasive nonindigenous plants: A hypothesis. *Journal of Ecology* 83, 887–889.

Blumenthal D, et al. (2009) Synergy between pathogen release and resource availability in plant invasion. *Proceedings of the National Academy of Sciences USA* 106, 7899–7904.

Blüthgen N, et al. (2006) Measuring specialization in species interaction networks. *BMC Ecology* 6, 9.

Boccaletti S, et al. (2006) Complex networks: Structure and dynamics. *Physics Reports* 424, 175–308.

Bossdorf O, et al. (2005) Phenotypic and genetic differentiation between native and introduced plant populations. *Oecologia* 144, 1–11.

Bramon MB, et al. (2020) Untangling the seasonal dynamics of plant-pollinator communities. *Nature Communications* 11, 4086.

Broido AD, Clauset A (2019) Scale-free networks are rare. *Nature Communications* 10, 1017.

Brown G, Sanders JW (1981) Lognormal genesis. *Journal of Applied Probability* 18, 542–547.

Bruno R, et al. (2005) Mesh networks: commodity multihop ad hoc networks. *IEEE Communications Magazine* 43, 123–131.

Cadotte MW, et al. (2006) Ecological patterns and biological invasions: Using regional species inventories in macroecology. *Biological Invasions* 8, 809–821.

Cagnolo L, et al. (2011) Network topology: Patterns and mechanisms in plant-herbivore and host-parasitoid food webs. *Journal of Animal Ecology* 80, 342–351.

Callaway RM, Aschehoug ET (2000) Invasive plants versus their new and old neighbors: a mechanism for exotic invasion. *Science* 290, 521–523.

Callaway RM, Ridenour WM (2004) Novel weapons: Invasive success and the evolution of increased competitive ability. *Frontiers in Ecology and the Environment* 2, 436–443.

Camacho J, et al. (2002) Robust patterns in food web structure. *Physical Review Letters* 88, 228102.

Campbell C, et al. (2015) Plant–pollinator community network response to species invasion depends on both invader and community characteristics. *Oikos* 124, 406–413.

CaraDonna PJ, Waser NM (2020) Temporal flexibility in the structure of plant–pollinator interaction networks. *Oikos* 129, 1369–1380.

Carpenter D, Cappuccino N (2005) Herbivory, time since introduction and the invasiveness of exotic plants. *Journal of Ecology* 93, 315–321.

Carroll SP, et al. (2005) And the beak shall inherit: Evolution in response to invasion. *Ecology Letters* 8, 944–951.

Catford JA, et al. (2009) Reducing redundancy in invasion ecology by integrating hypotheses into a single theoretical framework. *Diversity and Distributions* 15, 22–40.

Cattin MF, et al. (2004) Phylogenetic constraints and adaptation explain food-web structure. *Nature* 427, 835–839.

Chase JM (2003) Community assembly: When should history matter? *Oecologia* 136, 489–498.

Chase JM, et al. (2018) Embracing scale-dependence to achieve a deeper under-standing of biodiversity and its change across communities. *Ecology Letters* 21, 1737–1751.

Chesson P (2000) Mechanisms of maintenance of species diversity. *Annual Review of Ecology and Systematics* 31, 343–366.

Clark JS (2012) The coherence problem with the unified neutral theory of biodiver-sity. *Trends in Ecology & Evolution* 27, 198–202.

Clements FE (1916) *Plant Succession: An Analysis of the Development of Vegetation.* Washington, DC: Carnegie Institution of Washington.

Cohen JE, Briand F (1984) Trophic links of community food webs. *Proceedings of the National Academy of Sciences USA* 81, 4105–4109.

Cohen JE, Newman CM (1985) A stochastic theory of community food webs I: Models and aggregated data. *Proceedings of the Royal Society B: Biological Sciences* 224, 421–448.

Cohen JE, et al. (1990) *Community Food Webs: Data and Theory.* Berlin: Springer-Verlag.

Colautti R, et al. (2014) Quantifying the invasiveness of species. *NeoBiota* 21, 7.

Colautti RI, et al. (2004) Is invasion success explained by the enemy release hypothesis? *Ecology Letters* 7, 721–733.

Colautti RI, et al. (2017) Invasions and extinctions through the looking glass of evolutionary ecology. *Philosophical Transactions of the Royal Society B: Biological Sciences* 372, 20160031.

Condit R (1998) *Tropical Forest Census Plots.* Berlin: Springer.

Connell JH (1990) Apparent versus "Real" competition in plants. In Grace J, Tilman D (eds.) *Perspectives on Plant Competition*, pp. 9–26. London: Academic Press.

Connor EF, Simberloff D (1979) The assembly of species communities: Chance or competition? *Ecology* 60, 1132–1140.

Connor EF, Simberloff D (1983) Interspecific competition and species co-occurrence patterns on islands: Null models and the evaluation of evidence. *Oikos* 41, 455–465.

Cornwell WK, Schwilk DW, Ackerly DD (2006) A trait-based test for habitat filtering: Convex hull volume. *Ecology* 87, 1465–1471.

Costello CJ, Solow AR (2003) On the pattern of discovery of introduced species. *Proceedings of the National Academy of Sciences USA* 100, 3321–3323.

Cottingham K, Brown B, Lennon J (2001) Biodiversity may regulate the temporal variability of ecological systems. *Ecology Letters* 4, 72–85.

Coutts SR, Helmstedt KJ, Bennett JR (2018) Invasion lags: The stories we tell ourselves and our inability to infer process from pattern. *Diversity and Distributions* 24, 244–251.

Cuddington K, Hastings A (2016) Autocorrelated environmental variation and the establishment of invasive species. *Oikos* 125, 1027–1034.

Dale MRT, Fortin M-J (2021) *Quantitative Analysis of Ecological Networks.* Cambridge: Cambridge University Press.

D'Andrea R, Gibbs T, O'Dwyer JP (2020) Emergent neutrality in consumer-resource dynamics. *PLoS Computational Biology* 16, e1008102.

Davies KF, et al. (2005) Spatial heterogeneity explains the scale dependence of the native–exotic diversity relationship. *Ecology* 86, 1602–1610.

Delmas E, et al. (2019) Analysing ecological networks of species interactions. *Biological Reviews* 94, 16–36.

Diamond JM (1975) Assembly of species communities. In Diamond JM, Cody ML (eds.) *Ecology and Evolution of Communities*, pp. 342–344. Boston: Harvard University Press.

Dormann CF, Strauß R (2014) A method for detecting modules in quantitative bipartite networks. *Methods in Ecology and Evolution* 5, 90–98.

Dornelas M, et al. (2014) Assemblage time series reveal biodiversity change but not systematic loss. *Science* 344, 296–299.

Drake JA (1990) The mechanics of community assembly and succession. *Journal of Theoretical Biology* 147, 213–233.

Drossel B, et al. (2001) The influence of predator–prey population dynamics on the long-term evolution of food web structure. *Journal of Theoretical Biology* 208, 91–107.

Duncan R, Williams P (2002) Darwin's naturalization hypothesis challenged. *Nature* 417, 608–609.

Dunne JA, et al. (2002) Network structure and biodiversity loss in food webs: Robustness increases with connectance. *Ecology Letters* 5, 558–567.

Dupont YL, Olesen JM (2012) Stability of modular structure in temporal cumulative plant–flower-visitor networks. *Ecological Complexity* 11, 84–90.

Egas M, et al. (2005) Evolution of specialization and ecological character displacement of herbivores along a gradient of plant quality. *Evolution* 59, 507–520.

Egler FE (1954) Vegetation science concepts I: Initial floristic composition, a factor in old-field vegetation development with 2 figs. *Vegetatio Acta Geobot* 4, 412–417.

Ehrlich PR, Raven PH (1964) Butterflies and plants: A study in coevolution. *Evolution* 18, 586–608.

Elton CS (1958) *The Ecology of Invasions by Animals and Plants.* London: Metheun.

Epanchin-Niell RS, Hastings A (2010) Controlling established invaders: Integrating economics and spread dynamics to determine optimal management. *Ecology Letters* 13, 528–541.

Eppinga MB, et al. (2006) Accumulation of local pathogens: A new hypothesis to explain exotic plant invasions. *Oikos* 114, 168–176.

Erdős P, Rényi A (1959) On Random Graphs. I. *Publicationes Mathematicae* 6, 290–297.

Euler L (1736) Solutio problematis ad geometriam situs pertinenti. *Commentarii Academiae Scientiarum Petropolitanae* 8, 128–140.

Fakami T (2015) Historical contingency in community assembly: Integrating niches, species pools, and priority effects. *Annual Review of Ecology, Evolution, and Systematics* 46, 1–23.

Fargione JE, Tilman D (2005) Diversity decreases invasion via both sampling and complementarity effects. *Ecology Letters* 8, 604–611.

Farji-Brener AG, Amador-Vargas S (2014) Hierarchy of hypotheses or cascade of predictions? A comment on Heger et al. (2013). *Ambio* 43, 1112–1114.

Ferrer-Paris JR, et al. (2013) Congruence and diversity of butterfly-host plant associations at higher taxonomic levels. *PLoS ONE* 8, e63570.

Fontaine C, et al. (2011) The ecological and evolutionary implications of merging different types of networks. *Ecology Letters* 14, 1170–1181.

Fortuna MA, et al. (2010) Nestedness versus modularity in ecological networks: Two sides of the same coin? *Journal of Animal Ecology* 79, 811–817.

Fossette S, et al. (2012) Does prey size matter? Novel observations of feeding in the leatherback turtle (*Dermochelys coriacea*) allow a test of predator–prey size relationships. *Biology Letters* 8, 351–354.

Fricke EC, Svenning J (2020) Accelerating homogenization of the global plant–frugivore meta-network. *Nature* 585, 74–78.

Frost CM, et al. (2019) Using network theory to understand and predict biological invasions. *Trends in Ecology & Evolution* 34, 831–843.

Gaertner M, et al. (2009). Impacts of alien plant invasions on species richness in Mediterranean-type ecosystems: A meta-analysis. *Progress in Physical Geography* 33, 319–338.

Gaertner M, et al. (2014) Invasive plants as drivers of regime shifts: Identifying high-priority invaders that alter feedback relationships. *Diversity and Distributions* 20, 733–744.

Galiana N, et al. (2014) Invasions cause biodiversity loss and community simplification in vertebrate food webs. *Oikos* 123, 721–728.

Galiana N, et al. (2018) The spatial scaling of species interaction networks. *Nature Ecology & Evolution* 2, 782–790.

García-Callejas D, et al. (2018) Multiple interactions networks: Towards more realistic descriptions of the web of life. *Oikos* 127, 5–22.

Gaston KJ (2003) *The Structure and Dynamics of Geographic Ranges*. Oxford: Oxford University Press.

Gause GF (1934) Experimental analysis of Vito Volterra's mathematical theory of the struggle for existence. *Science* 79, 16–17.

Gauzens B, et al. (2015) Trophic groups and modules: Two levels of group detection in food webs. *Journal of the Royal Society Interface* 12, 20141176.

Genini J, et al. (2012) Mistletoes play different roles in a modular host–parasite network. *Biotropica* 44, 171–178.

Getz WM, et al. (2018) Making ecological models adequate. *Ecology Letters* 21, 153–166.

Gilljam D, Curtsdotter A, Ebenman B (2015) Adaptive rewiring aggravates the effects of species loss in ecosystems. *Nature Communications* 6, 8412.

Gilpin M, Hanski I (eds.) (1991) *Metapopulation Dynamics: Empirical and Theoretical Investigations*. London: Academic Press.

Girvan M, Newman MEJ (2002) Community structure in social and biological networks. *Proceedings of the National Academy of Sciences USA* 99, 7821–7826.

Gleason HA (1927) Further views on the succession concept. *Ecology* 8, 299–326.

Gotelli NJ (2000) Null model analysis of species co-occurrence patterns. *Ecology* 81, 2606–2621.

Gotelli NJ, Graves GR (1996) *Null Models in Ecology*. Washington, DC: Smithsonian Institution Press.

Götzenberger L, et al. (2012) Ecological assembly rules in plant communities: Approaches, patterns and prospects. *Biological Reviews* 87, 111–127.

Graham SP, et al. (2009) Nestedness of ectoparasite-vertebrate host networks. *PLoS ONE* 4, e7873.

Gravel D, et al. (2006), Reconciling niche and neutrality: the continuum hypothesis. *Ecology Letters* 9, 399–409.

Guimarães Jr PR, et al. (2011) Evolution and coevolution in mutualistic networks. *Ecology Letters* 14, 877–885.

Guimarães PR, Guimarães P (2006) Improving the analyses of nestedness for large sets of matrices. *Environmental Modelling & Software* 21, 1512–1513.

Guimerà R, Amaral LAN (2005) Cartography of complex networks: Modules and universal roles. *Journal of Statistical Mechanics: Theory and Experiment* 2, P02001.

Haak DM, et al. (2017) Coupling ecological and social network models to assess 'transmission' and 'contagion' of an aquatic invasive species. *Journal of Environmental Management* 190, 243–251.

Hansen BB, et al. (2020) The Moran effect revisited: Spatial population synchrony under global warming. *Ecography* 43, 1591–1602.

Hanski I (1982) Dynamics of regional distribution: The core and satellite species hypothesis. *Oikos* 38, 210–221.

Harrison JA, et al. (1997) *The Atlas of Southern African Birds*. Johannesburg: BirdLife South Africa.

Havens K (1992) Scale and structure in natural food webs. *Science* 257, 1107–1109.

Hayes KR, Barry SC (2008) Are there any consistent predictors of invasion success? *Biological Invasions* 10, 483–506.

Heger T, et al. (2019) Towards an integrative, eco-evolutionary understanding of ecological novelty: Studying and communicating interlinked effects of global change. *BioScience* 69, 888–899.

Hesse H (1943) *Das Glasperlenspiel*. New York: Henry Holt and Company.

Hierro JL, et al. (2005) A biogeographical approach to plant invasions: The importance of studying exotics in their introduced and native range. *Journal of Ecology* 93, 5–15.

Hillebrand H, et al. (2018) Biodiversity change is uncoupled from species richness trends: Consequences for conservation and monitoring. *Journal of Applied Ecology* 55, 169–184.

HilleRisLambers J, et al. (2012) Rethinking community assembly through the lens of coexistence theory. *Annual Review of Ecology, Evolution, and Systematics* 43, 227–248.

Hinz HL, Schwarzlaender M (2004) Comparing invasive plants from their native and exotic range: What can we learn for biological control? *Weed Technology* 18, 1533–1541.

Hobbs RJ, et al. (2014), Managing the whole landscape: Historical, hybrid, and novel ecosystems. *Frontiers in Ecology and the Environment* 12, 557–564.

Hocutt CH, et al. (1978) Fishes of the Greenbrier River, West Virginia, with drainage history of the Central Appalachians. *Journal of Biogeography* 5, 59–80.

Hokkanen HMT, Pimentel D (1989) New associations in biological control: Theory and practice. *Canadian Entomologist* 121, 829–840.

Holland JN, DeAngelis DL (2010) A consumer-resource approach to the density-dependent population dynamics of mutualism. *Ecology* 91, 1286–1295.

Holland JN, Okuyama T, DeAngelis DL (2006) Comment on 'Asymmetric coevolutionary networks facilitate biodiversity maintenance'. *Science* 313, 1887.

Holling CS (1959) Some characteristics of simple types of predation and parasitism. *Canadian Entomologist* 91, 385–398.

Hubbell SP (2001) *The Unified Neutral Theory of Biodiversity and Biogeography*. Princeton: Princeton University Press.

Hubbell SP, Condit R, Foster RB (2005) *Barro Colorado Forest Census Plot Data*. Panama City: Center for Tropical Forest Science.

Hubbell SP, et al. (1999) Light-gap disturbances, recruitment limitation, and tree diversity in a Neotropical forest. *Science* 283, 554–557.

Hui C (2012) Scale effect and bimodality in the frequency distribution of species occupancy. *Community Ecology* 13, 30–35.

Hui C (2021) Introduced species shape insular mutualistic networks. *Proceedings of the National Academy of Sciences USA* 118, e2026396118.

Hui C, McGeoch MA (2014) Zeta diversity as a concept and metric that unifies incidence-based biodiversity patterns. *The American Naturalist* 184, 684–694.

Hui C, Richardson DM (2017) *Invasion Dynamics*. Oxford: Oxford University Press.

Hui C, Richardson DM (2019) Network invasion as an open dynamical system: Response to Rossberg and Barabás. *Trends in Ecology & Evolution* 34, 386–387.

Hui C, et al. (2006) A spatially explicit approach to estimating species occupancy and spatial correlation. *Journal of Animal Ecology* 75, 140–147.

Hui C, et al. (2010) Measures perceptions and scaling patterns of aggregated species distributions. *Ecography* 33, 95–102.

Hui C, et al. (2011) Macroecology meets invasion ecology: Linking the native distributions of Australian acacias to invasiveness. *Diversity and Distributions* 17, 872–883.

Hui C, et al. (2012) Flexible dispersal strategies in native and non-native ranges: Environmental quality and the 'good-stay, bad-disperse' rule. *Ecography* 35, 1024–1032.

Hui C, et al. (2013) Increasing functional modularity with residence time in the co-distribution of native and introduced vascular plants. *Nature Communications* 4, 2454.

Hui C, et al. (2014) Macroecology meets invasion ecology: Performance of Australian acacias and eucalypts around the world revealed by features of their native ranges. *Biological Invasions* 16, 565–576.

Hui C, et al. (2015) Adaptive diversification in coevolutionary systems. In Pontarotti P (ed.) *Evolutionary Biology: Biodiversification from Genotype to Phenotype*, pp. 167–186. Cham: Springer.

Hui C, et al. (2016) Defining invasiveness and invasibility in ecological networks. *Biological Invasions* 18, 971–983.

Hui C, et al. (2017) Ranking of invasive spread through urban green areas in the world's 100 most populous cities. *Biological Invasions* 19, 3527–3539.

Hui C, et al. (2018) *Ecological and Evolutionary Modelling.* Cham: Springer.

Hui C, et al. (2020) The role of biotic interactions in invasion ecology: Theories and hypotheses. In Traveset A, Richardson DM (eds.) *Plant Invasions: The Role of Biotic Interactions,* pp. 26–44. Wallingford: CAB International.

Hulme PE (2009) Trade, transport and trouble: Managing invasive species pathways in an era of globalization. *Journal of Applied Ecology* 46, 10–18.

Janzen DH (1985) On ecological fitting. *Oikos* 45, 308–310.

Jaramillo P, et al. (2013) *CDF Checklist of Galápagos Vascular Plants: Charles Darwin Foundation Galápagos Species Checklist.* Ecuador: Charles Darwin Foundation.

Jarosz AM, Davelos AL (1995) Effects of disease in wild plant populations and the evolution of pathogen aggressiveness. *New Phytologist* 129, 371–387.

Jeschke JM (2014) General hypotheses in invasion ecology. *Diversity and Distribution* 20, 1229–1234.

Jeschke JM, Strayer DL (2006) Determinants of vertebrate invasion success in Europe and North America. *Global Change Biology* 12, 1608–1619.

Jeschke M, Heger T (eds)(2018) *Invasion Biology: Hypotheses and Evidence.* Oxfordshire: CAB International.

Jonsson LM, et al. (2001) Context dependent effects of ectomycorrhizal species richness on tree seedling productivity. *Oikos* 93, 353–364.

Jordano P (1987) Patterns of mutualistic interactions in pollination and seed dispersal: Connectance, dependence asymmetries, and coevolution. *The American Naturalist* 129, 657–677.

Jordano P, et al. (2003) Invariant properties in coevolutionary networks of plant–animal interactions. *Ecology Letters* 6, 69–81.

Joshi J, Vrieling K (2005) The enemy release and EICA hypothesis revisited: Incorporating the fundamental difference between specialist and generalist herbivores. *Ecology Letters* 8, 704–714.

Kaiser-Bunbury CN, et al. (2010) The robustness of pollination networks to the loss of species and interactions: A quantitative approach incorporating pollinator behaviour. *Ecology Letters* 13, 442–452.

Kauffman SA (2019) *A World Beyond Physics: The Emergence and Evolution of Life.* New York: Oxford University Press.

Keane RM, Crawley MJ (2002) Exotic plant invasions and the enemy release hypothesis. *Trends in Ecology & Evolution* 17, 164–170.

Kimbrell T, Holt RD (2005) Individual behaviour, space and predator evolution promote persistence in a two-patch system with predator switching. *Evolutionary Ecology Research* 7, 53–71.

King GE, Howeth JG (2019) Propagule pressure and native community connectivity interact to influence invasion success in metacommunities. *Oikos* 128, 1549–1564.

Klironomos JN (2003) Variation in plant response to native and exotic arbuscular mycorrhizal fungi. *Ecology* 84, 2292–2301.

Knevel IC, et al. (2004) Release from native root herbivores and biotic resistance by soil pathogens in a new habitat both affect the alien *Ammophila arenaria* in South Africa. *Oecologia* 141, 502–510.

Kondoh M (2003) Foraging adaptation and the relationship between food-web complexity and stability. *Science* 299, 1388–1391.

Kondoh M, Kato S, Sakato Y (2010) Food webs are built up with nested subwebs. *Ecology* 91, 3123–3130.

Kraft NJB, et al. (2008) Functional traits and niche-based tree community assembly in an Amazonian forest. *Science* 322, 580–582.

Krasnov BR, et al. (2012) Phylogenetic signal in module composition and species connectivity in compartmentalized host-parasite networks. *The American Naturalist* 179, 501–511.

Křivan V, et al. (2001) Alternative food, switching predators, and the persistence of predator-prey systems. *The American Naturalist* 157, 512–524.

Kueffer C, et al. (2013) Integrative invasion science: Model systems, multi-site studies, focused meta-analysis and invasion syndromes. *New Phytologist* 200, 615–633.

Kuussaari M, et al. (2009) Extinction debt: A challenge for biodiversity conservation. *Trends in Ecology & Evolution* 24, 564–571.

Laliberté E, Legendre P (2010) A distance-based framework for measuring functional diversity from multiple traits. *Ecology* 91, 299–305.

Landi P, Piccardi C (2014) Community analysis in directed networks: In-, out-, and pseudocommunities. *Physical Review E* 89, 012814.

Landi P, et al. (2018) Complexity and stability of ecological networks: A review of the theory. *Population Ecology* 60, 319–345.

Latombe G, et al. (2015) Beyond the continuum: A multi-dimensional phase space for neutral–niche community assembly. *Proceedings of the Royal Society B: Biological Sciences* 282, 20152417.

Latombe G, et al. (2018) Drivers of species turnover vary with species commonness for native and alien plants with different residence times. *Ecology* 99, 2763–2775.

Latombe G, et al. (2019) A four-component classification of uncertainties in biological invasions: Implications for management. *Ecosphere* 10, e02669.

Latombe G, et al. (2020) The effect of cross-boundary management on the trajectory to commonness in biological invasions. *NeoBiota* 62, 241–267.

Latombe G, et al. (2021) Mechanistic reconciliation of community and invasion ecology. *Ecosphere* 12, e03359.

Le Roux JJ (2020) Molecular ecology of plant-microbial interactions during invasions: progress and challenges. In Traveset A, Richardson DM (eds.) *Plant Invasions: The Role of Biotic Interactions*, pp. 340–362. Wallingford: CAB International.

Le Roux JJ, et al. (2017) Co-introduction vs ecological fitting as pathways to the establishment of effective mutualisms during biological invasions. *New Phytologist* 215, 1354–1360.

Le Roux JJ, et al. (2020) Biotic interactions as mediators of biological invasions: Insights from South Africa. In van Wilgen BW, et al. (eds.) *Biological Invasions in South Africa.* Cham: Springer.

Legault G, Fox JW, Melbourne BA (2019) Demographic stochasticity alters expected outcomes in experimental and simulated non-neutral communities. *Oikos* 128, 1704–1715.

Lehman CL, Tilman D (2000) Biodiversity, stability, and productivity in competitive communities. *The American Naturalist* 156, 534–552.

Leibold MA, Chase JM (2017) *Metacommunity Ecology*. Princeton: Princeton University Press.

Levine JM (2000) Species diversity and biological invasions: Relating local process to community pattern. *Science* 288, 852–854.

Levine JM, et al. (2004) A meta-analysis of biotic resistance to exotic plant invasions. *Ecology Letters* 7, 975–989.

Loeuille N, Loreau M (2005) Evolutionary emergence of size-structured food webs. *Proceedings of the National Academy of Sciences USA* 102, 5761–5766.

Lonsdale WM (1999) Global patterns of plant invasions and the concept of invasibility. *Ecology* 80, 1522–1536.

Lopezaraiza–Mikel ME, et al. (2007) The impact of an alien plant on a native plant–pollinator network: An experimental approach. *Ecology Letters* 10, 539–550.

Lurgi M, et al. (2014) Network complexity and species traits mediate the effects of biological invasions on dynamic food webs. *Frontiers in Ecology and Evolution* 2, 1–11.

MacArthur R, Levins R (1967) The limiting similarity, convergence, and divergence of coexisting species. *The American Naturalist* 101, 377–385.

MacArthur RH (1972) Strong, or weak, interactions. *Transactions of the Connecticut Academy of Arts and Sciences* 44, 177–188.

MacDougall AS, Gilbert B, Levine JM (2009) Plant invasions and the niche. *Journal of Ecology* 97, 609–615.

Mack RN (1996) Predicting the identity and fate of plant invaders: Emergent and emerging approaches. *Biological Conservation* 72, 107–121.

Maron JL, et al. (2004) Rapid evolution of an invasive plant. *Ecological Monographs* 74, 261–280.

Maron JL, Vilà M (2001) When do herbivores affect plant invasion? Evidence for the natural enemies and biotic resistance hypotheses. *Oikos* 95, 361–373.

Martinez ND (1992) Constant connectance in community food webs. *The American Naturalist* 139, 1208–1218.

May RM (1972) Will a large complex system be stable? *Nature* 238, 413–414.

McCullagh P, Nelder JA (1983) *Generalized Linear Models*. London: Chapman and Hall.

McGeoch MA, et al. (2019) Measuring continuous compositional change using decline and decay in zeta diversity. *Ecology* 100, e02832.

McGill BJ, et al. (2006) Rebuilding community ecology from functional traits. *Trends in Ecology & Evolution* 21, 178–185.

McGill BJ, et al. (2007) Species abundance distributions: Moving beyond single prediction theories to integration within an ecological framework. *Ecology Letters* 10, 995–1015.

McGrannachan CM, McGeoch MA (2019) Multispecies plant invasion increases function but reduces variability across an understorey metacommunity. *Biological Invasions* 21, 1115–1129.

McKane AJ (2004) Evolving complex food webs. *European Physical Journal B* 38, 287–295.

Mello MAR, et al. (2011) The modularity of seed dispersal: Differences in structure and robustness between bat–and bird–fruit networks. *Oecologia* 167, 131–140.

Memmott J (1999) The structure of a plant-pollinator food web. *Ecology Letters* 2, 276–280.

Memmott J, Waser NM (2002) Integration of alien plants into a native flower–pollinator visitation web. *Proceedings of the Royal Society B: Biological Sciences* 269, 2395–2399.

Meskens C, et al. (2011) Host plant taxonomy and phenotype influence the structure of a Neotropical host plant-hispine beetle food web. *Ecological Entomology* 36, 480–489.

Miele V, et al. (2019) Non-trophic interactions strengthen the diversity: Functioning relationship in an ecological bioenergetic network model. *PLoS Computational Biology* 15, e1007269.

Milgram S (1967) The small world problem. *Psychology Today* 2, 60–67.

Minoarivelo HO, et al. (2014) Detecting phylogenetic signal in mutualistic inter-action networks using a Markov process model. *Oikos* 123, 1250–1260.

Minoarivelo HO, Hui C (2016a) Trait-mediated interaction leads to structural emergence in mutualistic networks. *Evolutionary Ecology* 30, 105–121.

Minoarivelo HO, Hui C (2016b) Invading a mutualistic network: To be or not to be similar. *Ecology and Evolution* 6, 4981–4996.

Minoarivelo HO, Hui C (2018) Alternative assembly processes from trait-mediated co-evolution in mutualistic communities. *Journal of Theoretical Biology* 454, 146–153.

Mitchell C, Power A (2003) Release of invasive plants from fungal and viral pathogens. *Nature* 421, 625–627.

Mitchell CE and Power AG (2006) Disease dynamics in plant communities. In Collinge SK, Ray C (eds.) *Disease Ecology: Community Structure and Pathogen Dynamics*, pp. 58–72. Oxford: Oxford University Press.

Mlinarić A, et al. (2017) Dealing with the positive publication bias: Why you should really publish your negative results. *Biochemia Medica* 27, 447–452.

Molontay R, Nagy M (2019) Two decades of network science: As seen through the co-authorship network of network scientists. Proceedings of the 2019 IEEE/ACM International Conference on Advances in Social Networks Analysis and Mining (ASONAM '19). New York: Association for Computing Machinery, pp. 578–583.

Montoya JM, Solé RV (2002) Small world patterns in food webs. *Journal of Theoretical Biology* 214, 405–412.

Moore J, Hunt WH (1988) Resource compartmentation and the stability of real ecosystems. *Nature* 333, 261–263.

Mora BB, et al. (2020) Untangling the seasonal dynamics of plant-pollinator com-munities. *Nature Communications* 11, 4086.

Morales CL, Traveset A (2009) A meta-analysis of impacts of alien vs. native plants on pollinator visitation and reproductive success of co-flowering native plants. *Ecology Letters* 12, 716–728.

Morton RD, Law R (1997) Regional species pools and the assembly of local ecological communities. *Journal of Theoretical Biology* 187, 321–331.

Mougi A, Kondoh M (2016) Food-web complexity, meta-community complexity and community stability. *Scientific Reports* 6, 24478.

Müller-Schärer H, et al. (2004) Evolution in invasive plants: Implications for biological control. *Trends in Ecology & Evolution* 19, 417–422.

Murdoch WW (1969) Switching in general predators: Experiments on predator specificity and stability of prey populations. *Ecological Monographs* 39, 335–354.

Neutel A, et al. (2002) Stability in real food webs: Weak links in long loops. *Science* 296, 1120–1123.

Newman MEJ (2003) The structure and function of complex networks. *SIAM Review* 45, 167–256.

Newman MEJ, Girvan M (2004) Finding and evaluating community structure in networks. *Physical Review E* 69, 026113.

Nnakenyi CA, et al. (2019) Fine-tuning the nested structure of pollination networks by adaptive interaction switching, biogeography and sampling effect in the Galápagos Islands. *Oikos* 128, 1413–1423.

Novella-Fernandez R, et al. (2019) Interaction strength in plant–pollinator networks: Are we using the right measure? *PLoS ONE* 14, e0225930.

Novoa A, et al. (2020) Invasion syndromes: A systematic approach for predicting biological invasions and facilitating effective management. *Biological Invasions* 22, 1801–1820.

Nowak M, Sigmund KA (1993) Strategy of win-stay, lose-shift that outperforms tit-for-tat in the Prisoner's Dilemma game. *Nature* 364, 56–58.

Nuismer SL, et al. (2010) When is correlation coevolution? *The American Naturalist* 175, 525–537.

Nuwagaba S, Hui C (2015) The architecture of antagonistic networks: Node degree distribution compartmentalization and nestedness. *Computational Ecology and Software* 5, 317–327.

Nuwagaba S, et al. (2015) A hybrid behavioural rule of adaptation and drift explains the emergent architecture of antagonistic networks. *Proceedings of the Royal Society B: Biological Sciences* 282, 20150320.

Nuwagaba S, et al. (2017) Robustness of rigid and adaptive networks to species loss. *PLoS ONE* 12, e0189086.

Okuyama T, Holland JN (2008) Network structural properties mediate the stability of mutualistic communities. *Ecology Letters* 11, 208–216.

Olesen JM, et al. (2007) The modularity of pollination networks. *Proceedings of the National Academy of Sciences USA* 104, 19891–19896.

Olesen JM, Jordano P (2002) Geographic patterns in plant–pollinator mutualistic networks. *Ecology* 83, 2416–2424.

Padrón B, et al. (2009) Impact of alien plant invaders on pollination networks in two archipelagos. *PLoS ONE* 4, e6275.

Paine R (1992) Food-web analysis through field measurement of per capita interaction strength. *Nature* 355, 73–75.

Parker IM, Gilbert GS (2004) The evolutionary ecology of novel plant-pathogen interactions. *Annual Review of Ecology, Evolution, and Systematics* 35, 675–700.

Parker JD, et al. (2020) Biotic resistance to plant invasions. In Traveset A, Richardson DM (eds.) *Plant Invasions: The Role of Biotic Interactions*, pp. 177–191. Wallingford: CAB International.

Parker JD, Hay ME (2005) Biotic resistance to plant invasions? Native herbivores prefer non-native plants. *Ecology Letters* 8, 959–967.

Patterson BD, et al. (2009) Nested distributions of bat flies (Diptera: Streblidae) on Neotropical bats: Artifact and specificity in host-parasite studies. *Ecography* 32, 481–487.

Pauchard A, et al. (2018) Biodiversity assessments: Origin matters. *PLoS Biology* 16, e2006686.

Pearson DE, et al. (2018) Community assembly theory as a framework for biological invasions. *Trends in Ecology & Evolution* 33, 313–325.

Perkins LB, Nowak RS (2013) Invasion syndromes: Hypotheses on relationships among invasive species attributes and characteristics of invaded sites. *Journal of Arid Land* 5, 275–283.

Piazzon M, et al. (2011) Are nested networks more robust to disturbance? A test using epiphyte-tree, comensalistic networks. *PLoS ONE* 6, e19637.

Pigliucci M (2010) Genotype–phenotype mapping and the end of the 'genes as blueprint' metaphor. *Philosophical Transactions of the Royal Society B: Biological Sciences* 365, 557–566.

Pires MM, et al. (2011) The nested assembly of individual-resource networks. *Journal of Animal Ecology* 80, 896–903.

Raimundo RLG, et al. (2018) Adaptive networks for restoration ecology. *Trends in Ecology & Evolution* 33, 664–675.

Ramirez KS, et al. (2018) Network analyses can advance above-belowground ecology. *Trends in Plant Science* 23, 759–768.

Ramos-Jiliberto R, et al. (2012) Topological plasticity increases robustness of mutualistic networks. *Journal of Animal Ecology* 81, 896–904.

Ramsay-Newton C, et al. (2017) Species, community, and ecosystem-level responses following the invasion of the red alga *Dasysiphonia japonica* to the western North Atlantic Ocean. *Biological Invasions* 19, 537–547.

Rezende E, et al. (2007) Non-random coextinctions in phylogenetically structured mutualistic networks. *Nature* 448, 925–928.

Richardson DM, et al. (2000a) Naturalization and invasion of alien plants: concepts and definitions. *Diversity and Distributions* 6, 93–107.

Richardson DM, et al. (2000b) Plant invasions: The role of mutualisms. *Biological Reviews* 75, 65–93.

Richardson DM, et al. (2005) Species richness of alien plants in South Africa: Environmental correlates and the relationship with indigenous plant species richness. *ÉcoScience* 12, 391–402.

Richardson DM, et al. (2004) Using natural experiments in the study of alien tree invasions: Opportunities and limitations. In Gordon MS, Bartol SM (eds.), *Experimental Approaches to Conservation Biology*, pp. 180–201. Berkeley: University of California Press.

Richardson DM, Pyšek P (2006) Plant invasions: Merging the concepts of species invasiveness and community invasibility. *Progress in Physical Geography* 30, 409–431.

Richardson DM, Pyšek P (2012) Naturalization of introduced plants: Ecological drivers of biogeographical patterns. *New Phytologist* 196, 383–396.

Rodríguez-Echeverría S, et al. (2011) Jack-of-all-trades and master of many? How does associated rhizobial diversity influence the colonization success of Australian Acacia species? *Diversity and Distributions* 17, 946–957.

Romanuk TN, et al. (2009) Predicting invasion success in complex ecological networks. *Philosophical Transactions of the Royal Society B: Biological Sciences* 364, 1743–1754.

Rosindell J, et al. (2012) The case for ecological neutral theory. *Trends in Ecology & Evolution* 27, 203–208.

Rossberg AG, et al. (2010) How trophic interaction strength depends on traits. *Theoretical Ecology* 3, 13–24.

Rosvall M, Bergstrom CT (2007) An information-theoretic framework for resolving community structure in complex networks. *Proceedings of the National Academy of Sciences USA* 104, 7327–7331.

Rouget M, et al. (2015) Plant invasions as a biogeographical assay: Vegetation biomes constrain the distribution of invasive alien species assemblages. *South African Journal of Botany* 101, 24–31.

Rouget M, et al. (2016), Invasion debt: Quantifying future biological invasions. *Diversity and Distributions* 22, 445–456.

Saavedra S, et al. (2009) A simple model of bipartite cooperation for ecological and organizational networks. *Nature* 457, 463–466.

Sanchez A (2015) Fidelity and promiscuity in an ant-plant mutualism: A Case Study of Triplaris and Pseudomyrmex. *PLoS ONE* 10, e0143535.

Sax DF, et al. (2005) *Species Invasions: Insights into Ecology, Evolution and Biogeography.* Sunderland: Sinauer Associates Incorporated.

Scheffer M, van Nes EH (2006) Self-organized similarity, the evolutionary emergence of groups of similar species. *Proceedings of the National Academy of Sciences USA* 103, 6230–6235.

Schelling M, Hui C (2015) modMax: Community structure detection via modularity maximization. R package version 1.0. https://cranr-projectorg

Schemske DW, Horvitz CC (1984) Variation among floral visitors in pollination ability: A precondition for mutualism specialization. *Science* 225, 519–521.

Schlaepfer MA (2018) On the importance of monitoring and valuingall forms of biodiversity. *PLoS Biology* 16, e3000039.

Shea K, Chesson P (2002) Community ecology theory as a framework for biological invasions. *Trends in Ecology & Evolution* 17, 17–176.

Simberloff D (2009) The role of propagule pressure in biological invasions. *Annual Review of Ecology, Evolution, and Systematics* 40, 81–102.

Simberloff D, Von Holle B (1999) Positive interactions of nonindigenous species: Invasional meltdown? *Biological Invasions* 1, 21–32.

Smith-Ramesh LM, et al. (2017) Global synthesis suggests that food web connectance correlates to invasion resistance. *Global Change Biology* 23, 465–473.

Soininen J, et al. (2007) The distance decay of similarity in ecological communities. *Ecography* 30, 3–12.

Staniczenko PPA, et al. (2017) Linking macroecology and community ecology: Refining predictions of species distributions using biotic interaction networks. *Ecology Letters* 20, 693–707.

Staniczenko PPA, et al. (2010) Structural dynamics and robustness of food webs. *Ecology Letters* 13, 891–899.

Stastny M, et al. (2005) Do vigour of introduced populations and escape from specialist herbivores contribute to invasiveness? *Journal of Ecology* 93, 27–37.

Stephens DW, Krebs JR (1986) *Foraging Theory*. Princeton: Princeton University Press.

Stohlgren TJ, et al. (2003) The rich get richer: Patterns of plant invasions in the United States. *Frontiers in Ecology and the Environment* 1, 11–14.

Stone L, Roberts A (1990) The checkerboard score and species distributions. *Oecologia* 85, 74–79.

Stouffer DB, et al. (2014) How exotic plants integrate into pollination networks. *Journal of Ecology* 102, 1442–1450.

Stubbs WJ, Wilson BJ (2004) Evidence for limiting similarity in a sand dune community. *Journal of Ecology* 92, 557–567.

Summerhayes VS, Elton CS (1923) Contributions to the ecology of Spitsbergen and Bear Island. *Journal of Ecology* 11, 214–216.

Sutherland JP, Karlson RH (1977) Development and stability of the fouling community at Beaufort, North Carolina. *Ecological Monographs* 47, 425–446.

Suweis S, et al. (2013) Emergence of structural and dynamical properties of ecological mutualistic networks. *Nature* 500, 449–452.

Tansley AG (1935) The use and abuse of vegetational concepts and terms. *Ecology* 16, 284–307.

Thébault E, Fontaine C (2010) Stability of ecological communities and the architecture of mutualistic and trophic networks. *Science* 329, 853–856.

Tilman D (1982) *Resource Competition and Community Structure*. Princeton: Princeton University Press.

Tilman D (2004) Niche tradeoffs, neutrality, and community structure: A stochastic theory of resource competition, invasion, and community assembly. *Proceedings of the National Academy of Sciences USA* 101, 10854–10861.

Tilman D, et al. (1994) Habitat destruction and the extinction debt. *Nature* 371, 65–66.

Timi JT, Poulin R (2008) Different methods, different results: Temporal trends in the study of nested subset patterns in parasite communities. *Parasitology* 135, 131–138.

Tomasetto F, et al. (2019) Resolving the invasion paradox: Pervasive scale and study dependence in the native-alien species richness relationship. *Ecology Letters* 22, 1038–1046.

Torchin ME, Mitchell CE (2004) Parasites, pathogens, and invasions by plants and animals. *Frontiers in Ecology and the Environment* 2, 183–190.

Townsend CR, et al. (1998) Disturbance, resource supply and food-web architecture in streams. *Ecology Letters* 1, 200–209.

Traveset A, et al. (2013) Invaders of pollination networks in the Galápagos Islands: Emergence of novel communities. *Proceedings of the Royal Society B: Biological Sciences* 280, 20123040.

Traveset A, et al. (2015) Bird–flower visitation networks in the Galápagos unveil a widespread interaction release. *Nature Communications* 6, 6376.

Traveset A, Richardson DM (2014) Mutualistic interactions and biological invasions. *Annual Review of Ecology, Evolution, and Systematics* 45, 89–113.

Traveset A, Richardson DM (eds.) (2020) *Plant Invasions: The Role of Biotic Interactions*. Wallingford: CAB International.

Tylianakis J, et al. (2007) Habitat modification alters the structure of tropical host–parasitoid food webs. *Nature* 445, 202–205.

Ulanowicz RE (1997) Limitations on the connectivity of ecosystem flow networks. In Rinaldo A, Marani A (eds.) *Biological Models*, pp. 125–143. Venice: Instituto Veneto de Scienze, Lettere ed Arti.

Vacher C, Piou D, Desprez-Loustau ML (2008) Architecture of an antagonistic tree/fungus network: The asymmetric influence of past evolutionary history. *PLoS ONE* 3: e1740.

Valdovinos FS (2019) Mutualistic networks: Moving closer to a predictive theory. *Ecology Letters* 22, 1517–1534.

Valdovinos FS, et al. (2010) Consequences of adaptive behaviour for the structure and dynamics of food webs. *Ecology Letters* 13, 1546–1559.

Valdovinos FS, et al. (2018) Species traits and network structure predict the success and impacts of pollinator invasions. *Nature Communications* 9, 2153.

van Baalen M, et al. (2001) Alternative food, switching predators, and the persistence of predatory-prey systems. *The American Naturalist* 157, 512–524.

van der Heijden M, et al. (1998) Mycorrhizal fungal diversity determines plant biodiversity, ecosystem variability and productivity. *Nature* 396, 69–72.

van der Putten WH, et al. (2005) Invasive plants and their escape from root herbivory: A worldwide comparison of the root-feeding nematode communities of the dune grass *Ammophila arenaria* in natural and introduced ranges. *Biological Invasions* 7, 733–746.

van Ruijven J, et al. (2003) Diversity reduces invasibility in experimental plant communities: The role of plant species. *Ecology Letters* 6, 910–918.

Vázquez DP, Aizen MA (2003) Null model analyses of specialization in plant–pollinator interactions. *Ecology* 84, 2493–2501.

Vellend M (2016) *The Theory of Ecological Communities*. Princeton: Princeton University Press.

Vermaat JE, et al. (2009) Major dimensions in food-web structure properties. *Ecology* 90, 278–282.

Vilà M, et al. (2009) Invasive plant integration into native plant–pollinator networks across Europe. *Proceedings of the Royal Society B: Biological Sciences* 276, 3887–3893.

Vilà M, et al. (2011) Ecological impacts of invasive alien plants: A meta-analysis of their effects on species, communities and ecosystems. *Ecology Letters* 14, 702–708.

Vivanco JM, et al. (2004) Biogeographical variation in community response to root allelochemistry: Novel weapons and exotic invasion. *Ecology Letters* 7, 285–292.

Vizentin-Bugoni J, et al. (2019) Structure, spatial dynamics, and stability of novel seed dispersal mutualistic networks in Hawai'i. *Science* 364, 78–82.

Waddell EH, et al. (2020) Trait filtering during exotic plant invasion of tropical rainforest remnants along a disturbance gradient. *Functional Ecology* 34, 2584–2597.

Wallace AR (1889) *Darwinism*. London: MacMillan.

Waser NM, et al. (1996) Generalization in pollination systems, and why it matters. *Ecology* 77, 1043–1060.

Watkins A, Wilson JB (2003) Local texture convergence: A new approach to seeking assembly rules. *Oikos* 102, 525–532.

Watts D, Strogatz S (1998) Collective dynamics of 'small-world' networks. *Nature* 393, 440–442.

Whitney KD, Gabler CA (2008) Rapid evolution in introduced species, 'invasive traits' and recipient communities: Challenges for predicting invasive potential. *Diversity and Distributions* 14, 569–580.

Whittall J, Hodges S (2007) Pollinator shifts drive increasingly long nectar spurs in columbine flowers. *Nature* 447, 706–709.

Williams R, Martinez N (2000) Simple rules yield complex food webs. *Nature* 404, 180–183.

Wilson EO, Simberloff DS (1969) Experimental zoogeography of islands: Defaunation and monitoring techniques. *Ecology* 50, 267–278.

Wolfe BE, Husband BC, Klironomos JN (2005) Effects of a belowground mutualism on an aboveground mutualism. *Ecology Letters* 8, 218–223.

Wootton JT, Emmerson M (2005) Measurement of interaction strength in nature. *Annual Review of Ecology, Evolution, and Systematics* 36, 419–444.

Zhang F, Hui C (2014) Recent experience-driven behaviour optimizes foraging. *Animal Behaviour* 88, 13–19.

Zhang F, et al. (2011) An interaction switch predicts the nested architecture of mutualistic networks. *Ecology Letters* 14, 797–803.

4 · Regimes and Panarchy

It is not so much the organism or the species that evolves, but the entire system, species and environment. The two are inseparable.

Lotka (1925)

4.1 Open Adaptive Systems

Before diving into a discussion of open adaptive systems, we need to revisit the definition of an ecological network. Material covered in Chapters 2 and 3 showed that ecological networks are webs of co-evolving and co-fitting interactions among species residing in an ecosystem. Such networks subjected to regular incursions of new members in the form of biological invasions are a good example of Open Adaptive Systems (OASs). OASs are different from Clements' (1916) superorganism metaphor that was further developed and scaled up into the concept of Lovelock's (1972) Gaia theory, which posits that organisms interact to form a synergistic and self-regulating complex system. The reason for considering an ecological network (or its embedded ecological community) a system, rather than an organism or an organisation (*sensu* Keller 2005), lies with the type of its boundaries. A system can have either permeable or closed boundaries, while an organism cannot survive with a closed boundary. More importantly, a system has more flexible and tenuous boundaries, the positions of which are often set by the beholder. Boundaries drawn around sampling areas based on what we call an ecological community or an ecosystem are largely subjective. In contrast, the boundary of an organism is clear-cut and plays important physiological and metabolic roles. The value of a system's boundary, albeit usually subjectively defined, is to identify and differentiate its residents from alien visitors, thereby providing the foundation for labelling entities for management purposes. In contrast, the organic boundary is inseparable from the organism; they belong to an irreducible whole.

The fallacy of considering an ecological community as a superorganism lies with the potential risk of upscaling the unit of selection; natural selection works on each population, not an entire community. To this end, Gleason (1939) and Whittaker (1962) crushed the notion of considering an ecological community as a superorganism. However, the individualistic pieces left behind after this debate have led many to wonder whether there is such a thing as community integrity, or even a community, and whether biotic interactions should be discarded together with the detritus of a largely obsolete concept. Tansley (1935) provided a remedy for this problem by calling an ecological community an ecological system, or ecosystem in short. Levin (1999) and others have further refined this notion to advocate considering an ecosystem as a complex adaptive system, to emphasise the complexity emerging from the intertwined interactions within and processes over multiple scales (Levin 1992; McGill 2010), as well as the adaptability that emerges from population dynamics in response to these interactions. This is not an isolated incidence but part of the collective rise of the science devoted to exploring complex systems that is extending its tentacles across many research fronts (Figure 4.1; e.g., Anderson 1972; Kauffman 1993; Bar-Yam 1997; Ulanowicz 2002). To this end, ecological interaction networks have emerged as an attractive framework for capturing the essence of complex adaptive systems.

A complex adaptive system is a dynamic system comprising multiple interacting parts that respond, adaptively and collectively, to perturbations, often reactively but sometimes actively or proactively. The concept of complex adaptive systems traditionally falls under the umbrella of Dynamical Systems in mathematics but its influence has grown quickly in many disciplines. Structural complexity can emerge from the interaction and connectivity of its parts and exchange of information, materials and energies between its boundaries (Gell-Mann 1995). Natural systems, such as brains, immune systems, animal flocks, ecosystems and societies, and artificial systems, such as parallel and distributed computing systems, artificial intelligence, neural networks, and evolutionary programs, have become the typical models for discussion under the banner of complex adaptive systems. As an effective representation of an ecosystem, an ecological network, therefore, is a complex adaptive system. As explained in Chapter 2, ecological networks exhibit key features of such a system: complexity and adaptability (Levin 1999; Solé and Goodwin 2000; Parrott and Meyer 2012). These key features have been further refined in systems science to differentiate complex adaptive systems from

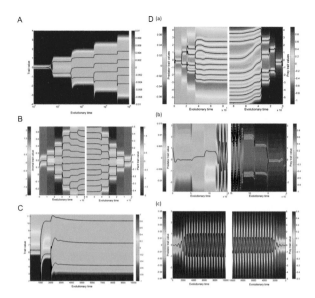

Figure 2.13 Illustrations of trait evolution from co-evolving biotic interactions. Black curves represent traits (vertical axis) in the system over evolutionary time (horizontal axis); background colour represents the invasion fitness. A: Trait diversification under resource competition, starting from a monomorphic population and gradually diversifying into five different morphs. B: Trait diversification under mutualistic co-evolution, starting from a monomorphic population and gradually diversifying into twelve different morphs on each side of the trophic. C: Trait diversification in a food web, starting from a monomorphic heterotroph population and diversifying into different morphs. D: Trait evolution under antagonistic co-evolution, representing trait diversification from antagonistic co-evolution (a) and Red Queen co-evolutionary cycle of prey and predator in polymorphic (b) and monomorphic (c) systems populations. See model details and parameters in Hui et al. (2018), reproduced with permission.

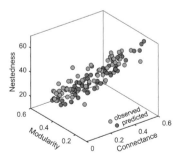

Figure 3.8 Observed and predicted relationships among connectance, modularity and nestedness for 60 bipartite antagonistic networks. For each network, simulations were run until equilibrium was reached based on a simple Lotka–Volterra model with Holling's type II functional response and implementing adaptive interaction switching. From Nuwagaba et al. (2017), under CC-BY Licence.

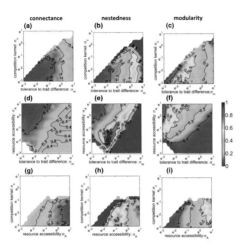

Figure 3.31 Quantitative connectance (first column), weighted nestedness (second column) and quantitative modularity (third column) of emerged bipartite mutualistic networks as a function of model parameters. In each plot we varied two parameters and kept the third parameter fixed. The white area in each plot represents an unfeasible parameter zone where the system becomes monomorphic with no network metrics calculated. From Minoarivelo and Hui (2016a), reprinted with permission.

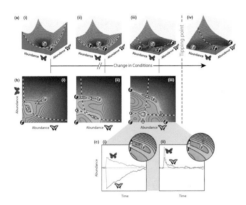

Figure 4.16 Stability properties for a small network of two pollinators (shown) and two plants (not shown). (a) Attraction basins (valleys) of alternative stable states (balls) are separated by thresholds (dashed curves). Initially, the only alternative to pristine state 1 is fully collapsed state 2 (a.i). When conditions change, two additional, partially collapsed states appear (states 3 and 4). The initial, pristine state loses resilience after state 3 appears (a.ii and a.iii). Eventually, the threshold towards state 3 approaches the pristine state so closely that a critical transition towards this state becomes inevitable (a.iii and a.iv). (b) Alternative stable states, saddle points (yellow dots) and hilltops (grey dots) are surrounded by areas in which the landscape's slope, and thus the rate at which abundances change, is nearly zero (indicated in orange). Higher speeds are found further away from these points. The direction of slowest recovery changes substantially before future state 3 appears (yellow arrow, b.i and b.ii). After state 3 appears, the system slows down in the direction of the saddle point on the approaching threshold (b.ii and b.iii). (c) Slow recovery from a perturbation towards the saddle point (c.i) as opposed to the much faster recovery from an equally large perturbation in another direction (c.ii). From Lever et al. (2020), under CC-BY Licence.

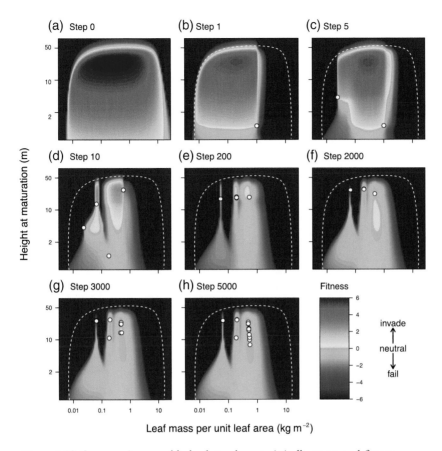

Figure 4.19 Community assembly leads to characteristically structured fitness landscapes that exhibit peaks, valleys and plateaus. a–h show fitness landscapes at different stages of community assembly. Colouring across trait space indicates the invasion fitness of rare species competing with the resident species (circles). Dashed curves in each panel delimit trait combinations that are not viable even in the absence of any competition. Community assembly starts with a species randomly selected across the trait space (step 1). At each subsequent step, the abundances of resident species are updated according to their fitness, and new invaders are added around existing residents and across trait space. Invasion is successful only in regions with positive invasion fitness (coloured yellow to red), whereas species that successfully invade but then find themselves in regions of negative invasion fitness (coloured blue to cyan; e.g., lowest circle in d) are ultimately driven to extinction. A ridge-shaped fitness plateau (green region in h) forms naturally in the course of the community assembly, after a high diversity of late-successional shade-tolerant species establish in regions of positive fitness. From Falster et al. (2017), reprinted with permission.

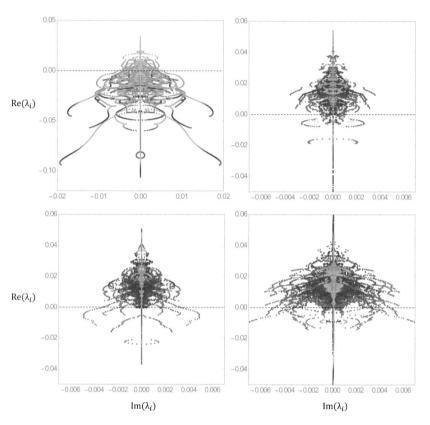

Figure 4.23 Dynamics of eigenvalues of the system Jacobian matrix during the assembly of an ecological network. The ecological network follows the standard Lotka–Volterra model (Eq.(2.6)) and the intrinsic growth rate of each species r_i follows a normal distribution with zero mean and standard deviation of 0.02; interaction strength α_{ij} for $i \neq j$ follows a normal distribution with zero mean and standard deviation 0.1; $\alpha_{ii} = k_i = 1$. Note, the complex plane is rotated to visualise the real part vertically. Top left: 100 species; the rest: five species initially followed by two species per unit time (top right and bottom left) and by four species per unit time (bottom right). Eigenvalues are overlaid on each other with yellow-green colours indicating early succession time and red colours late succession time. Note, the dense distributions have blocked some eigenvalues in the visualisation.

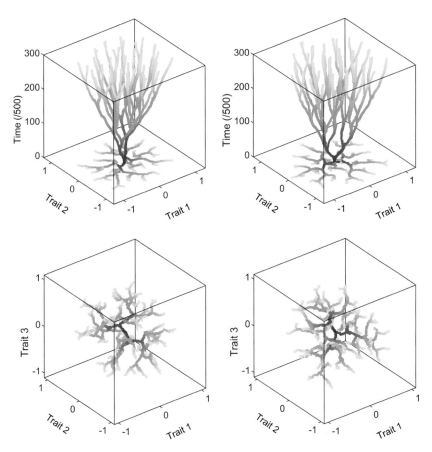

Figure 5.2 Alternative trait evolution dynamics in an ecological network where resident species are engaging in resource competition with each other. Top, a species is represented by two trait values; bottom, by three trait values. Simulations on each row were run twice (left and right) under the same set of parameters and initial conditions, while alternative evolutionary pathways emerged from the uncertainty in the directions of sequential evolutionary branching. Redrawn from Zhang et al. (2021).

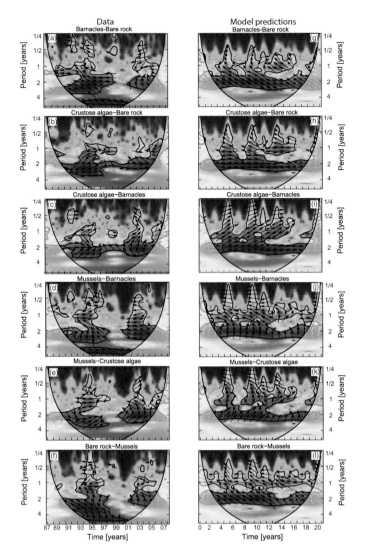

Figure 5.3 Cross-wavelet spectra of all species pairs of a rocky intertidal community in the Cape Rodney-Okakari Point Marine Reserve on the North Island of New Zealand. Cross-wavelet spectra of (a–f) the observed time series and (g–l) the model predictions. The spectra show how common periodicities in the fluctuations of two species (y axis) change over time (x axis). Colour indicates cross-wavelet power (from low power in blue to high power in red), which measures to what extent the fluctuations of the two species are related. Black contour lines enclose significant regions, with > 95% confidence that cross-wavelet power exceeds red noise. Arrows indicate phase angles between the fluctuations of the two species. Arrows pointing right represent in-phase oscillations (0°), and arrows pointing upward indicate that the first species lags the second species by a quarter period (90°). Shaded areas on both sides represent the cone of influence, where edge effects may distort the results. From Benincà et al. (2015), reprinted with permission.

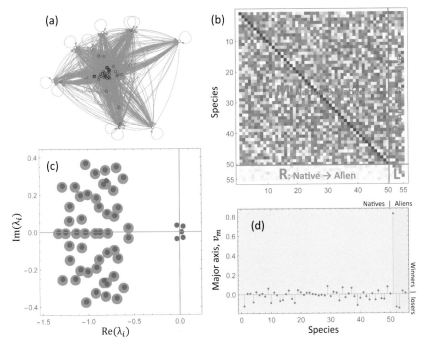

Figure 5.15 Dynamic responses of an ecological network to biological invasions. (a) The graph diagram of an ecological network of biotic interactions (50 species with interaction strengths randomly assigned from a normal distribution with zero mean and standard deviation $\sigma = 0.1$, with diagonal elements $-\mu = -1$). The network is illustrated by high-dimensional embedding using Wolfram Mathematica, with open black dots nodes and green directed lines pairwise interactions. (b) The adjacency matrix of the invaded network, with the orange lines dividing the matrix into four blocks: representing, respectively, the original native–native adjacency matrix M (top left; dimension: 50 by 50), invasion impacts on resident species (matrix P, top right; dimension: 5 by 50), biotic resistance to invasions (matrix R, bottom left, with $\sigma = 0.01$ for illustration; dimension: 50 by 5) and alien–alien interactions (matrix L, bottom right, with diagonal zeros for illustration; dimension: 5 by 5). The colours of matrix elements correspond to interaction strength; darker shading denotes stronger interactions. (c) Visualisation of all eigenvalues in the original network (50 green disks) and invaded network (55 red dots). Each eigenvalue is composed of a real part $\mathrm{Re}(\lambda_i)$ and an imaginary part $\mathrm{Im}(\lambda_i)$; the two parts of an eigenvalue allow us to locate it in the complex plane. Note, the real part of the lead eigenvalue for the original network (the rightmost green disk) is negative and becomes positive for the invaded network (the rightmost red dot), where the alien species define the lead eigenvalue because of its low abundance. (d) The major axis (i.e., the eigenvector of the lead eigenvalue) of the joint matrix (note, v_m in the figure is the same as v_1 in the text here), revealing how each species dynamically responds to the invasion (species 1 to 50 are the original resident species; species 51 to 55 marked with red circles are the newly introduced species). Note, introduced species 51 became highly invasive; species 52 and 53 experienced invasion failures; species 54 and 55 established but are yet not invasive. Some resident species responded positively, others negatively, and many resident species are insensitive to the invasion. From Hui and Richardson (2019), reprinted with permission.

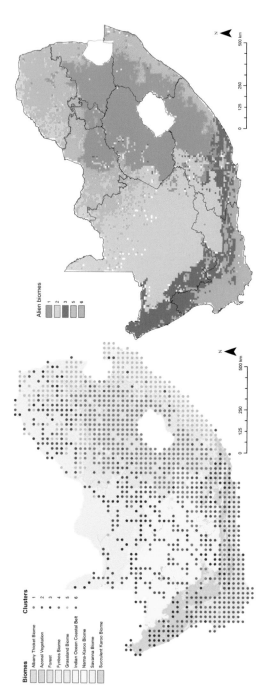

Figure 6.1 (Left) Clusters of invasive alien plant species in relation to biomes defined on the basis of natural vegetation (shading). Clusters were derived based on presence/absence of invasive alien species (shown as circles). (Right) The potential distribution of alien biomes. Environmental determinants of the distribution of invasive alien species clusters were identified from a classification tree and were mapped over the full range of environmental conditions in South Africa. From Rouget et al. (2015), reprinted with permission.

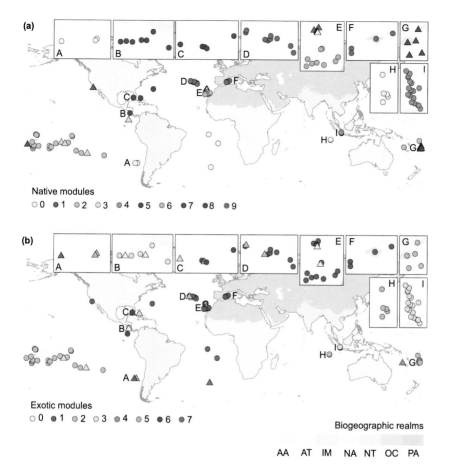

Figure 6.17 Distribution of (a) native and (b) alien modules for islands (i.e., distinct clusters of islands with a certain assemblage of species) derived from a modularity analysis. Colours indicate the module assigned by the modularity analysis, while the symbols show whether the conditional inference tree grouped that particular island in a node where the majority of islands have the same module (o) or a different one (Δ). Module 0 means that the island does not have ant species. The colours of the land masses show the biogeographical realms (AA, Australasia; AT, Afrotropical; IM, Indo-Malay; NA, Nearctic; NT, Neotropical; OC, Oceania, PA, Palaearctic; Olson et al., 2001). The insets represent an expanded view of areas where symbols overlap. From Roura-Pascual et al. (2016), reprinted with permission.

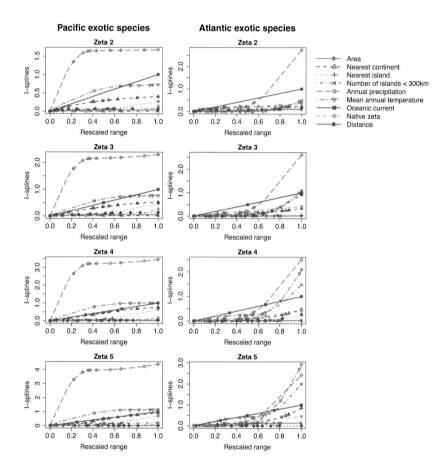

Figure 6.18 I-splines computed with the multi-site generalised dissimilarity model of zeta diversity for alien species of the Pacific and Atlantic island groups, from ζ_2 to ζ_5 for Simpson zeta diversity. The horizontal axes represent the original variables, rescaled between 0 and 1 for comparison (environmental variables are in red, geographical ones are in blue and the biotic native zeta variable is in green). The relative amplitude of each spline in a given panel therefore represents the relative importance of the corresponding variable to explain zeta diversity for a specific order. For each variable, the symbols are located at the percentiles, providing information on the distribution of values. From Latombe et al. (2019), under CC-BY Licence.

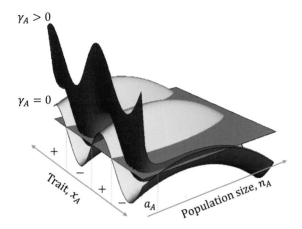

$\gamma_A > 0$

$\gamma_A = 0$

+

−

Trait, x_A

+

−

a_A

Population size, n_A

Figure 7.4 A detailed landscape of invasion fitness, illustrating the purpose of the concepts encompassed in the generic model proposed. Invasiveness of a non-native species ($\dot{n}_A/n_A$, vertical axis) is determined by its trait (x_A) and population size (n_A) as the horizontal plane, as well as propagule pressure (γ_A). The blue surface represents invasiveness for one-off introduction ($\gamma_A = 0$), while the yellow surface represents invasiveness with a constant rate of propagule influx ($\gamma_A > 0$). Green plane represents zero fitness (for reference purpose). For one-off introduction ($\gamma_A = 0$), invasion fitness (blue surface) divides the trait axis into a number of positive (+) and negative (-) performance pockets when the initial population size n_A is small. A non-native species possessing a trait value located in the positive pockets in the trait space will successfully invade, while other traits fail. For a particular non-native trait, its invasiveness also depends on the initial population size: for instance, when $n_A < a_A$ the non-native population could suffer from positive density dependence (Allee effect) and fail to establish, but can establish when $n_A > a_A$; ultimately, demographic and invasiveness will be constrained by negative density dependence (thus the blue and yellow curves eventually bend downwards when the population size n_A becomes too large (e.g., exceeding the carrying capacity)). From Hui et al. (2021), under CC-BY Licence.

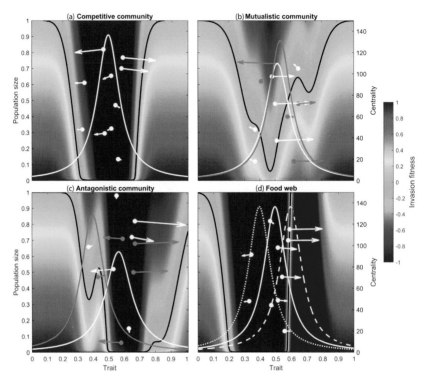

Figure 7.6 Four examples of eco-evolutionary dynamics on demographic fitness landscapes (see Figure 7.1). (a) A competitive community; (b) a bipartite mutualistic network between animals and plants with within-guild competition, facing the introduction of a non-native plant species; (c) a bipartite antagonistic network with within-guild competition, facing the introduction of a non-native resource species; (d) a food web with interspecific competition. White/grey dots represent trait and population size of resident species (white for plants and grey for animals in (b); white for resources and grey for consumers in (c)). Arrows represent the joint ecological dynamics (projection along the population size vertical axis) and evolutionary dynamics (projection along the trait horizontal axis) of resident species. For (b) and (c), grey dots and arrows indicate the other functional guild relative to the non-native species. The background colour represents the demographic performance of an incoming non-native ($\dot{n}_A/n_A$), with black contour lines representing zero invasion fitness. White bell-shaped lines represent within-guild non-native trait centrality (measured by Eq. (7.5)), while grey bell-shaped lines represent centrality for exploiting non-native mutualism (b) and non-native resource (c). In (d) dotted line represents centrality for consuming the non-native resource, while dashed line represents the centrality for the non-native to consume resident resource species. From Hui et al. (2021), under CC-BY Licence.

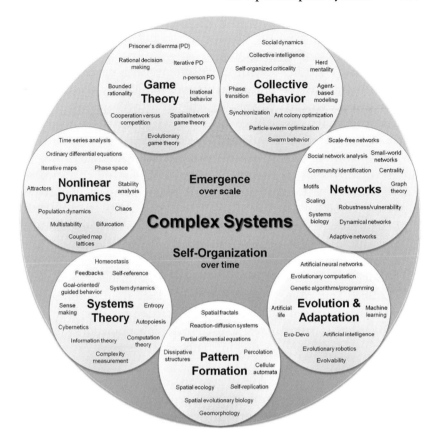

Figure 4.1 A visual, organisational map of complex systems broken into seven sub-groups. Created by Hiroki Sayama under CC BY licence.

simple rigid systems (Chan 2001): distributed control; connectivity; co-evolution; initial condition sensitivity; emergent order; far from equilibrium; and the state of paradox. We have elaborated the corresponding seven features associated with biological invasions of an ecological network (Hui and Richardson 2017):

1. No single centralised mechanism governs the invasibility of recipient ecosystems, and no single trait or syndrome of traits determines the invasiveness of alien species. The resistance from resident species to invasion is not simply the sum of the effect of each species; species richness is therefore usually not a good indicator of biotic resistance.

2. Impacts from biological invasion are not only felt by the species that interact directly with the invasive species but also by other species through mediated interactions.
3. The invasiveness and performance of an invader is mediated not only by its own traits and propagule pressure but also by the context of recipient ecosystems and traits/densities of resident species.
4. The community assemblage of an invaded ecosystem might not simply converge to a modified climax; alternative system regimes may result from priority effects and invasional meltdown.
5. Invasion patterns are retrospectively observable, and characteristics of successful invasions are identifiable and invasion impacts can be explained.
6. Besides characteristics of invaders and recipient ecosystems, introduction rate, propagule pressure and disturbance are also critical determinants of invasion performance and the features of novel ecosystems.
7. Although characteristics of high invasion performance can be clearly identified and strongly supported by evidence from historical invasions or comparison with other areas, prediction based on such strong knowledge is rather poor, a paradoxical phenomenon known as spontaneous self-organisation in complexity science.

In the past two decades, ecological networks of biotic interactions among species, whether native or alien, have become a powerful modelling proxy for capturing ecosystem complexity and transition (Bascompte and Jordano 2007; Hui and Richardson 2017). In particular, the dynamics of invading species and the response of ecological networks are largely self-organised, through a multitude of interactions and feedback loops, many of them facilitated by humans. However, with most ecosystems facing constant biological invasions and with alien species making up an ever-growing portion of resident species, the features of openness and transition deserve further attention (Dornelas and Madin 2020). Local and regional boundaries become blurred when considering this highly transformed global novel ecosystem (Figure 4.2; Fricke and Svenning 2020). Consequently, we coined the term Open Adaptive System (OAS) to denote a specific type of complex adaptive system that has all the key ingredients at play when considering an invaded ecological network (Hui and Richardson 2019a,b). Chapter 3 described a number of dynamic network models that can explain the emergence of structure in ecological networks and how such network structures respond to biological invasions. However, the structures and the changes caused by biological invasions are evidently interrelated, context dependent and diverse. To make sense of these complex and interweaved patterns, we need to dive more deeply into the multifaceted concept of OAS stability to understand the diverse patterns that emerge from ecological networks under invasions. We first introduce the concept of stability and then address a series of issues in erecting our final argument of complexity emergence through persistent transition at criticality.

4.2 Multifaceted System Stability

Do certain network structures promote stability in a complex adaptive system? Is system stability incompatible with changes and transitions? To address these questions, we need more clarity on how stability is defined. Intuitively, a system that can be easily tipped over from small perturbations or disturbance is not stable (imagine a stilt performer), while a system that can withstand some levels of shocks can be defined as stable (picture a cement road block). Inconsistency in the scientific and policy literature on the precise meaning of stability and disturbance has created much confusion (Donohue et al. 2016); Grimm and Wissel (1997) call the concept of ecological stability the tower of Babel. The stability of an ecological network can be assessed by monitoring how its dynamic behaviour responds to perturbations (Yodzis 1981). Given the multifaceted nature of network structures (Chapter 3), it is clear that the stability of a network must also be multifaceted (Figure 4.3; Logofet 2005; Borrelli et al. 2015). For an ecological network, the concept of stability has two components: demographic and structural; both influence how the network responds to perturbations but the consequences are measured in different ways (Hui et al. 2018; Frost et al. 2019). Biological invasions challenge both components of stability.

The definition of demographic stability is based on the effects of small fluctuations (in an ecological network these typically refer to changes in population sizes), normally caused by small perturbations to a system when it sits close to a steady state (an equilibrium). If the fluctuations diminish gradually after the forcing caused by the perturbation, this equilibrium of the system is said to be locally asymptotically stable (May 1973; for a review, see Mascolo 2019). In other words, the dynamic regime of this equilibrium is stable, or the system is considered stable when operating near this equilibrium. Clearly, the stability of a system can only be discussed under certain conditions or within certain bounds. A system can always be crushed by sufficiently strong forcing.

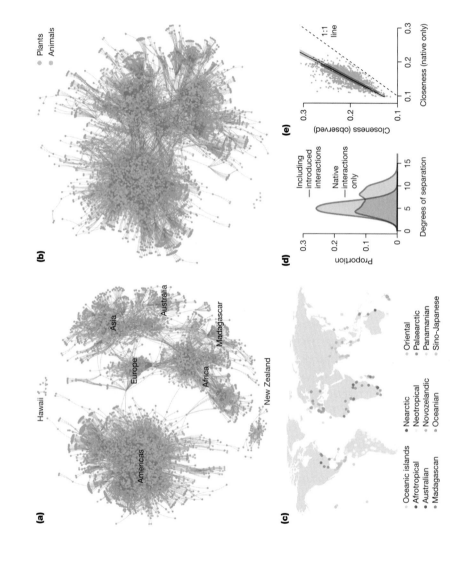

(a)

Hawaii

Asia

Europe

Australia

Americas

Africa

Madagascar

New Zealand

(b)

Plants
Animals

(c)

Oceanic islands
Afrotropical
Australian
Madagascan
Nearctic
Neotropical
Novozelandic
Oceanian
Oriental
Palaearctic
Panamanian
Sino-Japanese

(d)

Including introduced interactions

Native interactions only

Proportion

0.3
0.2
0.1
0

0   5   10   15
Degrees of separation

(e)

1:1 line

Closeness (observed)

0.3
0.2
0.1

0.1   0.2   0.3
Closeness (native only)

For an ecological network, its dynamics can be described in vector format (see Chapter 2) as $\dot{n} = f(n)$. Let $\bar{n}$ be the population densities of residing species at an equilibrium (i.e., $f(\bar{n}) = 0$); note that there could be multiple equilibria. When a small perturbation ($\varepsilon(t)$) to the network affects this equilibrium, the realised population densities from this perturbation become $n(t) = \bar{n} + \varepsilon(t)$ and the dynamics of this perturbation can be assessed at this equilibrium as: $\dot{\varepsilon} = \dot{n} = f(n)$. Ignoring some negligible detail, we could use the Taylor series to approximate the dynamics at the equilibrium (Chapter 2; section 2.2). To know whether the system fluctuation will diminish or increase over time, a linear approximation will suffice: $\dot{\varepsilon} = J\varepsilon(t)$, with J representing the Jacobian matrix of the population dynamics in the ecological network, evaluated at the equilibrium $\bar{n}$. The dynamic regime and the local asymptotic stability of the system at this equilibrium depends on the spectra (the empirical eigenvalue distribution) of the Jacobian matrix.

In particular, the Jacobian matrix represents the adjacency matrix (i.e., interaction strength matrix) of an ecological network (Figure 4.4A), with the diagonal elements (a_{ii}) representing the strength of self-regulation and the non-diagonal elements (a_{ij} for $i \neq j$) representing the interaction strength from the species on the column to the species on the row. An eigenvalue λ_i of matrix J is the root to the characteristic equation $Jv_i = \lambda_i v_i$, with v_i its corresponding right-hand eigenvector. Eigenvalues can be complex conjugates that possess a real part, $\text{Re}(\lambda_i)$ and an imaginary part, $\text{Im}(\lambda_i)$ (Figure 4.4B). Two metrics of eigenvalues are closely related to the dynamic regime of a system (e.g., Figure 4.5 for a two-dimensional system): the determinant $\det(J) = \prod_i \lambda_i$ and the trace $\text{tr}(J) = \sum_i a_{ii} = \sum_i \lambda_i$. The eigenvalues are normally located with a circle (or an ellipse) centred on

---

Figure 4.2 Altered interaction biogeography in the global plant–frugivore meta-network. **a, b,** Meta-networks including native interactions only (**a**) or both native and introduced interactions (**b**). Each node represents a plant or animal species, and links represent interactions observed at any study location. **c,** Interaction data derived from 410 spatially or temporally distinct networks. **d,** The degree of separation among species pairs in the meta-network is greater when only interactions among native species are considered than when interactions with introduced species are included. **e,** Closeness centrality for each species is higher in the observed meta-network that includes introduced interactions than in the native-only meta-network. Points represent species; line and shading show standardised major-axis regression fit and confidence interval. From Fricke and Svenning (2020), reprinted with permission.

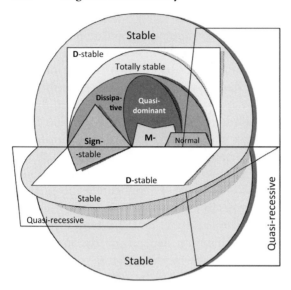

Figure 4.3 Different concepts of system stability illustrated in a 3D Venn diagram. Let a community be depicted by a Jacobian matrix. The Venn diagram contains three half-planes: the upper vertical one represents Jacobian matrices with no zero entries on the main diagonal, the horizontal one with all the diagonal entries non-positive and at least one zero, the lower vertical half-plane with at least one positive diagonal element. Lyapunov stability requires the real parts of all eigenvalues of the matrix to be negative (green petals). From Borrelli et al. (2015); originally adapted from Logofet (2005), reprinted with permission.

$(\mathrm{tr}(\boldsymbol{J})/S, 0)$ in the complex plane and with a specific radius (especially for a large network; Figure 4.4C; see details in the next section).

The distribution of eigenvalues can fully capture the dynamic regime of a system (Perko 2001). The non-zero imaginary part of an eigenvalue $\mathrm{Im}(\lambda_i) \neq 0$ signals an oscillating behaviour along its corresponding eigenvector in species densities from a perturbation; large absolute values imply faster oscillations while zero values indicate no oscillations. A positive real part of an eigenvalue $\mathrm{Re}(\lambda_i) > 0$ suggests that the perturbation will be amplified over time along its corresponding eigenvector (at a rate of $\exp(\mathrm{Re}(\lambda_i)t)$, while a negative real part indicates the exponential damping of the perturbation. We call the eigenvalue that has the maximal real part the leading eigenvalue of the network Jacobian, denoted as λ_m and by definition $\mathrm{Re}(\lambda_m) \geq \mathrm{Re}(\lambda_i)$ (there can be more than one leading eigenvalue such as the two conjugated complex numbers in Figure 4.4B). To summarise the response of an ecological

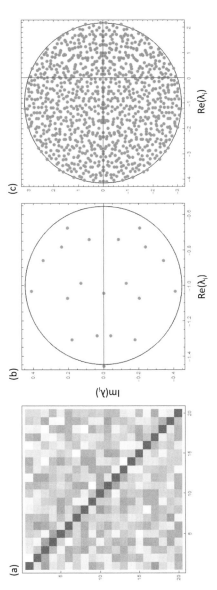

Figure 4.4 Jacobian matrix of a linear dynamic system and its eigenvalues. (a) The Jacobian matrix J (representing interaction strength) of a linear system $\dot{n} = J(n - \bar{n})$, of S ($= 20$) species interacting with each other (20×20 in dimension), with the diagonal elements assigned -1 and non-diagonal elements randomly generated from a normal distribution with zero mean and standard deviation $\sigma = 0.1$. Levels of grey represent values of interaction strength (level of grey represents magnitude). (b) The distribution of the eigenvalues (λ_i) of matrix J from (a) in the complex plane, with the horizontal axis representing the real part of the eigenvalues ($\text{Re}(\lambda_i)$) and the vertical axis the imaginary part of the eigenvalues ($\text{Im}(\lambda_i)$). (c) The eigenvalues of a randomly generated matrix of 1000×1000 in dimension, representing 1,000 species randomly interacting with each other. The black circles in (b) and (c) are centred at $(-1, 0)$ with a radius of $\sigma\sqrt{S-1}$.

Poincaré Diagram: Classification of Phase Portaits in the $(\det A, \operatorname{Tr} A)$-plane

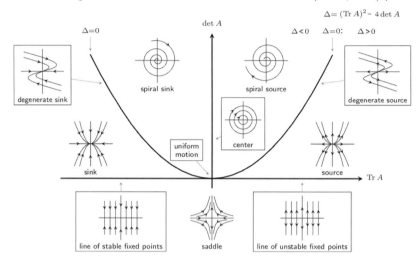

Figure 4.5 Trace-determinant plane for classifying phase portraits (dynamic trajectories of state variables) in 2D linear dynamical systems. Created by Freesodas under CC-BY Licence.

network to small perturbations, we therefore only need to check the sign of the real part of its leading eigenvalue(s) of the Jacobian matrix at the focal equilibrium point. If $\operatorname{Re}(\lambda_m) > 0$, the perturbation will be amplified along its corresponding eigenvector (the major axis) and the system is unstable. If $\operatorname{Re}(\lambda_m) < 0$, the perturbation dies out and the system returns to its equilibrium (i.e., locally asymptotically stable). As the damping rate of perturbation is $\exp(\operatorname{Re}(\lambda_m)t)$ and $\operatorname{Re}(\lambda_m) < 0$ for a stable system, resilience can be measured as

$$\operatorname{Res} = \ln|\operatorname{Re}(\lambda_m)|. \tag{4.1}$$

In some resilient systems, fluctuations can grow temporally, seemingly pushing the system away from an equilibrium, but these eventually diminish around the equilibrium, normally reflecting the transient dynamics and delayed response of the system to perturbations. In such cases, reactivity is often used to measure the maximum instantaneous rate at which a perturbation runs away from an equilibrium (Chen and Cohen 2001),

$$R = \max_{\|\varepsilon\| \neq 0}\left[\left(\frac{1}{\|\varepsilon\|}\frac{d\|\varepsilon\|}{dt}\right)\Big|_{t=0}\right], \tag{4.2}$$

where the norm is the Euclidean distance $\|\varepsilon\| = \left(\sum_i \varepsilon_i^2\right)^{1/2}$. Species can coexist stably in a resilient network with specified densities according to the equilibrium. In these and many other metrics that address demographic stability, perturbations are only considered to affect the population sizes or densities of resident species, while the underlying network structures are assumed to remain unchanged.

Structural stability, on the other hand, addresses the effects of adding or removing species on network topology. For instance, in many simulated ecological networks, a number of species (S_{ini}) are initially randomly assembled and allowed to interact with each other according to specified rules (e.g., Figure 3.26), while the number of species (S_{eqn}) that have survived and which can coexist after the system has reached an equilibrium is often just a small portion of the initial species. Accordingly, the metric of persistence can be simply defined as,

$$P = S_{eqn}/S_{ini}. \tag{4.3}$$

A large score of persistence often signals the presence of coexisting mechanisms that can accommodate even randomly assembled species. Robustness is another structural stability and accounts for the ability of a network to defy secondary extinctions from the removal of some species, also known as deletion stability (Pimm 1979; Paine 1980). In practice, it can be measured as the fraction or number of species to be removed that can result in 50 per cent subsequent species loss in a network (Dunne et al. 2002). Species removal can be random or targeted (e.g., first remove species with lower or higher node degrees, or from lower or higher trophic levels), or it can be defined for different fractions of secondary species loss (e.g., Nuwagaba et al. 2017). Non–random species removal is necessary when assessing robustness as the sequence of extinctions resulted from biological invasions is often not random (e.g., specialists are more likely to experience secondary extinctions than generalists). Invasibility is related to the addition stability and describes the response of a network to the introduction of new species (Hui and Richardson 2017).

Instead of species removal or addition, a more subtle structural change in an ecological network happens in the system parameters such as the per-capita interaction strengths, depicting random or adaptive changes in the payoffs of specific life-history strategies and behaviours, as well as interaction rewiring. If the system regime (e.g., the equilibrium can be a limit cycle or a stable attractor) does not change from such subtle changes in system parameters, it lacks the chance for any bifurcation (a drastic change of dynamic regimes from minute shifts in system parameters) and is therefore structurally stable (Solé and Valls 1992; Rohr et al. 2014). Co-evolution, ecological fitting and biological invasions can all impose changes to system parameters of the recipient ecosystem and profoundly affect the network's adaptability and structural stability.

Each assembly process has its own characteristic scale and pace (Figure 3.1); emergent patterns and the associated stability of an OAS can often be divided into shorter-term fast-paced features and longer-term slow-paced ones. The demographic stability of an ecological network often reflects the response of fast-paced processes to perturbations, in the form of population fluctuations and interaction rewiring, while structural stability normally captures the response of slow-paced processes to perturbations (e.g., adaptation and trait evolution at the pace of generations). Moreover, the feedback between such fast and slow processes can add extra layers to the structural stability of eco-evolutionary systems (see Section 2.6). For instance, the canonical equation of adaptive dynamics for a single trait (Eq. (2.17); Dieckmann and Law 1996) can be further expanded to represent multispecies interactions with each also represented by multiple traits. Let there be s interacting species, each represented by k traits. This means that we can use a column vector $\boldsymbol{x}_i = \langle x_{i,j} \rangle$ to represent the trait composition of species i, with $x_{i,j}$ its value for the j-th trait. For the entire ecological network, the combined trait vector $\boldsymbol{x} = \langle \boldsymbol{x}_i^T \rangle^T$ with a total of $s \cdot k$ entries, thus, captures the combined strategies of all resident species. The canonical equation for this multispecies network with each also possessing multidimensional traits can be given as the following (Dieckmann and Law 1996; Durinx et al. 2008; Metz and de Kovel 2013; Zhang et al. 2021),

$$\dot{\boldsymbol{x}}_i = \boldsymbol{M}_i(\boldsymbol{x}) \left(\nabla_{\boldsymbol{x}_i'} f_i\left(\boldsymbol{x}_i', \boldsymbol{x}\right) \Big|_{\boldsymbol{x}_i' = \boldsymbol{x}_i} \right), \tag{4.4}$$

where the mutational matrix $\boldsymbol{M}_i(\boldsymbol{x})$ is k by k in dimension and can be further elaborated as $1/2\mu_i(\boldsymbol{x}_i)\hat{n}_i(\boldsymbol{x})\boldsymbol{C}_i(\boldsymbol{x}_i)$, with $\boldsymbol{C}_i(\boldsymbol{x}_i)$ the covariance matrix of mutational trait increments (Dieckmann and Law 1996; Metz and de Kovel 2013). As the mutations are assumed to be directionless and allometric in nature, the covariance mutational matrix is symmetric and positive (semi-) definite (meaning $\boldsymbol{v}^T \boldsymbol{C}_i \boldsymbol{v} \geq 0$ for any non-zero column vector $\boldsymbol{v}$). The term in the brackets represents the selection gradient and is defined as the nabla operator ∇ (the partial derivatives) of the invasion fitness vector f_i (Section 2.6 for examples).

At an evolutionary singularity of a trait coalition of all species in the network (when selection gradients for all species diminish), we could evaluate the stability of Eq. (4.4) at this singularity using its combined Jacobian matrix (J_C, $s \cdot k$ by $s \cdot k$ in dimension). Again, if all eigenvalues of J_C have negative real parts, the singularity is convergence stable; otherwise, if the real part of one eigenvalue is positive, the singularity is convergence unstable (Figure 2.11). A singularity that cannot be invaded by any mutants within its vicinity in the trait space is evolutionarily stable, while a singularity that is both evolutionarily stable [un-invadable] and convergence stable [traits evolve towards the singularity] is termed continuously stable (Eshel et al. 1997). Disruptiveness, a measure of evolutionary instability, can be computed as the average strength, over all species, of disruptive selection in the system (Brännström et al. 2011), with the strength of disruptive selection for a particular species measured as the curvature of its invasion fitness. Convergence stability and evolutionary stability are the two extra structural stabilities of system adaptability that have emerged as a result of the consideration of evolutionary processes in ecological networks.

An ecological network facing biological invasions can lose any or all facets of its system stability. For instance, biological invasions could cause large demographic impacts to resident species and trigger the recipient ecological network to lose its demographic stability (Simberloff 2001; Gaertner et al. 2014; Colautti et al. 2017). By adding new alien species and imposing novel interactions with resident species, biological invasions can also, inevitably, challenge addition stability (invasibility). Should a biological invasion lead to the local extinction or extirpation of resident species, the deletion stability (robustness) needs to be assessed. With alien species and resident species co-evolving and forming novel interactions, it can modify the realised niches and selection pressures facing each residing species and alter the path of community assembly, in consequence affecting convergence stability and evolutionary stability. We explore these aspects next and also in subsequent chapters.

4.3 Stability Criteria

A fundamental quest in biology is to elucidate the relationship between the structure and function of an organism, or the notion that structure determines function. Adopting ecological networks as the model of an ecosystem, Chapter 3 discussed some of the many inter-related structures that can emerge from the co-evolving and co-fitting network. It also considered how biological invasions can tentatively alter these emerged structures. These network structures reflect different topological and architectural features of the same underlying interaction relationships between species, captured by the interaction matrix that can be considered a proxy of the system Jacobian matrix. Network stability, as introduced previously, deals with the multifaceted ways in which a network responds to perturbations and can also be measured in different ways based on the system Jacobian matrix. These two properties of an OAS – structural complexity and stability – are therefore interlinked as the two sides of a coin, while their relationships have been a primary thread through the development of modern ecology (see review, Landi et al. 2018).

Intuition, based on ecological theory, tells us that complex communities and networks should be more stable than simple ones; in other words, network complexity implies redundancy and therefore enhances system resilience and robustness. MacArthur (1955) argued that 'a large number of paths [links of biotic interactions] through each species is necessary to reduce the effects of overpopulation of one species', and thus that stability increases with the node degree of species. Based on his observations of natural communities, Elton (1958) argued that 'simple communities were more easily upset than richer ones'. Simple systems are thus more susceptible to destructive oscillations in populations, and more vulnerable to invasions; this notion has become known as the diversity-invasibility hypothesis (Fridley 2011). Such ecological intuition of a positive relationship between system complexity and stability was, nevertheless, flawed and was challenged mathematically in the seminal work of Robert May (1972, 1973) on the stability criterion of modelled ecosystems. In particular, May depicted network dynamics using linear differential equations, $\dot{n} = A(n - \bar{n})$ near an equilibrium ($\bar{n}$) and explored its local asymptotic stability by examining the community matrix $A = \langle a_{ij} \rangle$ (similar to the system Jacobian matrix in a generic nonlinear system). The non-diagonal entries, a_{ij}, representing the strength of interspecific interactions, were randomly assigned from a normal distribution with zero mean and standard deviation σ (i.e., the average strength of interspecific interactions), while the diagonal entries of the matrix, representing self-regulation from intraspecific interactions, were set to -1 (Figure 4.4A). In addition, interspecific interactions were assigned zero (the absence of an interaction) with probability $1 - C$. Consequently, in such a modelled ecosystem network, complexity is measured by three network structures: species richness (S), the connectance of the community matrix (C) and the average interaction strength (σ). May then asked the question – what specific constraints or network structures of this modelled ecosystem could ensure system stability and resilience against small perturbations? The answer, surprisingly, goes against ecological intuition.

The community interaction matrix assigned in May's linear dynamic network is a random square matrix. To answer his question, May followed two advances in dynamical systems and algebra. First, we mentioned in the Preface the n-body problem challenge from Oscar II, King of Sweden, in 1887. The prize was given to Henri Poincaré, who developed the qualitative theory of differential equations as early as 1881 (Figure 4.5). Over the years, interest in this topic has grown into a subdiscipline of mathematics known as stability theory, and different stability criteria have been developed for specific systems (Bhatia and Szegö 1970). For such linear dynamical systems, the eigenvalues of the Jacobian matrix determine the system stability (Lyapunov 1892), and May's question thus becomes how eigenvalues of a random matrix distribute in the complex plane (Figure 4.4). Second, the spectra of random matrices began featuring in mainstream statistics in the early twentieth century (Wishart 1928). In 1931 Gershgorin developed his circle theorem, which showed that the eigenvalues of a square matrix lie within at least one so-called Gershgorin circle with the i-th circle centred at $(a_{ii}, 0)$ and the radius $\sum_{j \neq i} |a_{ij}|$. In 1955 Wigner introduced the random matrix approach to nuclear physics. While modelling the Hamiltonian of atomic nuclei, Wigner (1958) derived the random matrix theory regarding the eigenvalues of symmetric matrices, known as the semicircle law. Recently, Tao et al. (2010) proved the universality of an expanded random matrix theory, known as the circular law. In short, the random matrix theory states that the eigenvalues of a large random matrix, with its entries following identical independent random distributions with zero mean and unit variance, will cover a unit round disk evenly in the complex plane.

Applying these random matrix theories, May (1972, 1973) found that the eigenvalues of the modelled community matrix are distributed within a disk in the complex plane centred at $(-\mu, 0)$ with a radius of $\sigma\sqrt{C(S-1)}$ ($\mu = 1$ indicates the absolute strength of self-regulation; see an example in Figure 4.4 for $C = 1$). This means that the real part of the leading eigenvalue must be less than the right end of the circle in the complex plane, $Re(\lambda_m) \leq \sigma\sqrt{C(S-1)} - \mu$. As a stable system requires $Re(\lambda_m) < 0$, May thus derived his stability criterion governing the stability of a modelled ecosystem: an ecological network near its equilibrium almost certain of its stability if,

$$\sigma\sqrt{C(S-1)} < \mu. \tag{4.5}$$

This simple stability criterion suggests that complex networks with a large number of species which are highly connected via strong interactions are destined to be **un**stable. Complexity, therefore, counterintuitively, begets instability, although the interpretation of this is also intuitive to many. Like Gause's law, which posits that interspecific competition needs to be weaker than intraspecific regulation to allow two competing species to coexist, May's stability criterion defines the prerequisites for the stable coexistence of resident species in complex ecological networks. It argues that, as with Gause's law, for stable persistence and coexistence the average pressure from interspecific interactions experienced by a species in the network ($\sigma\sqrt{C(S-1)}$) needs to be weaker than its intraspecific self-regulation (μ). Community stability, thus, is fundamentally linked to species persistence and coexistence.

As a result of its oversimplification of ecological reality, May's stability criterion has been widely challenged (e.g., Jacquet et al. 2016). These challenges have produced many milestone contributions in the complexity-stability literature. For instance, De Angelis (1975) argued for the inclusion of non-linear functional response (Holling 1959) as an essential ingredient for modelled ecosystems, while Nunney (1980) showed that doing this can generate a positive complexity-stability relationship in food webs. Dambacher et al. (2003) further aligned stability measures with our intuition and introduced a weighted measure to show that complex networks do not always beget instability. Several milestones were achieved in the 1980s. First, Hastings (1984) extended May's way of assigning interaction strengths randomly with zero mean ($E(a_{ij}) = 0$ for $i \neq j$) for non-diagonal elements to a non-zero mean ($E(a_{ij}) \equiv \omega \neq 0$) and showed the drastic change in the stability criterion,

$$\omega C(S-1) < \mu. \tag{4.6}$$

This implies two separate constraints behind the radius of the circle surrounding the eigenvalues. Nevertheless, we see that the left-hand side of Eq. (4.6) still represents the average strength of interspecific interactions experienced by a species in the ecological network, so the generality and validity of Gause's law hold. Second, Sommers et al. (1988) developed an elliptic law, showing the eigenvalues of a square matrix with its elements assigned randomly with zero mean and unit variance, but imposing a correlation (ρ) between off-diagonal element pairs (between a_{ij} and a_{ji}). This is a reasonable consideration as the strengths of reciprocal interactions are often highly correlated. For instance, the gain of a predator species from consuming a particular prey species, say a_{ij}, is essentially the same as the loss of the prey species weighted by a conversion factor. Sommers et al. (1988) showed that the eigenvalues distribute evenly over an ellipse in the complex plane, centred at $(0, 0)$ in the complex plane, with the semi-major axis of length $1 + \rho$ along the real axis and the semi-minor axis of length $1 - \rho$ along the imaginary axis.

Recent studies have revisited these milestones to fine-tune May's stability criterion even further by considering the role of different types of biotic interactions and their combinations on network stability. In particular, Allesina and Tang (2012) discovered that different types of biotic interactions can constrain the distribution of eigenvalues into ellipses and therefore impose different stability criteria. For instance, for a network with a mixture of competition and mutualism, the eigenvalues are distributed in an ellipse centred at $(-\mu, 0)$ and the horizontal semi-major axis of length $\sigma\sqrt{C(S-1)}(1 + 2/\pi)$ and the vertical semi-minor axis of length $\sigma\sqrt{C(S-1)}(1 - 2/\pi)$. The stability criterion of a predator–prey network with zero mean interspecific interaction strength becomes, $\sigma\sqrt{C(S-1)} < \mu/(1 - \beta^2)$, where $\beta = E(|a_{ij}|)/\sigma$. For a network with only mutualistic interactions, the stability becomes $\sigma C(S-1) < \mu/\beta$. Tang et al. (2014) explored the role of the correlation between off-diagonal matrix elements a_{ij} and a_{ji} and showed two alternative stability criteria in a food web, $\omega C(S-1) < \mu$ (eq.(4.6)) or $\sigma\sqrt{(S-1)(1+\rho)} - C\omega < \mu$. Grilli et al. (2016) considered a network comprising two equal sized functional guilds with within-guild and between-guild interaction strengths following different random distributions.

Following this line of thinking of a structured multilayer network, a generic stability criterion has been derived (Nnakenyi et al. 2021). Here, let us consider the spectral distribution of a structured community of S species comprising k functional guilds with functional guild i containing s_i species (Figure 4.6; therefore, $S = \sum_{i=1}^{k} s_i$). This forms a community

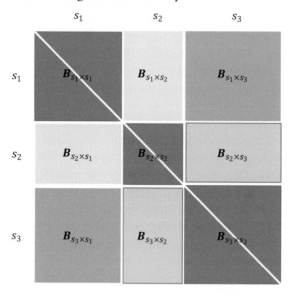

Figure 4.6 An illustration of the interaction matrix of an ecological network with three functional guilds. The matrix is therefore composed of 3-by-3 blocks, with the three diagonal blocks representing intra-guild interactions and the non-diagonal blocks representing cross-guild interactions. The diagonal elements (white line across diagonal blocks) represent self-regulation strength of each species. Each block has its own distribution and correlation of interaction strengths and connectance. See text for the generic stability criterion.

matrix $A_{S \times S}$ with block structures $B_{s_i \times s_j}$ representing the interactions between species of function guild i and functional guild j; $B_{s_i \times s_i}$ represents within-guild interactions between species of the same functional guild i. Let the diagonal elements of block $B_{s_i \times s_i}$ follow a distribution of mean $\mu_{diag,i}$ and standard deviation $\sigma_{diag,i}$, while the non-diagonal elements follow a distribution of $\mu_{i,i}$ and standard deviation $\sigma_{i,i}$ and correlation between elements a_{ij} and a_{ji} of the non-diagonal elements $\rho_{i,i}$, and connectance $C_{i,i}$. Let the elements of the off-diagonal blocks $B_{s_i \times s_j}$ $(i \neq j)$ follow a distribution of mean $\mu_{i,j}$, standard deviation $\sigma_{i,j}$, correlation between symmetric elements $\rho_{i,j}$ and connectance $C_{i,j}$. We can have the following bound on the real part of the leading eigenvalue (Nnakenyi et al. 2021),

$$\lim_{S \to \infty} \mathrm{Re}(\lambda_m) \leq \max(r_1, r_2) - \mathrm{E}(a_{ii}), \qquad (4.7)$$

where

$$E(a_{ii}) = \sum_{j=1}^{k} \mu_{diag,j} \frac{s_j}{S}$$

$$r_1 = (S-1) \sum_{i=1}^{k} \sum_{j=1}^{k} C_{i,j} \mu_{i,j} \frac{s_i s_j}{S^2}$$

$$r_2 = \sqrt{(S-1) \sum_{i=1}^{k} \sum_{j=1}^{k} C_{i,j} \sigma_{i,j}^2 \frac{s_i s_j}{S^2} \left(1 + \sum_{i=1}^{k} \sum_{j=1}^{k} \rho_{ij} \frac{s_i s_j}{S^2}\right)}$$

The stability criterion for such a generic block-structured multilayer network is, therefore, $\max(r_1, r_2) < E(a_{ii})$.

To resolve the oversimplification of ecological networks as a linear dynamical system, many researchers have attempted to develop more realistic models that include more plausible network structures of non-linear (McCann et al. 1998; Krause et al. 2003; Okuyama and Holland 2008) and non-random (Neutel et al. 2002; Suweis et al. 2015; van Altena et al. 2016) interactions. These have helped to further elucidate the conceptual layers of network stability and complexity. May's stability criterion, in particular, has been extended to consider density dependence in the formulating interaction strength (McCann 2012; Stone 2018) and the effect of dispersal of propagules between communities on network stability (Mougi and Kondoh 2016; Gravel et al. 2016; see Chapter 6). Furthermore, interaction promiscuity and switching can have a profound influence on the complexity–stability relationship (Montoya et al. 2006). By incorporating interaction switching (i.e., with foraging adaptation), Kondoh (2003) showed that the classic negative complexity–stability relationship in static food webs does not necessarily hold. In mutualistic networks, allowing interaction switching can enhance the robustness of interaction strength against species loss over static networks (Kaiser-Bunbury et al. 2010). Considering ecological fitting via adaptive interaction switching, Langat et al. (2021) also found that May's stability criterion can be further relaxed by reducing the effective strength of interspecific interactions, $\theta\sigma\sqrt{C(S-1)} < \mu$, with $\theta = C\sqrt{\pi(1/2 - C/3)}$, thus facilitating network stability. Taken together, what May's stability criterion and its diverse variants have achieved is to complete the jigsaw puzzle by relating network structural complexity to its stability, in particular to system resilience based on the position of the leading eigenvalue of the system Jacobian.

We can make (tentative) sense of the diverse network structures that emerge from community assembly and biological invasions. For example, May's stability criterion highlights four factors that could disrupt network stability and thus initiate assemblage reshuffling: stronger interaction strength, higher species richness, more generalists (fostering a higher level of connectance) and weaker intraspecific self-regulation. These four factors are largely congruent with the factors that facilitate the establishment and impact of alien species in recipient ecosystems. In particular, propagule pressure (Ricciardi et al. 2011), trait distinctiveness (Daehler 2001; Moles et al. 2012), resource availability (Guo et al. 2015), interaction strength (Williamson 1996; Lonsdale 1999), system connectivity (Ives and Carpenter 2007) and disturbance (Davis et al. 2000) clearly contribute substantially to the susceptibility of an ecosystem to invasion (Richardson and Pyšek 2006).

First, large propagule pressure of the invader can increase the standard deviation of interaction strength and network connectance, thereby increasing invasibility (Ricciardi et al. 2011). More generalist invaders and invaders with distinctive traits will increase interaction strength, making the system more susceptible to invasion (Moles et al. 2012; Minoarivelo and Hui 2016). Second, high levels of species saturation in a community (meaning that all or most niches are occupied and most resources are used) can increase intraspecific density regulation, leading to strong resistance to opportunistic invasions (Guo et al. 2015). Third, a high level of network connectance has been predicted to enhance the resistance of food webs to invasion (Romanuk et al. 2009), although this view is contested (Baiser et al. 2010; Lurgi et al. 2014). Network complexity, measured as network size and connectivity (number of interactions), can also enhance network resilience (Okuyama and Holland 2008). Finally, disturbance and fluctuating resources can dismantle self-regulation, thereby allowing species to capitalize on limiting resources and become established (Davis et al. 2000; Horak et al. 2013; Cuddington and Hastings 2016). All these factors will enhance network invasibility, although caution is needed when generalising results from such simple systems (Evans et al. 2013; Rohr et al. 2014; Valverde et al. 2017; Maynard et al. 2018). We predict that more realistic findings will emerge when using recently updated structure-function relationships for specific networks (e.g., Grilli et al. 2016; Romanuk et al. 2017; Valdovinos et al. 2018).

Although network architectures, such as nestedness and modularity, are not explicitly included in the stability criteria, they are nevertheless strongly related to network size and connectance and therefore strong predictors of network stability (Figure 4.7; also see Chapter 3). However, these empirical observations do not allow us to determine whether these network features have triggered or facilitated the invasion, or are simply a

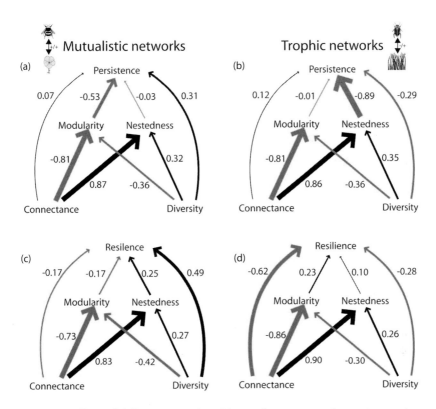

Figure 4.7 Effects of different network architectural patterns on the persistence (a and b) and resilience (c and d) of mutualistic [(a) and (c)] and trophic [(b) and (d)] networks. Arrow thickness is scaled to standardized coefficients from path analysis and illustrates relative effect strengths. Negative effects are represented in grey and positive effects in black. The effects of connectance and diversity are split between direct effects and indirect effects through changes in modularity and nestedness. The strength of the indirect effects is calculated by the product of the coefficients along the path. For example, in (a), connectance has a direct effect of strength 0.07, an indirect effect through modularity of strength 0.43 (−0.81 × −0.53) and an indirect effect through nestedness of strength −0.026 (0.87 × −0.03), which lead to an overall connectance effect of 0.47. From Thébault and Fontaine (2010), reprinted with permission.

consequence of the invasion (Traveset and Richardson 2006). On the one hand, a highly nested network suggests a strong hierarchical architecture. This is potentially the result of sorting of species by multiple ecological filters which creates unbalanced energy/material flow from specialists to generalists and opportunity niches for introduced specialists that can more efficiently exploit resources than resident generalists. In other words, network asymmetry creates opportunity niches for specialists and thus enhances invasibility. For instance, Thébault and Fontaine (2010) demonstrated that a highly connected and nested architecture promotes stability in mutualistic networks (also, Okuyama and Holland 2008; Bastolla et al. 2009; Thébault and Fontaine 2010; Zhang et al. 2011; Rohr et al. 2014). Invaded pollination networks are often more nested and (obviously) contain more species than uninvaded networks (Padrón et al. 2009; Traveset et al. 2013; Stouffer et al. 2014). However, when interaction strength is considered in more detail, stability becomes negatively affected by both connectance and nestedness (Allesina and Tang 2012; Vieira and Almeida-Neto 2015).

On the other hand, a highly compartmentalised network is formed by clearly bounded modules which could have niches and habitats that are spatially or temporally partitioned. This nonetheless provides opportunities for invaders that are generalist enough to explore two or more modules. Species possessing traits with high plasticity or tolerance, and those with complex life cycles (through ontogenetic niche shift), could thus invade highly compartmentalised networks. For instance, modularity (i.e., the structure of compartmentalisation in networks) has been observed to be lower in invaded pollination networks and food webs than in uninvaded ones (Albrecht et al. 2014; Lurgi et al. 2014). Extremely low levels of nestedness are, therefore, associated with high levels of modularity and thus destabilize the community due to the detrimental effect of modularity on the stability of mutualistic networks (Thébault and Fontaine 2010; Campbell et al. 2012). Inconsistency of the correlation between network architecture and stability seems to be attributable to the confusion in choosing appropriate measures of network stability, as each measure only specifies one particular facet of stability and thus often leads to contradictions when interpreted as general stability for comparison (Vallina and Quéré 2011).

We must emphasise that, as elucidated in Section 4.2, demographic stability is only one of many facets of system stability; discussing only this facet masks other important issues related to other dimensions of system stability, such as feasibility (Rohr et al. 2014) and network invasibility

(Hui et al. 2016). When considering the adaptive network (Eq. (4.4)), for instance, newly emerged concepts of stability can also be explained by decomposing the Jacobian matrix J_C of the canonical equation, Eq. (4.4) (Zhang et al. 2021),

$$J_C = MJ_S = M(H + F + C), \qquad (4.8)$$

where M is the mutational matrix; J_S is the Jacobian of the selection gradient (Leimar 2005, 2009). The matrix M is a block diagonal with blocks $M_i(x)$. The stability criterion for such an adaptive network has been developed based on this matrix decomposition (Figure 4.8). In particular, the matrix H captures the local shape of invasion fitness around the singularity and is a block diagonal matrix with the block for species i its Hessian matrix H_i of invasion fitness $f_i(x'_i, x)$, with its element $H_{i,kl} = \partial^2 f_i/(\partial x'_{i,k} \partial x'_{i,l})$. The F matrix is also a block diagonal matrix with the block for species i, F_i, capturing the effects of frequency-dependent selection (compare Metz and Geritz 2016), with its element of its block $F_{i,kl} = \partial^2 f_i/(\partial x'_{i,k} \partial x_{i,l})$. If $v^T F_i v < 0$ for a unit vector v

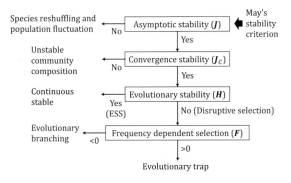

Figure 4.8 A summary of the eco-evolutionary regime scenarios of an adaptive ecological network and its dependent matrix. The decomposition of the Jacobian matrix J_C demonstrates that the Hessian matrix (H), frequency-dependent selection matrix (F), co-evolutionary matrix (C) and mutational correlation matrix (M), together, determine whether or not a singularity is convergence stable. Evolutionary trap can be induced by a positive definite or indefinite Hessian matrix (H) to ensure evolutionary instability and a positive definite or indefinite frequency-dependent selection (matrix F) to ensure the lack of locally protected dimorphisms. Such constraints on the matrices H and F make it impossible, in the absence of co-evolution (C) and mutational correlations (M), for all eigenvalues of the Jacobian J_C to have negative real parts. Drawn based on Zhang et al. (2021).

indicating a direction in the trait space, we say that species i is experience negative frequency-dependent selection along this trait direction at the vicinity of the singularity; $\boldsymbol{v}^T \boldsymbol{F}_i \boldsymbol{v} > 0$ defines positive frequency-dependent selection; $\boldsymbol{v}^T \boldsymbol{F}_i \boldsymbol{v} = 0$ frequency-independent selection. The $\boldsymbol{C}$ matrix comes from the co-evolution of the species in the community (due to interspecific interactions) and is a block matrix with zero sub-matrices for the diagonal blocks and other non-diagonal blocks $\boldsymbol{C}_{ij}$, with its element $C_{ij,kl} = \partial^2 f_i / (\partial x'_{i,k} \partial x_{j,l})$. This decomposition of the Jacobian differentiates between the various evolutionary forces and shows how these four forces jointly (i.e., the eigenvalues of matrix $\boldsymbol{J}_C$) determine the convergence stability of a singularity, while its evolutionary stability is solely determined by the eigenvalues of matrix $\boldsymbol{H}$.

Although network structures and stability are constrained by the stability criterion and are therefore related, the relationship often appears weak and complex (Minoarivelo and Hui 2016). Hierarchical clustering analyses revealed three clusters of metrics that can coalesce depending on the level of propagule pressure (Figure 4.9). Apart from the two clusters formed by most network structural metrics and most network stability metrics, respectively, the third cluster contains nestedness, invasion impact and the average invasiveness of invaders. Nestedness is weakly related to both the other network structural metrics and the cluster of network stability metrics, and is negatively correlated with resilience and robustness (Allesina and Tang 2012; Campbell et al. 2012; Minoarivelo and Hui 2016). Interestingly, there are very strong positive relationships among resilience, invasibility and disruptiveness. Although invasibility and disruptiveness measure network instability rather than network stability, they are nonetheless positively correlated with robustness and resilience. This is understandable because the ability of a network to quickly adapt in the face of perturbations also confers on it the ability to accommodate the invader and make it more invasive. Measurement of invasion impact, as the reduction in the total abundance of all resident species as a result of invasion, is the least related to the other network stability metrics. The clustering of network metrics reveal the multifaceted and interrelated network features, while the positive relationships between stability metrics (resilience, robustness) and instability metrics (invasibility, invasiveness, disruptiveness, impact) highlight the necessity to use multiple network metrics from different clusters for a cohesive view on network functioning (Ives and Carpenter 2007; Minoarivelo and Hui 2016).

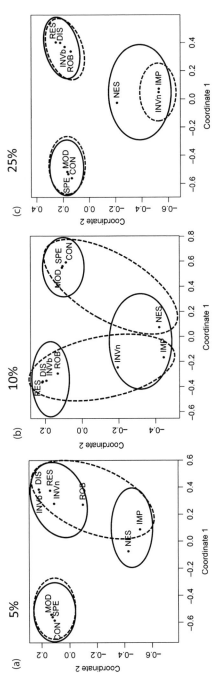

Figure 4.9 Visualization of the relationship between all network metrics in a two-dimensional space using a multi-dimensional scaling analysis, for different initial propagule size. Invader propagule pressure is respectively (a) 5%, (b) 10% and (c) 25% of the average native population density. Clusters formed by a k-mean clustering analysis are shown by the solid circles and those formed by a hierarchical clustering analysis are shown by the broken circles. CON: connectance; SPE: specialisation; MOD: modularity; NES: nestedness; IMP: impact; ROB: robustness; INVb: invasibility; INVn: average invasiveness; RES: resilience; DIS: disruptiveness. From Minoarivelo & Hui (2016), under CC–BY Licence.

4.4 Collapses and Panarchy

Successful invasions can only happen if some alien species can grow in abundance from small founder populations. In other words, invasions can succeed only if both the standing resident species and newly introduced alien ones form an unstable mixed network. The outcome of invasions is thus directly related to the loss of stability in invaded networks; being unstable makes a network susceptible to invasion. The synonyms of losing system stability are regime collapses and shifts. System collapses are precursors for drastic transition and transformation, often followed by reorganisation and rebound. It is the Aufheburg of Hegel, progressing in an unending dialectic spiral, from the thesis of a system regime, to the antithesis of its collapse, followed by the synthesis of a new alternative regime. If the ecological interaction network is the spirit of an ecosystem, then in Hegel's own words, 'the wounds of the Spirit heal, and leave no scars behind'. Such systems collapse and reorganise, like the tail-biting Oroborous and the phoenix rising from its own ash. To emphasise the lack of predetermined rules in a functioning complex network, Gunderson and Holling (2001) ditched the word 'hierarchy' (meaning sacred rules) in favour of the term 'panarchy' to describe a framework of nature's rules (Pan is the Greek god of nature). Panarchy describes the transformation of a social-ecological system undergoing an adaptive cycle of exploitation, conservation, release and reorganisation (Figure 4.10; Holling 2001). This suggests that an invaded ecosystem could pass through four phases – from early exploitation by invasive species (r) to the transformation of ecosystem

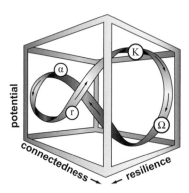

Figure 4.10 Panarchy, the adaptive cycle of complex socio-ecological systems. Four letters – r, K, α and Ω – indicate the ecosystem function of exploitation, conservation, release and reorganisation. From Holling (2001), reprinted with permission.

resources (K), then from the release (Ω) of current ecosystem regime to the reorganisation (α) into a novel ecosystem.

Violation of the forces that connect species in an ecological network, as described in the stability criteria, can trigger system collapse. In other words, the stability criteria provide us with a feasibility criterion for system-level selection (Rohr et al. 2014). Unstable systems, with their stability criteria that hold their resident species together being violated, are in transition towards alternative regimes. This means that unstable ecosystems are often transient and less likely to be observed in nature. The potential ways of violating the inequality of a stability criterion can also be conjectured (Eq. (4.5)), as discussed earlier, by invasions with higher propagule pressure (increasing σ and C, reducing μ), unique invader traits (increasing σ), more other invasive species (increasing S and C as invasional meltdown), more resources and stronger disturbance (reducing μ). Biological invasions and resulting local extinctions are therefore the strongest manifestations of network instability.

Biological invasions into an ecological network often lead to complex and delayed reactions as reflected by the decline and extirpation of original resident species. For instance, increased connectance from biological invasions often precedes extinctions and changes in network topology (Albrecht et al. 2014). This is because invasions manifest much more quickly than extinctions (Baiser et al. 2010; Downey and Richardson 2016), and most invasions are, therefore, observed to only cause changes in the abundances of residing species (Gallardo et al. 2016). Moreover, the incursions and resulting extinctions often experience time lags, as invasion and extinction debts (Tilman et al. 1994; Essl et al. 2015; Rouget et al. 2016), largely as a result of the particular feature of a complex network − its interaction feedback loops and pathways. This means that tensions can accumulate and be stored through the complex feedback loops and released in sudden bursts of pushes or pulls (Figueiredo et al. 2019). Indeed, compared to a rigid system, a complex adaptive system (often with highly connected parts, such as the plant–frugivore meta-networks created by invasions; Fricke and Svenning 2020) can collapse drastically with the loss of stability (Figure 4.11; Scheffer et al. 2012). This highlights the importance of understanding the mechanisms and early warning signs of complex adaptive systems possessing alternative regimes and facing imminent system collapses or regime shifts resulting from the bombardment of non-native species (Shackleton et al. 2018).

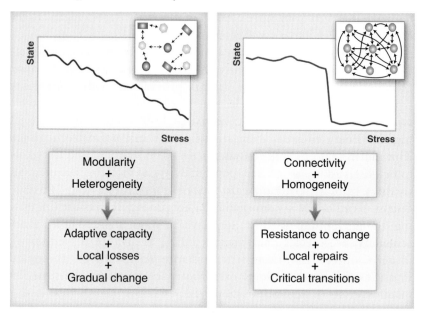

Figure 4.11 The connectivity and homogeneity of the units affect the way in which distributed systems with local alternative states respond to changing conditions. Networks in which the components differ (are heterogeneous) and where incomplete connectivity causes modularity tend to have adaptive capacity in that they adjust gradually to change. By contrast, in highly connected networks, local losses tend to be 'repaired' by subsidiary inputs from linked units until at a critical stress level the system collapses. The particular structure of connections also has important consequences for the robustness of networks, depending on the kind of interactions between the nodes of the network. From Scheffer et al. (2012), reprinted with permission.

Two issues are related to regime shifts or system collapses from biological invasions: the presence of alternative stable states and the processes that can push a system to traverse through these alternative stable states. First, to resolve the Clements–Gleason debate, Tansley (1935) proposed the polyclimax theory by positing the existence of alternative stable states in an ecosystem and the possibility of ecosystems traversing towards different climax states contingent on the priority effect of succession history. Whittaker's (1953) climax pattern theory further acknowledged the potential transition of communities between different climax states. These two succession theories were formalised into the hypothesis of alternative stable states in an ecological community by Lewontin (1969), with subsequent developments by Holling (1973),

Sutherland (1974), May (1977) and Scheffer et al. (2001). The presence of alternative stable states requires the system to be hysteretic, acting like a switch with a memory – a phenomenon known as bistability in mathematics. It requires positive feedbacks, especially through the coupling of fast and slow positive feedback loops (Brandman et al. 2005). To explore how interaction loops and chains in complex networks affect network stability and the formation of alternative stable states, we need to take a detour into another alleyway of history that sheds different light on the typical system complexity–stability debate.

In 1921, geneticist Sewell Wright developed path analysis to explore candidate causal structures among multiple measured entities within a system, based on their correlations. Path analysis was developed further into the powerful tools of structural equation modelling (Anderson and Rubin 1949; Koopmans and Hood 1953; Theil 1953; Freedman 1987) and signal-flow graphs (Shannon 1942; Mason 1953) for studying the role of feedbacks in complex dynamic networks. Inspired by this line of research, Richard Levins (1974) developed the loop analysis for modelling qualitatively specified systems. Loop analysis was immediately adopted by many researchers for studying how interaction topology in an ecological network (such as feedback loops, trophic cascade and interaction intransitivity) affects community stability (Levine 1976; Briand and McCauley 1978; Vandermeer 1980; Boucher et al. 1982; Lane 1986; Dambacher et al. 2003). Even though it has faced criticism largely because of the scarcity of data for parameterising such models (Orzack and Sober 1993), this weakness is rapidly being resolved as ecology embraces informatics and the data revolution. For instance, the role of interaction intransitivity (e.g., the rock-paper-scissors dominance of species A outcompeting B, B outcompeting C and C outcompeting A) has been extensively discussed as a potential mechanism to promote species persistence and coexistence in an ecological network (Huisman and Weissing 2001; Allesina and Levine 2011). Although modern coexistence theory developed largely with reference to pairwise interactions (Chesson 2000; Chesson and Kuang 2008), its scope can be expanded by integrating interaction loops to achieve a cohesive understanding of how multiple species coexist in complex ecological networks (Gallien et al. 2017; Godoy et al. 2017; Levine et al. 2017; Soliveres and Allan 2018; Yang and Hui 2021). Weak intransitive interactions can emerge naturally and increase with functional trait diversification, thereby imposing greater resistance to invasion but slowing the rate of evolution and network resilience against disturbance (Figure 4.12;

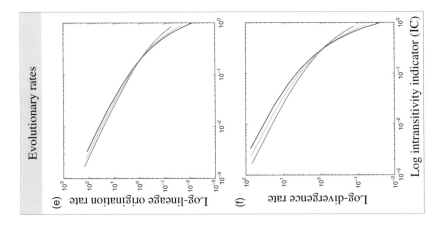

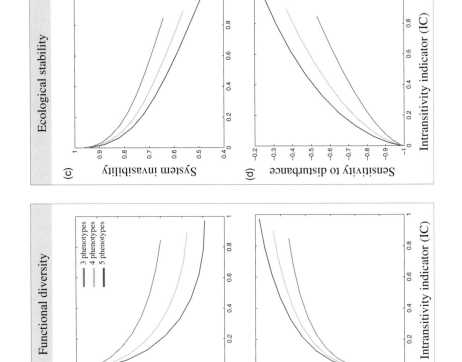

Gallien et al. 2018). Elucidating the role of interaction complexity, especially feedback loops, in network functioning and how it integrates aliens requires us to look into Levins' loop analysis.

Similar to the dynamical system approach used in Chapter 3 to describe an ecological network, loop analysis also describes a system by its system Jacobian matrix J at an equilibrium. However, instead of summarising the overall features of a resilient network as in May's stability criterion, Levins' loop analysis provides the means to gain insights on the role of internal network topology – interaction chains and feedback loops – in system stability. In a loop analysis, an ecological network can be visualised as a directed graph (Figure 4.13; Justus 2006). A path from node j to i is a series of directed edges that traverses through intermediate nodes but with each passing through not more than once, with the number of edges traversed known as the step length of the path. Such a path is also termed the Eulerian path to honour Leonhard Euler's solution to the Königsberg bridge problem in 1735 that has flourished into modern graph and network theory. For instance, in a fully connected network with three nodes, a two-step path from node 1 to 2 via intermediate node 3 is $a_{23}a_{31}$ (Figure 4.13); in a three-species network, this is equivalent to the indirect interaction strength from species 1 to 2 via species 3. A loop in this network is defined as a path from a node to itself; for instance a one-step loop of node 1 is a_{11}, a two-step loop of node 1 is $a_{12}a_{21}$ and a three-step loop of node 1 is $a_{13}a_{32}a_{21}$. Paths or loops that share no nodes are termed disjunct (e.g., a_{11} and $a_{32}a_{23}$ are two disjunct loops). Clearly, for a network with many species, the number of loops and paths increase exponentially. Levins (1975) defined the (entire) k-level feedback from a network of S species as (Justus 2006),

---

Figure 4.12 Relationships between the level of intransitivity and three main eco-evolutionary properties of the simulated system: (a–b) functional diversity, (c–d) ecological stability and (d–e) average evolutionary rates. The functional diversity is measured by (a) the abundance-weighted mean trait value and (b) the average trait difference between phenotypes. The ecological stability of the coexisting phenotypes is measured by (c) the invasibility of the system (as estimated with the mean invasion fitness of invaders) and its sensitivity to pulse disturbances (as estimated with the real part of the dominant eigenvalue of the system's Jacobian at equilibrium). The evolutionary rates are measured as (e) lineage origination rate (i.e., average rate of a new branching event) and (f) phenotype divergence rate (i.e., trait distance per unit of time); these relationships (e–f) are presented in a log–log scale to better visualise differences between numbers of phenotypes. Different lines represent three to five phenotypes, respectively. From Gallien et al. (2018), reprinted with permission.

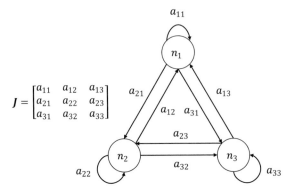

$$J = \begin{bmatrix} a_{11} & a_{12} & a_{13} \\ a_{21} & a_{22} & a_{23} \\ a_{31} & a_{32} & a_{33} \end{bmatrix}$$

Figure 4.13 A directed graph of the system Jacobian matrix of a 3-species network.

$$F_k(\boldsymbol{J}) = \sum_{i=1}^{k} (-1)^{i+1} \sum_{L(i,k) \in \mathbf{L}_{i,k}} L(i,k), \qquad (4.9)$$

where $L(i, k)$ is the product of k elements of the Jacobian matrix forming i number of disjunct loops, while $\mathbf{L}_{i,k}$ is the set of all such products in the Jacobian matrix. For instance in Figure 4.13, $F_1(\boldsymbol{J}) = \sum_{i=1}^{3} a_{ii}$ and $F_2(\boldsymbol{J}) = (-1)^2(a_{12}a_{21} + a_{23}a_{32} + a_{31}a_{13}) + (-1)^3(a_{11}a_{22} + a_{11}a_{33} + a_{22}a_{33})$, while $F_S(\boldsymbol{J}) = \det(\boldsymbol{J})$. The factor $(-1)^{i+1}$ indicates the role of loop length in regulating species coexistence, with a loop having an odd step length often stabilising coexistence through weakening negative density dependence, while those with an even length amplify perturbation and destabilise ecological networks (Hui et al. 2004; Allesina and Levine 2011; Vandermeer 2011; Gallien 2017).

Taken together, positive feedbacks amplify perturbations while negative feedbacks dampen perturbations; their combined effect determines system stability. The eigenvalues of the Jacobian matrix are the solution to the characteristic equation, $\det(\boldsymbol{J} - \lambda \boldsymbol{I}) = 0$, where $\boldsymbol{I}$ is the identity matrix with 1 for diagonal elements and the rest zeros. Levins (1975) showed that $\det(\boldsymbol{J} - \lambda \boldsymbol{I}) = \lambda^S - \sum_{k=1}^{S} F_k \lambda^{S-k}$, and ingeniously applied the Routh–Hurwitz stability criterion (Routh 1877; Hurwitz 1895) to express network stability as from the interplay of feedback loops of different lengths (see a review, Justus 2005): (i) negative feedback at every level ($F_k < 0$) and (ii) stronger feedback at lower levels than higher ones (e.g., $F_1 F_2 + F_3 > 0$). As the quantity of a_{ij} is often unknown but the sign can be qualitatively assigned as being positive, negative or zero based on the type of interactions, the quantitative Jacobian matrix can be simplified into a sign table where its stability can be quickly assessed using Levins' loop analysis. However, it is not clear how this sign stability is related to asymptotic stability (e.g., resilience). Quirk and Ruppert (1965) proved for a network with S species with its diagonal elements negative ($a_{ii} < 0$ for any i), the system is sign stable if and only if all paths $a_{ij}a_{ji} \leq 0$ for any $i \neq j$ and no loops in the system have three or more step lengths. The condition, however, appears too strong for any networks containing mutualistic interactions. More complicated cases have been developed to introduce the strength of interactions, not just their signs (Jeffries 1974; Logofet 1993).

In an ecological network, positive feedbacks abound, and many invasive species have been shown to establish more positive feedback loops that cause major impacts (Figure 4.14; Gaertner et al. 2012, 2014; Shackleton et al. 2018). Such novel positive feedbacks in the invaded ecosystems can be structured in three different ways (D'Antonio and Vitousek 1992), which can alter (i) resource supply (e.g., *Carpobrotus edulis*, *Mesembryanthemum crystallinum*, *Morella faya*), (ii) trophic structure

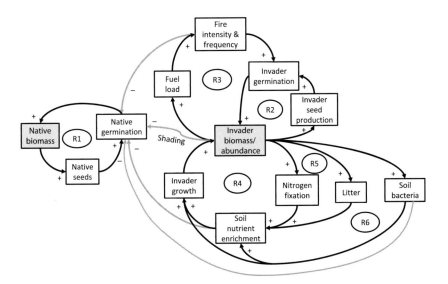

Figure 4.14 Causal loop diagram of five main feedback mechanisms implicated in plant invasions [R1: seed-biomass (native); R2: seed-biomass (invasive); R3: fire feedback; R4, 5 and 6: soil-nutrient-litter and soil biota feedback]. Note, feedbacks can also operate in the opposite direction (e.g., fire may be enhanced or suppressed). From Gaertner et al. (2014), reprinted with permission.

(e.g., through introducing the brown tree snake as a new top predator to Guam) and (iii) disturbance regimes [e.g., feral pigs *Sus scrofa* altering decomposition; cheatgrass (*Bromus tectorum*) in North American prairies; *Acacia* trees in South African fynbos altering fire regimes, thereby promoting their invasions]. In practice, invasion management can follow a full spectrum of interaction matrices, from fully qualitative to semi-quantitative to fully quantitative (Figure 4.15; Hui et al. 2016). For instance, checking the two-step interaction loops created from the invasion of the alien ladybird *Harmonia axyridis* in European agricultural systems and boreal forests (Figure 4.15), we can identify a two-step feedback loop between *H. axyridis* and the common green lacewing *Chrysoperla carnea* that could be responsible for the impact of invasive ladybirds in both these systems. This could be an effective tool for invasion management, and the approach warrants further attention.

Once the mechanisms giving rise to alternative stable states (in particular, those driven by positive feedback loops) are identified, system collapses and regime shifts are the factors and processes that can drive an ecosystem to traverse between the alternative basins of attraction. An ecological network and its alternative stable states have been typically visualised as a ball, located by the abundances of its residing species, over a hilly landscape. In a low-dimensional system (e.g., two interacting species), this landscape can be represented by field magnitude (Figure 2.5), while in complex ecological networks with many species this landscape can represent the sum of square of the population change rate (Figure 4.16; Lever et al. 2020),

$$v = \sqrt{\sum_{i=1}^{S} \left(\frac{dn_i}{dt}\right)^2}, \tag{4.10}$$

where the square root serves a similar role as the logarithm of the field magnitude in Figure 2.5, to reduce the skewness and enhance visualisation. The ball of an ecological network can be settled, in theory, at various places of different local shapes – a peak, a ditch, a saddle-shaped mountain pass, the ridge of a watershed – while only the ditch (a point or a valley) can withstand small perturbations (an attractor). However, it is worth pointing out that as the landscape reflects field magnitude, the system slows down at these resting places, regardless of their shapes. Perturbations can bring about both demographic changes ($n(t) = \bar{n} + \varepsilon(t)$) and structural changes (a_{ij}), pushing the ball through a co-dancing landscape (Figure 4.16). The response of ecosystems to a

(a)

	spp1	spp2	spp3	spp4	spp5
spp1	-0.96	-0.44	-0.11	0.58	-0.35
spp2	0.33	0.58	0.65	-0.13	-0.85
spp3	-0.23	-0.98	-0.24	-0.37	0.45
spp4	0.64	-0.44	0.36	-0.01	-0.43
spp5	-0.03	-0.46	0.40	-0.19	0.62

spp1	-10	-1	-0.1	1	-1
spp2	0.1	1	1	-0.1	-10
spp3	-0.1	-10	-0.1	-1	1
spp4	1	-1	1	-0.1	-1
spp5	-0.1	-0.1	1	-0.1	1

spp1	-1	-1	0	1	-1
spp2	0	1	1	0	-1
spp3	0	-1	0	-1	1
spp4	1	-1	1	0	-1
spp5	0	-1	1	0	1

(c)

	HA	AB	CS	CC	EB	PN	DC	AP
HA	-1	0	-0.1	-0.1	0	0	-0.1	10
AB	-10	-0.1	-0.1	-0.1	0	0	-0.1	10
CS	-0.1	0	-1	-0.1	0	0	-1	10
CC	-10	0	0	-0.1	0	0	0	10
EB	-10	0	0	-0.1	0	0	0	10
PN	0.1	0.1	0.1	0	0	0	0	10
DC	0	0	0	0	0	0	-0.1	0
AP	-10	-10	-10	-10	-10	-10	0	0

(d)

	HA	AB	AD	HS	CC	EB	PN	DC	PF	AP
HA	-1	0	0	0	-0.1	0	0	-0.1	-0.1	10
AB	-10	-0.1	0	0	-0.1	0	0	-0.1	-1	10
AD	-10	0	-0.1	0	-0.1	0	0	-0.1	-1	10
HS	-10	0	0	0	-0.1	0	0	-0.1	0	0.1
CC	-10	0	0	0	-0.1	0	0	0	0	10
EB	-10	0	0	0	0	0	0	0	0	10
PN	0.1	0.1	0.1	0	0	0	0	0	0	10
DC	0	0	0	0	0	0	0	-0.1	0	0
PF	-0.1	0	0	0	-0.1	0	0	-0.1	0	0
AP	-10	-10	-10	0	-10	-10	-10	0	0	0

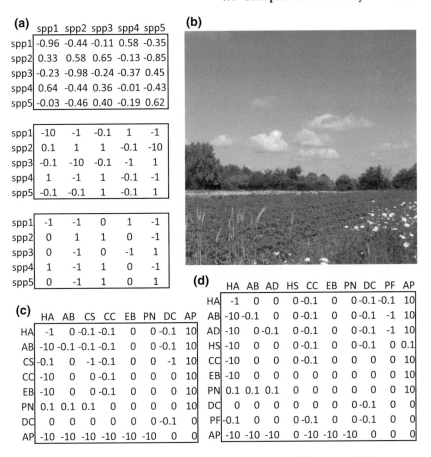

Figure 4.15 Interaction matrices for evaluating invasion performance. (a) An illustration of quantitative, semi-quantitative and qualitative interaction matrices (top, middle and bottom), with decreasing demands for data quality. (b) A picture of one European agricultural system with boreal habitat in background (Photo: H.E. Roy); both invaded by invasive alien ladybird *Harmonia axyridis*. (c, d) are semi-quantitative interaction matrices for agricultural and boreal systems in Europe, based on the literature and expert opinion. Acronyms: HA: *Harmonia axyridis*, AB: *Adalia bipunctata*, CS: *Coccinella septempunctata*, CC: *Chrysoperla carnea*, EB: *Episyrphus balteatus*, PN: *Pandora neoaphidis*, DC: *Dinocampus coccinellae*, AD: *Adalia decempunctata*, HS: *Halyzia sedecimguttata*, PF: Phorid fly, AP: aphids. The leading eigenvalues before the invasion (removing the entries related to *H. axyridis* in the matrix) is effectively zero for both the agricultural system and boreal forests, suggesting that both systems are at weak asymptotically stable. After invasion by *H. axyridis*, both systems become ecologically unstable, with the boreal forests more unstable than the agricultural system (leading eigenvalue: 5.51 vs. 4.12), suggesting a stronger impact of *H. axyridis* on the boreal forests from the perspective of stability. From Hui et al. (2016), reprinted with permission.

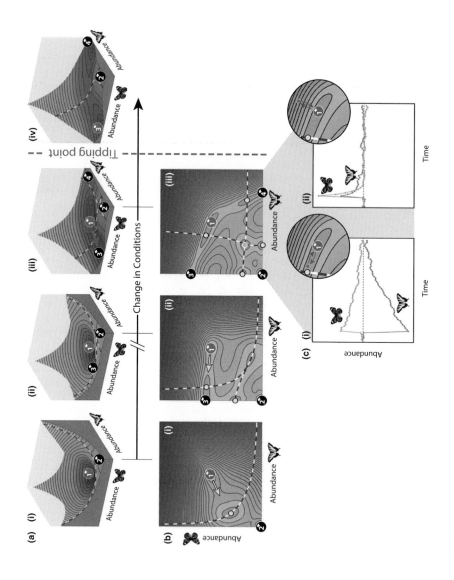

change in environmental conditions is determined by the relative strengths of positive and negative feedback loops in the networks of interactions between species or between species and their environment. Immediate negative feedbacks (e.g., due to intraspecific competition) have stabilising effects, while positive or 'reinforcing' feedbacks are destabilising and a necessary condition for the existence of alternative stable states (Thomas 1981; Gouzé 1998; Snoussi 1998; Lever et al. 2020). The impact from biological invasions through establishing positive feedbacks with key resources and resident species needs to accumulate first through positive feedbacks to push the system towards a tipping point at the boundary of the recipient ecosystem's current regime, in order to bring about regime shift. Forecasting the possibility of a regime shift can be assessed by measuring the strength of positive feedbacks between alien species and any resident species (and environments) and can be achieved by using early warning signals exhibited during network dynamics, largely from critical slowing down features when close to a tipping point (see Chapter 5).

With the coupling of eco-evolutionary feedbacks, an ecological network can exhibit a phenomenon known as evolutionary bistability – the existence of two or more alternative evolutionary or functional stable attractors of trait compositions, which appears rather common when trait evolution is under multiple opposing selective forces for adaptation. For instance, such eco-evolutionary couplings often create a strong priority

---

←

Figure 4.16 Stability properties for a small network of two pollinators (shown) and two plants (not shown). (a) Attraction basins (valleys) of alternative stable states (balls) are separated by thresholds (dashed curves). Initially, the only alternative to pristine state 1 is fully collapsed state 2 (a.i). When conditions change, two additional, partially collapsed states appear (states 3 and 4). The initial, pristine state loses resilience after state 3 appears (a.ii and a.iii). Eventually, the threshold towards state 3 approaches the pristine state so closely that a critical transition towards this state becomes inevitable (a.iii and a.iv). (b) Alternative stable states, saddle points (yellow dots) and hilltops (grey dots) are surrounded by areas in which the landscape's slope, and thus the rate at which abundances change, is nearly zero (indicated in orange). Higher speeds are found further away from these points. The direction of slowest recovery changes substantially before future state 3 appears (yellow arrow, b.i and b.ii). After state 3 appears, the system slows down in the direction of the saddle point on the approaching threshold (b.ii and b.iii). (c) Slow recovery from a perturbation towards the saddle point (c.i) as opposed to the much faster recovery from an equally large perturbation in another direction (c.ii). From Lever et al. (2020), under CC-BY Licence. *A black and white version of this figure will appear in some formats. For the colour version, please refer to the plate section.*

effect in mutualistic networks (May 1977; Young et al. 2001; Fukami 2005); they can steer the community succession to favour specific functional makeups depending on the initial trait composition of pioneer species (Dieckmann et al. 1995; Dercole et al. 2006). Biological invasions that add new distinct traits to a system can strongly affect the functional makeup of a system (e.g., trait dispersion in a community) (Ricciardi and Atkinson 2004), potentially triggering the reshuffling of trait compositions to alternative evolutionary regimes. This often leads to chaotic evolutionary scenarios, especially in the case of evolutionary bistability and bifurcation. For instance, in a bipartite mutualistic network where species are engaging in cross-guild mutualistic interactions and within-guild resource competition, based on the trait compositions the system can develop two alternative evolutionary regimes (Minoarivelo and Hui 2018) – one optimised for resource acquisition and competition within the function guild and the other fine-tuned for exploiting mutualistic gains (Figure 4.17). The first evolutionary regime selects for species with traits aligned for optimal resource acquisition, while the second evolutionary regime selects for species with cross-guild trait alignment and matching to enhance mutualistic benefit exchange. Subsequent co-evolution and diversification, and species packing, will be dictated by this priority effect in community assembly by fine-tuning the pre-determined evolutionary regimes.

As biological invasions often involve the introduction of species with totally new sets of traits, successful invasions thus often create a strong directional selection, and may force the system into an alternative evolutionary trajectory. Factors that could affect the strength of directional selection, such as reduced pollination success from phenological mismatches as a result of rising temperatures, may alter the evolutionary trajectory (Ferrière and Legendre 2013). As such, trait evolution in a community cannot depend solely on the evolutionary history of ancestral or pioneer species; resilience of the pre-determined evolutionary trajectory in the face of biotic perturbations also matters (Smallegange and Coulson 2013). Disruption of interactions as a result of invasion can have not only ecological impacts, but can also alter selective pressures that drive evolutionary responses in native species as well as the path of community succession (Traveset and Richardson 2011, 2014; Minoarivelo and Hui 2018). Moreover, perturbations could push the adaptive trait onto the trajectory of evolutionary suicide (negative fitness) or evolutionary trap (local fitness valley but cannot escape) (Rankin et al. 2007; Zhang et al. 2013). Besides such bistability, many evolutionary

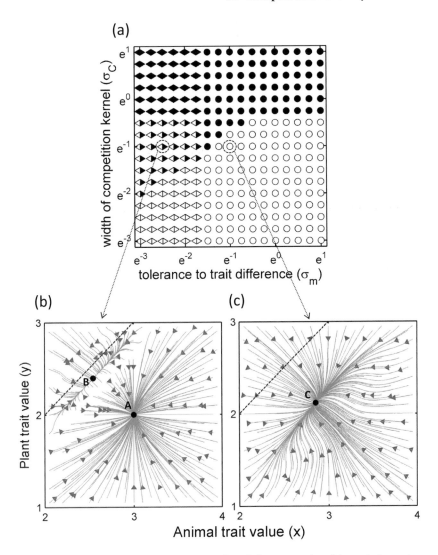

Figure 4.17 Bifurcation diagram and examples of phase portraits of the trait dynamics before the first branching event in a bipartite mutualistic network with species represented as adaptive traits. (a) Diamonds represent the existence of bistability, while circles represent the existence of a unique steady state. Circles and diamonds are either closed, indicating an ESS, or open, indicating a disruptive selection. The left and right sides of the diamonds correspond, respectively, to evolutionary regime I and II; (b) Species evolve towards either attractor A by optimizing utilization of within-guild resources (Evolutionary regime I) or attractor B by maximizing benefits from mutualism (Evolutionary regime II); (c) Species evolve towards a unique attractor. The broken lines in (b) and (c) represent the diagonal. From Minoarivelo and Hui (2018), reprinted with permission.

systems show a bifurcation of punctuated equilibria where slight changes in system parameters lead to a drastic jump of the evolutionary trajectory in trait space from one to another evolutionary attractor (Dercole et al. 2003; Ferrière and Legendre 2013); that is, an evolutionary regime shift. Assessing the current evolutionary regime in an ecological network and how novel traits from biological invasions alter selection force on each resident species has intrinsic value for invasion management as context dependent. This aspect is dealt with in Chapter 7.

4.5 Persistent Transition at Marginal Instability

Adaptive cycles of panarchy and regime shift often assume the existence of well-formed alternative stable states. This suggests that the regime residing in the basin of attraction is rather stable with enough room to absorb small perturbations. This is the case in most social systems and some ecosystems. Instead, we show here that in most open adaptive networks, community compositions are neither static (reflecting the standing regime) nor shifting towards a well-formed alternative regime, but are in constant transition due to the continuous incursions of aliens and extinctions of resident species (see Figure 1.2 and Figure 3.32 and related text), especially in many novel ecosystems. This means that in such an adaptive network with a relatively high rate of species flow, well-formed regimes, and therefore adaptive cycles, are often lacking. Instead, the system shows signs of *marginal instability*, which is also known as self-organised criticality (SOC). In one of the classics of modern physics, Per Bak, Chao Tang and Kurt Wiesenfeld (1988) used the example of ecological networks to introduce the concept of SOC:

For instance, ecological systems are organized such that the different species 'support' each other in a way which cannot be understood by studying the individual constituents in isolation. The same interdependence of species also makes the ecosystem very susceptible to small changes or 'noise'. However, the system cannot be too sensitive since then it could not have evolved into its present state in the first place. Owing to this balance we may say that such a system is 'critical'.

Ten years later, Richard Solé, Susanna Manrubia, Michael Benton, Stuart Kauffman and Per Bak introduced the SOC concept to ecology (Figure 4.18; Solé et al. 1999a); they began their perspective with a bold claim:

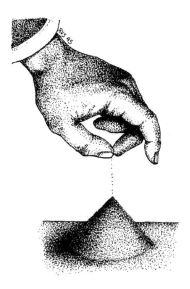

Figure 4.18 The sand pile is a simple example of how to reach a critical point. By slowly adding grains of sand, the system is driven to instability. When a critical slope is reached, adding one more grain can generate an avalanche involving s grains of sand. Mostly, only small avalanches are observed, but eventually very large avalanches will occur. The number of avalanches $N(s)$ involving s grains will follow a power law. From Solé et al. (1999a), reprinted with permission.

Fluctuations in ecological systems are known to involve a wide range of spatial and temporal scales, often displaying self-similar (fractal) properties. Recent theoretical approaches are trying to shed light on the nature of these complex dynamics. The results suggest that complexity in ecology and evolution comes from the network-like structure of multispecies communities that are close to instability.

This view was quickly contested by Fukami et al. (1999), who had two main concerns: (i) power law fluctuations in ecological systems might not be a better fit than exponential; (ii) such fluctuations might not result from internal processes of the system but reflect the behaviours of external forces. These two points seemed weak and were challenged in Solé et al.'s (1999b) rebuttal: (i) fact check on curve fitting under information criteria; (ii) on this point, the rebuttal is acute:

Current studies on the organization of biological systems recognize the presence of complex networks of interactions acting at different levels and of strong self-reinforcing processes among the hierarchy. This gives rise to invariant properties and to processes acting at different scales. An ecosystem is formed by many interacting parts, the relevant quantities characterizing it are rarely (if ever) independent and, as a result, the response of that system to an external perturbation will be typically nonlinear, in many cases unpredictable, and very often strongly dependent on its internal state. .... If the nonlinear response of a self-organized ecosystem (usually quantified through several dependent variables) distributes according to a power-law, the internal mechanism poising the system to the observed state is termed self-organized criticality (SOC). And, by definition, SOC requires the concomitant action of an external (slow) driving mechanism that maintains the system out of equilibrium: there is no evolution in its absence.

Of course, the driving force in our context is biological invasion. An ecological network facing the constant flux of species (invasions and extinctions) and with fine-tuning internal structures via co-evolving and fitting biotic interactions is ideal for demonstrating the SOC in complex adaptive systems. Added sand grains represent introduced species, and avalanches represent interaction cascades and species extinctions from biological invasions. The concept of SOC took off in many spheres of complex systems sciences (Figure 4.1) such as astrophysics (Aschwanden et al. 2016). The acceptance of this concept in ecology is, however, similar to the view of Fukami et al. (1999). While an ecological network facing constant biological invasions could exhibit some panarchy of drastic regime shifts, most fluctuations represent only pseudo-cycles at the criticality. To complete our proposal of open adaptive network emergence (Figure 3.36), we provide a list of evidence to demonstrate how an ecological network is a jammed complex system self-organised at marginal instability due to the flow from biological invasions and resulting species extinctions.

First, in an adaptive network, co-evolution and biological invasions can sequentially exploit available niches and eventually flatten the fitness landscape (Figure 4.19; also Figure 3.28b). This can sometimes include an intermediate stage of niche construction and expansion as a result of the establishment of novel interactions by alien species within the ecosystem, through either acquiring abiotic resources that are inaccessible to standing resident species or serving as resource species or mutualistic partners to resident species. With the establishment of alien species, the positive fitness landscape in a community that initially has few species can be sequentially filled up until no species with feasible traits are able to establish in the community. This means that not only the history and

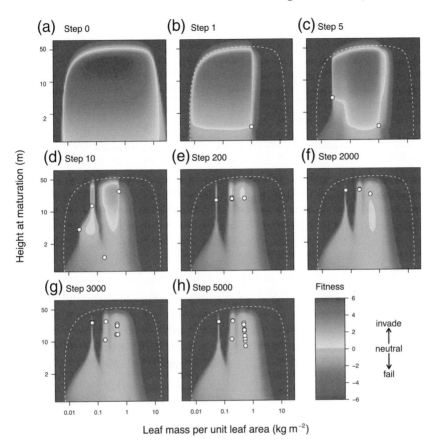

Figure 4.19 Community assembly leads to characteristically structured fitness landscapes that exhibit peaks, valleys and plateaus. a–h show fitness landscapes at different stages of community assembly. Colouring across trait space indicates the invasion fitness of rare species competing with the resident species (circles). Dashed curves in each panel delimit trait combinations that are not viable even in the absence of any competition. Community assembly starts with a species randomly selected across the trait space (step 1). At each subsequent step, the abundances of resident species are updated according to their fitness, and new invaders are added around existing residents and across trait space. Invasion is successful only in regions with positive invasion fitness (coloured yellow to red), whereas species that successfully invade but then find themselves in regions of negative invasion fitness (coloured blue to cyan; e.g., lowest circle in d) are ultimately driven to extinction. A ridge-shaped fitness plateau (green region in h) forms naturally in the course of the community assembly, after a high diversity of late-successional shade-tolerant species establish in regions of positive fitness. From Falster et al. (2017), reprinted with permission. *A black and white version of this figure will appear in some formats. For the colour version, please refer to the plate section.*

sequence of invasions affect the shape of the invasion fitness landscape but also the path leading to the eventual stage of a flattened fitness landscape. Note, the eventual fitness landscape can be flattened by different sets of trait compositions, accommodating different levels of richness and functional diversity (a scenario similar to packing your suitcase for travelling). This also does not imply that no new aliens will be able to establish, but the necessity of selection and extinction for some resident species (the undecided traveller needs to find space for additional items by discarding some already packed items). When exploring the population dynamics of species in this jammed ecosystem, the energy landscape is also flattened (Figure 4.20). The flattened energy landscape implies that species, in this jammed ecological network at marginal instability, are ecologically equivalent, although still with some levels of rather important differences. The energy landscape is flattened, but the terrain is not smooth, but rugged, with numerous small peaks. These small peaks allow the system to possess short-term local memory (regime) but still be capable of shifting from one small peak, adaptively or randomly, to another in response to further biological invasions and environmental change. Such changes need not be total collapses or drastic regime shifts, but can be akin to tentative sampling of multiple feasible states with mostly small but sometimes large network transitions. Open community assembly is, therefore, not towards a dominant peak (representing a climax or a dead end) as a result of the obvious topology of a flattened landscape; rather, community composition is persistently transitioning its residing species, allowing the community to function in a semi-fluid status, hopping over and learning the rugged landscape to cope with environmental crises and grasp opportunities.

Second, because most ecological networks are set at marginal instability, the complexity–stability relationships of real ecological networks are destined to be weak or lacking. In other words, an open adaptive network operates at its marginal instability regardless of the emerged network structures. For instance, Jacquet et al. (2016) found that in real food webs (Figure 4.21) most networks are unstable but with the real part of the leading eigenvalue ($Re(\lambda_{max})$) slightly positive, indicating marginal instability, and that the complexity–stability relationship in these empirical ecological networks is lacking. In addition, as part of network instability features, invasibility could also be only weakly correlated with the complexity and architecture of the ecological network of recipient ecosystems. Indeed, trivial changes in the overall network architecture following the introduction of an alien

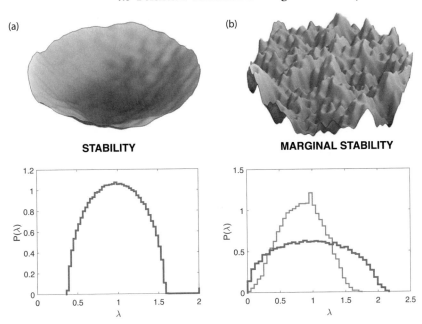

Figure 4.20 Possible scenarios for the energy landscape associated to Lotka–Volterra dynamics. (a) There is only a single equilibrium, i.e., a unique global and local minimum, as illustrated by the cartoon of the energy landscape. The corresponding density $P(\lambda)$ of eigenvalues of the stability matrix associated with a given minimum (the Hessian) has a strictly positive support and the number of species in the community is strictly smaller than the bound set from May's stability criterion. The curve is produced numerically from a standard Lotka–Volterra model. For a large number of species, $P(\lambda)$ is in this case a shifted Wigner semi-circle. (b) The energy landscape is rugged: there are many equilibria and local minima, as illustrated by the cartoon of the energy landscape. The corresponding density $P(\lambda)$ of eigenvalues of the stability matrix associated with a minimum has a support whose left edge touches zero, corresponding to marginal stability, and the number of possible surviving species saturates the bound of May's stability criterion. The dark grey curve is from a standard Lotka–Volterra model, with the light grey curve for a different functional response. In the former case $P(\lambda)$ is a shifted Wigner semi-circle, whereas in the latter it has a different shape. From Biroli et al. (2018), under CC-BY Licence.

species have been documented empirically (Padrón et al. 2009; Vilà et al. 2009) and were tentatively explained by the peripheral role of the invader in the network. In contrast, Albrecht et al. (2014) analysed 20 independent pairs of invaded and uninvaded pollination networks and found that the overall number of modules was not changed by invasion but that modules were more interconnected by

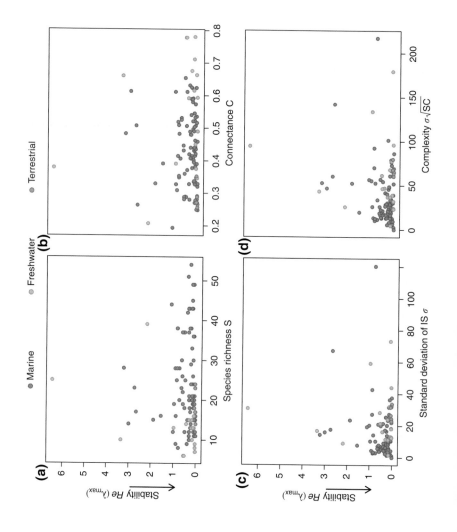

super-generalist invaders (also see Figure 4.2 on invaded global plant–frugivore networks; Fricke and Svenning 2020). This suggested the potential yet inconclusive role of super-generalist invaders in dismantling the compartmentalized structure in native pollination networks. However, this contrast does not necessarily support a strong relationship between network complexity and invasibility; rather, it merely highlights the necessity to differentiate network structures with alien species included versus those only considering native resident species. Nevertheless, adding alien species does increase richness and connectance in invaded ecological networks (at least temporally), which may push systems that were stable before invasion towards marginal instability. Importantly, lack of a strong complexity–stability relationship does not suggest that ecological networks are randomly structured. Rather, many realistic network structures emerge spontaneously and simultaneously while the entire network converges to marginal instability via co-evolution and ecological fitting (Figure 4.22; Zhang et al. 2011; Suweis et al. 2013; Nuwagaba et al. 2015).

Finally, what is the direction of system evolution and associated dynamics in such an open adaptive network driven by biological invasions? We have provided limited, but hopefully convincing, evidence to support our hypothesis that, driven by constant biological invasions, an open adaptive network evolves towards or operates at marginal instability. Two forces can be responsible for this system-level evolution. First, any randomly assembled large network tends to settle its leading eigenvalue towards the stability boundary ($\text{Re}(\lambda_m) = 0$; top left, Figure 4.23), but this dragging force weakens drastically when $\text{Re}(\lambda_m)$ approaches the stability boundary. An open ecological network experiences a thrust from the continuous stream of invasions that pushes the leading eigenvalue up to possess a positive real part (top right and bottom plots, Figure 4.23). The thrust can propel a persistent transformation of the system in response to new invasions. The thrust of biological

---

Figure 4.21 Food web stability related to complexity parameters in 116 food webs. (a) Number of species S (linear regression: P = 0.97, $R^2 < 10^{-5}$), (b) Connectance $C = (L/S2)$ where L is the number of links (P = 0.98, $R^2 < 10^{-6}$), (c) Standard deviation of interaction strengths σ (P = 0.1, $R^2 = 0.02$), (d) May's complexity measure $\sigma\sqrt{SC}$ (P = 0.02, $R^2 = 0.04$). Stability is measured as $\text{Re}(\lambda_{max})$ for marine, freshwater and terrestrial ecosystems (in different levels of grey). Food webs with eigenvalues close to zero are the most stable. All quantities are dimensionless. From Jacquet et al. (2016), under CC-BY Licence.

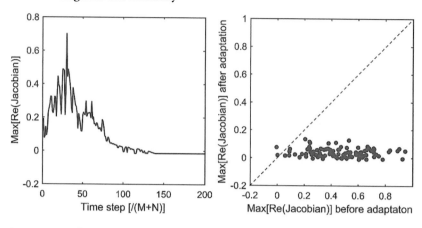

Figure 4.22 Effect of the adaptive process of interaction switching on the maximum real part of the eigenvalues of the Jacobian matrix (a metric of system resilience). The simulation is based on the binary interaction matrix of a stream food web in a pine plantation in New Zealand (Thompson and Townsend 2003). Positive values lead to instability and the system evolves towards marginal stability with the lead eigenvalue just above (or around) zero, with no intention to further strengthen the stability or resilience of the system. Points on the right panel represent 100 repeats of the simulation. From Nuwagaba et al. (2015), reprinted with permission.

invasions (destabilising force), compensated by the system drag (stabilising force), can create an invisible wave in the complex plane, carrying the invaded ecosystem, like a surfer on her board, moving at the wave front. At marginal instability, the surfer does not fall back as the drag force vanishes when the surfboard stops (Figure 4.24). She is, however, unable to be far ahead of the wave front because of the lack of thrust – she can only surf just slightly ahead of the wave front. At this point, the system inevitably undergoes persistent transition and species turnover, allowing successful alien species to invade; this causes rare underperforming resident species to be expelled, thereby giving rise to the spontaneously emergence of network structures (see Chapter 3 for network structure emergence and Chapter 5 for the dynamics of network transition and turnover). This is what we mean by persistent transition, which involves more than simply jumping basins of attraction.

At this point of marginal instability, the alternative regime of the system is yet to be formed; the surfer moves forward but her future path remains undecided. There are turns and bumps but the surfer is agile and tough. Clearly, biological invasions (and other means of species influxes) are crucial generators of the wave that propels the surfer. Could this wave

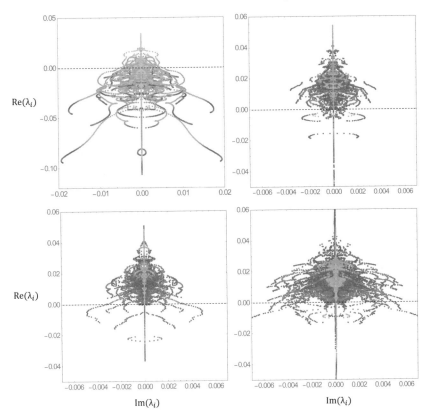

Figure 4.23 Dynamics of eigenvalues of the system Jacobian matrix during the assembly of an ecological network. The ecological network follows the standard Lotka–Volterra model (Eq.(2.6)) and the intrinsic growth rate of each species r_i follows a normal distribution with zero mean and standard deviation of 0.02; interaction strength α_{ij} for $i \neq j$ follows a normal distribution with zero mean and standard deviation 0.1; $\alpha_{ii} = k_i = 1$. Note, the complex plane is rotated to visualise the real part vertically. Top left: 100 species; the rest: five species initially followed by two species per unit time (top right and bottom left) and by four species per unit time (bottom right). Eigenvalues are overlaid on each other with yellow-green colours indicating early succession time and red colours late succession time. Note, the dense distributions have blocked some eigenvalues in the visualisation. *A black and white version of this figure will appear in some formats. For the colour version, please refer to the plate section.*

become a tsunami with the ongoing acceleration of global species translocation, thereby pushing the ecological network of local and regional ecosystems from this marginal instability to positions drastically far removed from equilibrium? Most studies to date suggest only persistent

SURF-SWIMMING, SANDWICH ISLANDS.
Page 419.

Figure 4.24 Surfing in gentle waves, a metaphor of ecological networks under persistent transition at criticality (see text). Engraving of *wahine* surfing after a 1829 drawing by Charles de Varigny. Image taken from 'Captain Cook's Voyages round the World', from the Mechanical Curator collection and released to Flickr Commons by the British Library in Public Domain.

transition. While many ecosystems are experiencing rapid transitions, the rates are largely comparable to secondary succession. Given the increasing awareness of the impacts of biological invasions, and enhanced international governance regarding the global translocation of alien species (Early et al. 2016), it is unlikely that current invasion pathways will be flooded by new alien species or that there will be a continuous acceleration in introduction rates, pushing ecosystems way beyond equilibrium. In the short to medium term we expect this ongoing global translocation to continue and largely maintain at the current rate (Seebens et al. 2020); this, of course, does not preclude occasional cases of acceleration or saturation for some invasion pathways. Together with the waves from other global change drivers, human-mediated introductions of alien species will continuously propel invaded ecological networks to transform and some will occasionally tumble and collapse (Shackleton et al. 2018), resembling the unpredictable avalanches in the sand pile model (Figure 4.18) and the occasional splash of eigenvalues (Figure 4.23). We discuss the network dynamics and interaction turnover from this persistent transition of ecological network at marginal instability in Chapter 5, but will revisit the potential of system transition far from equilibrium and associated jammed ecosystems in Chapter 7.

We end this chapter with a Polynesian chant known as *pohuehue* (beach morning glory, *Ipomoea pes-caprae*) recorded in Womack (2003) that describes the Polynesian surfing culture:

Arise! Arise! You great surfs from Kahiki,
The powerful curling waves. Arise with the *pohuehue*.
Well up, long raging surf.

References

Albrecht M, et al. (2014) Consequences of plant invasions on compartmentalization and species' roles in plant–pollinator networks. *Proceedings of the Royal Society B: Biological Sciences* 281, 20140773.

Allesina S, Levine JM (2011) A competitive network theory of species diversity. *Proceedings of the National Academy of Sciences USA* 108, 5638–5642.

Allesina S, Tang S (2012) Stability criteria for complex ecosystems. *Nature* 483, 205–208.

Anderson PW (1972) More is different. *Science* 177(4047), 393–396.

Anderson TW, Rubin H (1949) Estimation of the parameters of a single equation in a complete system of stochastic equation. *The Annals of Mathematical Statistics* 20, 46–63.

Aschwanden MJ, et al. (2016) 25 Years of self-organized criticality: Solar and astrophysics. *Space Science Reviews* 198, 47–166.

Baiser B, Russell GJ, Lockwood JL (2010) Connectance determines invasion success via trophic interactions in model food webs. *Oikos* 119, 1970–1976.

Bak P, Tan C, Wiesenfeld K (1988) Self-organized criticality. *Physical Review A* 38, 364–374.

Bar-Yam Y (1997) *About Complex Systems*. Reading: Addison-Wesley.

Bascompte J, Jordano P (2007) Plant-animal mutualistic networks: The architecture of biodiversity. *Annual Review of Ecology, Evolution, and Systematics* 38, 567–593.

Bastolla U, et al. (2009) The architecture of mutualistic networks minimizes competition and increases biodiversity. *Nature* 458, 1018–1020.

Bhatia NP, Szegö GP (1970) *Stability Theory of Dynamical Systems*. New York: Springer-Verlag.

Biroli G, Bunin G, Cammarota C (2018) Marginally stable equilibria in critical ecosystems. *New Journal of Physics* 20, 083051.

Borrelli JJ, et al. (2015) Selection on stability across ecological scales. *Trends in Ecology & Evolution* 30, 417–425.

Boucher DH, James S, Keeler KH (1982) The ecology of mutualism. *Annual Review of Ecology and Systematics* 13, 315–347.

Brandman O, et al. (2005) Interlinked fast and slow positive feedback loops drive reliable cell decisions. *Science* 310, 496–498.

Brännström Å, et al. (2011) Emergence and maintenance of biodiversity in an evolutionary food-web model. *Theoretical Ecology* 4, 467–478.

Briand F, McCauley E (1978) Cybernetic mechanisms in lake plankton systems: How to control undesirable algae. *Nature* 273, 228–230.

Campbell C, et al. (2012) Topology of plant-pollinator networks that are vulnerable to collapse from species extinction. *Physical Review E* 86, 021924.

Chan S (2001) Complex Adaptive Systems. ESD.83 Research Seminar in Engineering Systems October 31, 2001/November 6, 2001. http://web.mit.edu/esd.83/www/notebook/Complex%20Adaptive%20Systems.pdf (accessed 12 February 2021).

Chen X, Cohen JE (2001) Transient dynamics and food–web complexity in the Lotka–Volterra cascade model. *Proceedings of the Royal Society B: Biological Sciences* 268, 869–877.

Chesson P (2000) General theory of competitive coexistence in spatially-varying environments. *Theoretical Population Biology* 58, 211–237.

Chesson P, Kuang J (2008) The interaction between predation and competition. *Nature* 456, 235–238.

Clements FE (1916) Plant Succession: Analysis of the Development of Vegetation. *Washington: Carnegie Institution of Washington Publication Sciences*, 242, pp. 1–512.

Colautti RI, Alexander JM, Dlugosch KM, Keller SR, Sultan SE (2017) Invasions and extinctions through the looking glass of evolutionary ecology. *Philosophical Transactions of the Royal Society B: Biological Sciences* 372, 20160031.

Cuddington K, Hastings A (2016) Autocorrelated environmental variation and the establishment of invasive species. *Oikos* 125, 1027–1034.

D'Antonio CM, Vitousek PM (1992) Biological invasions by exotic grasses, the grass/fire cycle, and global change. *Annual Review of Ecology and Systematics* 23, 63–87.

Daehler CC (2001) Darwin's naturalization hypothesis revisited. *The American Naturalist* 158, 324–330.

Dambacher JM, Luh HW, Rossignol PA (2003) Qualitative stability and ambiguity in model ecosystems. *The American Naturalist* 161, 876–888.

Davis MA, Grime JP, Thompson K (2000) Fluctuating resources in plant communities: A general theory of invasibility. *Journal of Ecology* 88, 528–534.

De Angelis DL (1975) Stability and connectance in food web models. *Ecology* 56, 238–243.

Dercole F, et al. (2003) Numerical sliding bifurcation analysis: An application to a relay control system. *IEEE Transactions on Circuits and Systems I: Fundamental Theory and Applications* 50, 1058–1063.

Dercole F, et al. (2006) Coevolution of slow–fast populations: Evolutionary sliding, evolutionary pseudo-equilibria and complex Red Queen dynamics. *Proceedings of the Royal Society B: Biological Sciences* 273, 983–990.

Dieckmann U, Law R (1996) The dynamical theory of coevolution: A derivation from stochastic ecological processes. *Journal of Mathematical Biology* 34, 579–612.

Dieckmann U, Marrow P, Law R (1995) Evolutionary cycling in predator-prey interactions: Population dynamics and the red queen. *Journal of Theoretical Biology* 176, 91–102.

Donohue I, et al. (2016) Navigating the complexity of ecological stability. *Ecology Letters* 19, 1172–1185.

Dornelas M, Madin JS (2020) Novel communities are a risky business. *Science* 370, 164–165.

Downey PO, Richardson DM (2016) Alien plant invasions and native plant extinctions: A six-threshold framework. *AoB PLANTS* 8, plw047.

Dunne A, Williams RJ, Martinez ND (2002) Network structure and biodiversity loss in food webs: Robustness increases with connectance. *Ecology Letters* 5, 558–567.

Durinx M, Metz JAJ, Meszéna G (2008) Adaptive dynamics for physiologically structured population models. *Journal of Mathematical Biology* 56, 673–742.

Early R, et al. (2016) Global threats from invasive alien species in the twenty-first century and national response capacities. *Nature Communications* 7, 12485.

Elton CS (1958) *The Ecology of Invasions by Animals and Plants*. London: Metheun.

Eshel I, Motro U, Sansone E (1997) Continuous stability and evolutionary convergence. *Journal of Theoretical Biology* 185, 333–343.

Essl F, et al. (2015) Historical legacies accumulate to shape future biodiversity in an era of rapid global change. *Diversity and Distributions* 21, 534–547.

Evans DM, Pocock MJO, Memmott J (2013) The robustness of a network of ecological networks to habitat loss. *Ecology Letters* 16, 844–852.

Falster DS, et al. (2017) Multitrait successional forest dynamics enable diverse competitive coexistence. *Proceedings of the National Academy of Sciences USA* 114, E2719–E2728.

Ferriere R, Legendre S (2013) Eco-evolutionary feedbacks, adaptive dynamics and evolutionary rescue theory. *Philosophical Transactions of the Royal Society B: Biological Sciences* 368, 20120081.

Figueiredo L, et al. (2019) Understanding extinction debts: Spatio–temporal scales, mechanisms and a roadmap for future research. *Ecography* 42, 1973–1990.

Freedman DA (1987) As others see us: A case study in path analysis. *Journal of Educational Statistics* 12, 101–128.

Fricke EC, Svenning JC (2020) Accelerating homogenization of the global plant–frugivore meta-network. *Nature* 585, 74–78.

Fridley JD (2011). Biodiversity as a bulwark against invasion: Conceptual threads since Elton. In Richardson DM (ed.), *Fifty Years of Invasion Ecology: The Legacy of Charles Elton*, pp. 121–130. Oxford: Wiley-Blackwell.

Frost CM, et al. (2019) Using network theory to understand and predict biological invasions. *Trends in Ecology & Evolution* 34, 831–843.

Fukami T (2005) Integrating internal and external dispersal in metacommunity assembly: Preliminary theoretical analyses. *Ecological Research* 20, 623–631.

Fukami T, et al. (1999) Self-organized criticality in ecology and evolution. *Trends in Ecology & Evolution* 14, 321.

Gaertner M, et al. (2012) Biological invasions, resilience and restoration. In van Andel J, Aronson J (eds.), *Restoration Ecology: The New Frontier*, pp. 265–280. Oxford: Wiley-Blackwell.

Gaertner M, et al. (2014) Invasive plants as drivers of regime shifts: Identifying high-priority invaders that alter feedback relationships. *Diversity and Distributions* 20, 733–744.

Gallardo B, et al. (2016), Global ecological impacts of invasive species in aquatic ecosystems. *Global Change Biology* 22, 151–163.

Gallien L (2017) Intransitive competition and its effects on community functional diversity. *Oikos* 126, 615–623.

Gallien L, et al. (2017) The effects of intransitive competition on coexistence. *Ecology Letters* 20, 791–800.

Gallien L, et al. (2018) Emergence of weak-intransitive competition through adaptive diversification and eco-evolutionary feedbacks. *Journal of Ecology* 106, 877–889.

Gell-Mann M (1995) Complex adaptive systems. In Morowitz HJ, Singer JL (eds.), *The Mind, the Brain, and Complex Adaptive Systems, Sante Fe Institute Studies in the Science of Complexity*, pp. 11–24. Reading: Addison-Wesley.

Gleason HA (1939) The individualistic concept of the plant association. *The American Midland Naturalist* 21, 92–110.

Godoy O, et al. (2017) Intransitivity is infrequent and fails to promote annual plant coexistence without pairwise niche differences. *Ecology* 98, 1193–1200.

Gouzé JL (1998) Positive and negative circuits in dynamical systems. *Journal of Biological Systems* 06, 11–15.

Gravel D, Massol F, Leibold M (2016) Stability and complexity in model meta-ecosystems. *Nature Communications* 7, 12457.

Grilli J, Rogers T, Allesina S (2016) Modularity and stability in ecological communities. *Nature Communications* 7, 12031.

Grimm V, Wissel C (1997) Babel, or the ecological stability discussions: An inventory and analysis of terminology and a guide for avoiding confusion. *Oecologia* 109, 323–334.

Gunderson LH, Holling CS (eds.)(2001) *Panarchy: Understanding Transformations in Human and Natural Systems*. Washington DC: Island Press.

Guo Q, et al. (2015) A unified approach for quantifying invasibility and degree of invasion. *Ecology* 96, 2613–2621.

Hastings HM (1984) Stability of large systems. *Biosystems* 17, 171–177.

Holling C (2001) Understanding the complexity of economic, ecological, and social systems. *Ecosystems* 4, 390–405.

Holling CS (1959) Some characteristics of simple types of predation and parasitism. *The Canadian Entomologist* 91, 385–398.

Holling CS (1973) Resilience and stability of ecological systems. *Annual Review of Ecology and Systematics* 4, 1–23.

Horak J, et al. (2013) Changing roles of propagule, climate, and land use during extralimital colonization of a rose chafer beetle. *Naturwissenschaften* 100, 327–336.

Hui C, Richardson DM (2017) *Invasion Dynamics*. Oxford: Oxford University Press.

Hui C, Richardson DM (2019a) How to invade an ecological network. *Trends in Ecology & Evolution* 34, 121–131.

Hui C, Richardson DM (2019b) Network invasion as an open dynamical system: Response to Rossberg and Barabás. *Trends in Ecology & Evolution* 34, 386–387.

Hui C, et al. (2004) Metapopulation dynamics and distribution and environmental heterogeneity induced by niche construction. *Ecological Modelling* 177, 107–118.

Hui C, et al. (2016) Defining invasiveness and invasibility in ecological networks. *Biological Invasions* 18, 971–983.

Hui C, et al. (2018) *Ecological and Evolutionary Modelling*. Cham: Springer.

Huisman J, Weissing FJ (2001) Biological conditions for oscillations and chaos generated by multispecies competition. *Ecology* 82, 2682–2695.

Hurwitz A (1895) On the conditions under which an equation has only roots with negative real parts. *Mathematische Annalen* 46, 273–284.

Ives AR, Carpenter SR (2007) Stability and diversity of ecosystems. *Science* 317, 58–62.

Jacquet C, et al. (2016) No complexity–stability relationship in empirical ecosystems. *Nature Communications* 7, 12573.

Jeffries C (1974) Qualitative stability and digraphs in model ecosystems. *Ecology* 55, 1415–1419.

Justus J (2005) Qualitative scientific modeling and loop analysis. *Philosophy of Science* 72, 1272–1286.

Justus J (2006) Loop analysis and qualitative modeling: Limitations and merits. *Biology & Philosophy* 21, 647–666.

Kaiser-Bunbury CN, et al. (2010) The robustness of pollination networks to the loss of species and interactions: A quantitative approach incorporating pollinator behaviour. *Ecology Letters* 13, 442–452.

Kauffman SA (1993) *The Origins of Order: Self-organization and Selection in Evolution*. Oxford: Oxford University Press.

Keller EF (2005) Ecosystems, organisms, and machines. *BioScience* 55, 1069–1074.

Kondoh M (2003) Foraging adaptation and the relationship between food-web complexity and stability. *Science* 299, 1388–1391.

Koopmans T, Hood W (1953) The estimation of simultaneous linear economic relationships. In Hood W and Koopmans T (eds.), *Studies in Econometric Method*. New York: John Wiley.

Krause A, et al. (2003) Compartments revealed in food-web structure. *Nature* 426, 282–285.

Landi P, et al. (2018) Complexity and stability of ecological networks: A review of the theory. *Population Ecology* 60, 319–345.

Lane PA (1986) Preparing marine plankton data sets for loop analysis. Ecology. Supplementary Publication Source Document, (825B).

Langat GK, Minoarivelo HO, Hui C (2021) Adaptive interaction switching relaxes complexity-stability criteria. In prep.

Leimar O (2005) The evolution of phenotypic polymorphism: Randomized strategies versus evolutionary branching. *The American Naturalist* 165, 669–681.

Leimar O (2009) Multidimensional convergence stability. *Evolutionary Ecology Research* 11, 191–208.

Lever JJ, et al. (2020) Foreseeing the future of mutualistic communities beyond collapse. *Ecology Letters* 23, 2–15.

Levin SA (1992) The problem of pattern and scale in ecology. *Ecology* 73, 1943–1967.

Levin SA (1999) Towards a science of ecological management. *Conservation Ecology* 3(2), 6.

Levine J, et al. (2017) Beyond pairwise mechanisms of species coexistence in complex communities. *Nature* 546, 56–64.

Levine SH (1976) Competitive interactions in ecosystems. *The American Naturalist* 110, 903–910.

Levins R (1974) The qualitative analysis of partially specified systems. *Annals of the New York Academy of Sciences* 231, 123–138.

Levins R (1975) Evolution in communities near equilibirium. In Cody M, Diamond J (eds.), *Ecology and Evolution of Communities*, pp. 15–51. Cambridge: Belknap Press.

Lewontin RC (1969) The bases of conflict in biological explanation. *Journal of the History of Biology* 2, 35–45.

Logofet DO (1993) *Matrices and Graphs: Stability Problems in Mathematical Ecology.* Boca Raton: CRC Press.

Logofet DO (2005) Stronger-than-Lyapunov notions of matrix stability, or how "flowers" help solve problems in mathematical ecology. *Linear Algebra and its Applications* 398, 75–100.

Lonsdale WM (1999) Global patterns of plant invasions and the concept of invasibility. *Ecology* 80, 1522–1536.

Lotka AJ (1925) *Elements of Physical Biology.* Baltimore: Williams & Wilkins Company.

Lovelock JE (1972) Gaia as seen through the atmosphere. *Atmospheric Environment* 6, 579–580.

Lurgi M, et al. (2014) Network complexity and species traits mediate the effects of biological invasions on dynamic food webs. *Frontiers in Ecology and Evolution* 2, 36.

Lyapunov AM (1892) *The General Problem of the Stability of Motion.* Kharkov: Kharkov Mathematical Society.

MacArthur R (1955) Fluctuations of animal populations and a measure of community stability. *Ecology* 36, 533–536.

Mascolo I (2019) Recent developments in the dynamic stability of elastic structures. *Frontiers in Applied Mathematics and Statistics* 5, 51.

Mason SJ (1953) Feedback theory-some properties of signal flow graphs. *Proceedings of the IRE* 41, 1144–1156.

May R (1972) Will a large complex system be stable? *Nature* 238, 413–414.

May R (1977) Thresholds and breakpoints in ecosystems with a multiplicity of stable states. *Nature* 269, 471–477.

May RM (1973) Stability in randomly fluctuating versus deterministic environments. *The American Naturalist* 107, 621–650.

Maynard DS, Serván CA, Allesina S (2018) Network spandrels reflect ecological assembly. *Ecology Letters* 21, 324–334.

McCann K, Hastings A, Huxel G (1998) Weak trophic interactions and the balance of nature. *Nature* 395, 794–798.

McCann KS (2012) *Food Webs.* Princeton: Princeton University Press.

McGill BJ (2010) Matters of scale. *Science* 328, 575–576.

Metz JAJ, de Kovel CGF (2013) The canonical equation of adaptive dynamics for Mendelian diploids and haplo-diploids. *Interface Focus* 3, 20130025.

Metz JAJ, Geritz S (2016) Frequency dependence 3.0: An attempt at codifying the evolutionary ecology perspective. *Journal of Mathematical Biology* 72, 1011–1037.

Minoarivelo HO, Hui C (2016) Invading a mutualistic network: To be or not to be similar. *Ecology and Evolution* 6, 4981–4996.

Minoarivelo HO, Hui C (2018) Alternative assembly processes from trait-mediated co-evolution in mutualistic communities. *Journal of Theoretical Biology* 454, 146–153.

Moles AT, et al. (2012) Invasions: The trail behind, the path ahead, and a test of a disturbing idea. *Journal of Ecology* 100, 116–127.

Montoya J, Pimm S, Solé R (2006) Ecological networks and their fragility. *Nature* 442, 259–264.

Mougi A, Kondoh M (2016) Food-web complexity, meta-community complexity and community stability. *Scientific Reports* 6, 24478.

Neutel AM, Heesterbreek JAP, de Ruiter PC (2002) Stability in real food webs: Weak links in long loops. *Science* 296, 1120–1123.

Nnakenyi CA, Hui C, Minoarivelo HO, Dieckmann U (2021) Leading eigenvalue of block-structured random matrices. In prep.

Nunney L (1980) The stability of complex model ecosystems. *The American Naturalist* 115, 639–649.

Nuwagaba S, Zhang F, Hui C (2015) A hybrid behavioural rule of adaptation and drift explains the emergent architecture of antagonistic networks. *Proceedings of the Royal Society B: Biological Sciences* 282, 20150320.

Nuwagaba S, Zhang F, Hui C (2017) Robustness of rigid and adaptive networks to species loss. *PLoS ONE* 12, e0189086.

Okuyama T, Holland JN (2008) Network structural properties mediate the stability of mutualistic communities. *Ecology Letters* 11, 208–216.

Orzach SH, Sober E (1993) A critical assessment of Levins's The Strategy of Model Building in Population Biology (1966). *The Quarterly Review of Biology* 68, 533–546.

Padrón B, et al. (2009) Impact of alien plant invaders on pollination networks in two Archipelagos. *PLoS ONE* 4, e6275.

Paine RT (1980) Food webs: Linkage, interaction strength and community infrastructure. *Journal of Animal Ecology* 49, 667–685.

Parrott L, Meyer WS (2012) Future landscapes: Managing within complexity. *Frontiers in Ecology and the Environment* 10, 382–389.

Perko L (2001) *Differential Equations and Dynamical Systems*. New York: Springer.

Pimm SL (1979) Complexity and stability: Another look at MacArthur's Original Hypothesis. *Oikos* 33, 351–357.

Quirk J, Ruppert R (1965) Qualitative economics and the stability of equilibrium. *The Review of Economic Studies* 32, 311–326.

Rankin DJ, Bargum K, Kokko H (2007) The tragedy of the commons in evolutionary biology. *Trends in Ecology & Evolution* 22, 643–651.

Richardson DM, Pyšek P (2006) Plant invasions: Merging the concepts of species invasiveness and community invasibility. *Progress in Physical Geography* 30, 409–431.

Ricciardi A, Atkinson SK (2004) Distinctiveness magnifies the impact of biological invaders in aquatic ecosystems. *Ecology Letters* 7, 781–784.

Ricciardi A, Palmer ME, Yan ND (2011) Should biological invasions be managed as natural disasters? *BioScience* 61, 312–317.

Rohr RP, Saavedra S, Bascompte J (2014) On the structural stability of mutualistic systems. *Science* 345, 1253497.

Romanuk TN, et al. (2009) Predicting invasion success in complex ecological networks. *Philosophical Transactions of the Royal Society B: Biological Sciences* 364, 1743–1754.

Romanuk TN, et al. (2017) Robustness trade-offs in model food webs: Invasion probability decreases while invasion consequences increase with connectance. In Bohan DA, Dumbrell AJ and Massol F (eds.), *Advances in Ecological Research*, pp. 263–291. Academic Press.

Rouget M, et al. (2016) Invasion debt: Quantifying future biological invasions. *Diversity and Distributions* 22, 445–456.

Routh EJ (1877) *A Treatise on the Stability of a Given State of Motion, Particularly Steady Motion*. London: MacMillan.

Scheffer M, et al. (2001) Catastrophic shifts in ecosystems. *Nature* 413, 591–596.

Scheffer M, et al. (2012) Anticipating critical transitions. *Science* 338, 344–348.

Seebens H, et al. (2020) Projecting the continental accumulation of alien species through to 2050. *Global Change Biology* 27, 970–982.

Shackleton RT, et al. (2018) Social-ecological drivers and impacts of invasion-related regime shifts: Consequences for ecosystem services and human well-being. *Environmental Science and Policy* 89, 300–314.

Shannon CE (1942) The Theory and Design of Linear Differential Equation Machines. Technical report, National Defence Research Council.

Simberloff D (2001) Biological invasions: How are they affecting us, and what can we do about them? *Western North American Naturalist* 61, 308–315.

Smallegange IM, Couson T (2013) Towards a general, population-level understanding of eco-evolutionary change. *Trends in Ecology & Evolution* 28, 143–148.

Snoussi EH (1998) Necessary conditions for multistationarity and stable periodicity. *Journal of Biological Systems* 06, 3–9.

Solé R, et al. (1999a) Criticality and scaling in evolutionary ecology. *Trends in Ecology & Evolution* 14, 156–160.

Solé R, et al. (1999b) Reply from R.V. Solé, S.C. Manrubia, M.J. Benton, S. Kauffman and P. Bak. *Trends in Ecology & Evolution* 14, 321–322.

Solé R, Goodwin B (2000) *Signs of Life: How Complexity Pervades Biology*. New York: Basic Books.

Solé RB, Valls J (1992) On structural stability and chaos in biological systems. *Journal of Theoretical Biology* 155, 87–102.

Soliveres S, Allan E (2018) Everything you always wanted to know about intransitive competition but were afraid to ask. *Journal of Ecology* 106, 807–814.

Sommers HJ, et al. (1988) Spectrum of large random asymmetric matrices. *Physical Review Letters* 60, 1895–1898.

Stone L (2018) The feasibility and stability of large complex biological networks: A random matrix approach. *Scientific Reports* 8, 8246.

Stouffer DB, Cirtwill AR, Bascompte J (2014) How exotic plants integrate into pollination networks. *Journal of Ecology* 102, 1442–1450.

Sutherland JP (1974) Multiple stable points in natural communities. *The American Naturalist* 108, 859–873.

Suweis S, et al. (2013) Emergence of structural and dynamical properties of ecological mutualistic networks. *Nature* 500, 449–452.

Suweis S, et al. (2015) Effect of localization on the stability of mutualistic ecological networks. *Nature Communications* 6, 10179.

Tang S, Pawar S, Allesina S (2014) Correlation between interaction strengths drives stability in large ecological networks. *Ecology Letters* 17, 1094–1100.

Tansley AG (1935) The use and abuse of vegetational concepts and terms. *Ecology* 16, 284–307.

Tao T, Vu V, Krishnapur M (2010) Random matrices: Universality of ESDs and the circular law. *Annals of Probability* 38, 2023–2065.

Thébault E, Fontaine C (2010) Stability of ecological communities and the architecture of mutualistic and trophic networks. *Science* 329, 853–856.

Theil H (1953). *Repeated Least Squares Applied to Complete Equation Systems*. The Hague: Central Planning Bureau.

Thomas R (1981) On the relation between the logical structure of systems and their ability to generate multiple steady states or sustained oscillations. In Della DJ, Demongeot J, Lacolle B (eds.), *Numerical Methods in the Study of Critical Phenomena*, pp. 180–193. Heidelberg: Springer.

Thompson RM, Townsend CR (2003) Impacts on stream food webs of native and exotic forest: An intercontinental comparison. *Ecology* 84, 145–161.

Tilman D, et al. (1994) Habitat destruction and the extinction debt. *Nature* 371, 65–66.

Traveset A, et al. (2013) Invaders of pollination networks in the Galápagos Islands: Emergence of novel communities. *Proceedings of the Royal Society B: Biological Sciences* 280, 20123040.

Traveset A, Richardson DM (2006) Biological invasions as disruptors of plant reproductive mutualisms. *Trends in Ecology & Evolution* 21, 208–216.

Traveset A, Richardson DM (2011) Mutualisms: key drivers of invasions … key casualties of invasions. In Richardson DM (ed.), *Fifty Years of Invasion Ecology: The Legacy of Charles Elton*, pp. 143–160. Oxford: Wiley-Blackwell.

Traveset A, Richardson DM (2014) Mutualistic interactions and biological invasions. *Annual Review of Ecology, Evolution, and Systematics* 45, 89–113.

Ulanowicz RE (2002) The balance between adaptability and adaptation. *Biosystems* 64, 13–22.

Valdovinos FS, et al. (2018) Species traits and network structure predict the success and impacts of pollinator invasions. *Nature Communications* 9, 2153.

Vallina SM, Le Quéré C (2011) Stability of complex food webs: Resilience, resistance and the average interaction strength. *Journal of Theoretical Biology* 272, 160–173.

Valverde S, Elena SF, Solé R (2017) Spatially induced nestedness in a neutral model of phage-bacteria networks. *Virus Evolution* 3, vex021.

van Altena C, Hemerik L, de Ruiter PC (2016) Food web stability and weighted connectance: The complexity-stability debate revisited. *Theoretical Ecology* 9, 49–58.

Vandermeer J (1980) Indirect mutualism: Variations on a theme by Stephen Levine. *The American Naturalist* 116, 441–448.

Vandermeer J (2011) Intransitive loops in ecosystem models: From stable foci to heteroclinic cycles. *Ecological Complexity* 8, 92–97.

Vieira MC, Almeida-Neto M (2015) A simple stochastic model for complex coextinctions in mutualistic networks: Robustness decreases with connectance. *Ecology Letters* 18, 144–152.

Vilà M, et al. (2009) Invasive plant integration into native plant–pollinator networks across Europe. *Proceedings of the Royal Society B: Biological Sciences* 276, 3887–3893.

Whittaker RH (1953) A consideration of climax theory: The climax as a population and pattern. *Ecological Monographs* 23, 41–78.

Whittaker RH (1962) Classification of natural communities. *The Botanical Review* 28, 1–239.

Wigner EP (1958) On the distribution of the roots of certain symmetric matrices. *Annals of Mathematics* 67, 325–327.

Williamson M (1996) *Biological Invasions*. London: Chapman & Hall.

Wishart J (1928) The generalised product moment distribution in samples from a normal multivariate population. *Biometrika* 20A, 32–52.

Womack M (2003) *Sport as Symbol: Images of the Athlete in Art, Literature and Song*. North Carolina: McFarland & Company, Inc. Publishers.

Yang Y, Hui C (2021) How competitive intransitivity and niche overlap affect spatial coexistence. *Oikos* 130, 260–273.

Yodzis P (1981) The stability of real ecosystems. *Nature* 289, 674–676.

Young TP, Chase JM, Huddleston RT (2001) Succession and assembly as conceptual bases in community ecology and ecological restoration. *Restoration Ecology* 19, 5–19.

Zhang F, Hui C, Pauw A (2013) Adaptive divergence in Darwin's race: How coevolution can generate trait diversity in a pollination system. *Evolution* 67, 548–560.

Zhang F, Hui C, Terblanche JS (2011) An interaction switch predicts the nested architecture of mutualistic networks. *Ecology Letters* 14, 797–803.

Zhang F, Dieckmann U, Metz JAJ, Hui C (2021) Condition for evolutionary branching and trapping in multidimensional adaptive dynamics. In prep.

5 · *Network Transitions*

Communities have always changed in their composition and should continue to do so. A nostalgic longing for a lost Garden of Eden, which permeated the roots of the conservation movement, is not supported by what we know of the past and expect in the future.

Dornelas and Madin (2020)

5.1 The Forecasting Conundrum

Astrologists have predicted the occurrence of solar eclipses with increasing precision through the ages. Predicting celestial motions invokes the dynamics of a relatively simple and rigid system; it is straightforward and akin to identifying regularities in recurrent records. Discovering regularities, however, does not necessarily impart true comprehension. While we can speculate about the mechanisms and forces at work to fill gaps as we edge towards comprehension, such conjectured theories are often misleading. In early 2020, epidemiologists were confronted with a once-in-a-lifetime challenge: forecasting the number of infections of COVID-19 both regionally and globally. With little understanding of the viral transmission at the time, most forecasts failed miserably. Failed forecasts abound, especially for systems that are complex and adaptive; the bet between ecologist Paul Ehrlich and economist Julian Simon on the swings of metal price anticipated from socioeconomic impacts of overpopulation (Sabin 2013) is a good example. The forecasting conundrum is both typical and perplexing to ecologists and invasion scientists; hindsight is an exact science, while forecasting is no easier than catching the Cheshire Cat.

True comprehension of a system often empowers a revelation and results in a leap in forecasting precision. So, the Renaissance started when Nicolaus Copernicus shifted our planet out of the celestial centroid and placed the Sun, rather than the Earth, at the centre of the universe. Counterintuitively, however, a deep comprehension of how members of an ecological community are assembled and how the structure and function of embedded ecological networks emerge does not necessarily pave the way for strong predictions of the dynamics of the community under global change scenarios (Stokstad 2009). A review of all 18,076 articles published in three flagship ecology journals revealed alarming trends (Figure 5.1; Low-Décarie et al. 2014). Although increasing model complexity allows us to discern more and increasingly minute and intricate effects in ecological systems (captured by the increasing numbers of reported P values over the years), the explanatory power of these models, as indicated by the variance explained (R^2; different but related to predictability), has been declining steadily for decades. Conservation and the management of transformed ecosystems, the mainstay of ecology in the future, clearly need to address this forecasting conundrum. Although great strides have been made in invasion science in recent decades, the level of understanding has not progressed to the stage where robust predictions can be made. A number of factors can thwart predictability when forecasting future dynamics and responses of ecological networks to biological invasions.

First, and perhaps most importantly, the future of an open adaptive system depends on its current and past contexts and can unfold in different ways. 'Any replay of the tape of life', as Stephen Jay Gould (1989) writes in *Wonderful Life*, 'would lead evolution down a pathway radically different from the road actually taken'. Adaptive network models, discussed in Chapter 4, indeed produce alternative evolutionary pathways when run under exactly the same conditions (Figure 5.2), highlighting the emergence of such contingency during system evolution. With contingent interactions and components, ecosystems become computationally irreducible (Beckage et al. 2011); their dynamics can only be observed but are unpredictable, like a game: although the rules of engagement have been predetermined, it is impossible to predict precisely how a game of chess will unfold at each turn. As discussed at the end of Chapter 4, ecological networks facing biological invasions experience persistent transitions at criticality; this means that the alternative regimes and thus the direction and potential pathways of system evolution are yet to emerge. This contrasts with the typical view in the theory of transient dynamics that holds that perturbed systems traverse towards or between well-established alternative stable states (in the case of non-equilibrium dynamics towards fixed attractors). A surfer at a wave front (an open adaptive system at marginal instability; see Section 4.5) can follow highly flexible and contingent paths, although in hindsight only

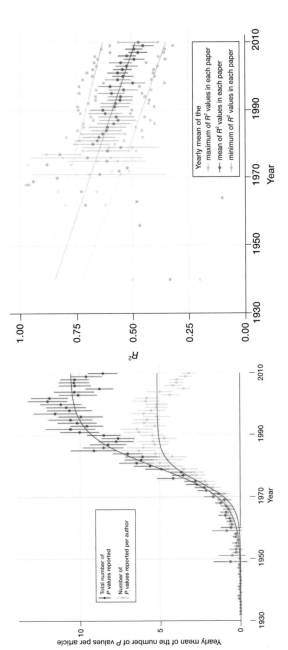

Figure 5.1 Top: Increase in mean number of P values reported in each publication per year in *Journal of Ecology* (established in 1913), *Journal of Animal Ecology* (established in 1932) and *Ecology* (established in 1920). The black data points are yearly means for the total number of P values, whereas the grey data points are means of the total number of P values standardised by the number of authors on a given paper. The error bars are 95% confidence intervals of the mean. Lines are best-fit logistic models. The increase in number of P values suggests an increase in the complexity of the research being reported. Bottom: maximum, mean and minimum R^2 values reported as yearly means. The trend lines are weighted least squares regressions ($R^2 = 0.62$ and slope of -0.005 per year for mean values). The error bars are 95% confidence intervals of the mean. The opacity of the point and trend line denotes the number of articles from which the R^2 values were extracted; fewer articles and fewer of these articles contained R^2 values in earlier years. From Low-Décarie et al. (2014), reprinted with permission.

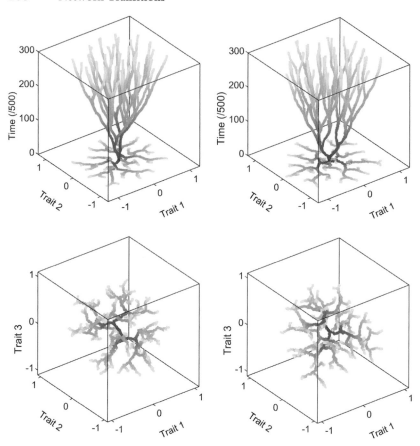

Figure 5.2 Alternative trait evolution dynamics in an ecological network where resident species are engaging in resource competition with each other. Top, a species is represented by two trait values; bottom, by three trait values. Simulations on each row were run twice (left and right) under the same set of parameters and initial conditions, while alternative evolutionary pathways emerged from the uncertainty in the directions of sequential evolutionary branching. Redrawn from Zhang et al. (2021). *A black and white version of this figure will appear in some formats. For the colour version, please refer to the plate section.*

one exact path is realised. Consequently, a system with high predictability often implies that its future can be predetermined, like celestial motions. Such a system is often rigid and leaves little room for management. This is certainly not the case for open adaptive systems such as socioecological systems and ecological networks whose components are co-evolving and undergoing co-fitting due to intrusions of new species, novel interactions and changing selection forces. Unpredictability – not following its fate – is therefore precisely the feature of an open adaptive system such as an ecological network facing invasions and experiencing persistent transitions.

Second, forecasting is often hypersensitive to observation and measurement errors. Ecological networks often comprise large numbers of species and numerous interactions, many of which experience intermittent fluctuations and flickering. While the complete parameterisation of a dynamic network model is itself challenging, high levels of model complexity do not necessarily guarantee high predictability (Figure 5.1). Indeed, there is no evidence that complex models are better than simple ones for forecasting system dynamics (Green and Armstrong 2015). As a result of the propagation of uncertainty, complex models often amplify prediction errors and result in prediction failures. Even for a simple system such as the discrete-time logistic equation, $n_{t+1} = m_t(1 - n_t)$, depicting density-dependent population dynamics, May (1976) demonstrated that prediction errors can accumulate dramatically, leading to the phenomenon of deterministic chaos when the intrinsic rate of growth (r) exceeds a certain threshold. A high rate of population growth is typical of highly successful invasive species (Sakai et al. 2001); the invasion dynamics of these species are, therefore, likely to exhibit deterministic chaos. Indeed, species interactions, coupled with seasonal fluctuations, can push many ecological networks to the edge of chaos (e.g., Figure 5.3; Benincà et al. 2015), making dynamic forecasting hypersensitive to measurement errors. As ecological networks facing biological invasions undergo persistent transitions, the precise and prompt measurement of state variables – species abundances and interaction strengths – is inherently challenging since these are representative of the system's behaviour only for a short period. Adding to such hypersensitivity, lag phases of invasion dynamics and associated invasion and extinction debts of an invaded ecosystem make it even more challenging to monitor the status quo and trace introduction and invasion histories (e.g., Hirsch et al. 2017). The combination of model complexity, measurement errors and deterministic chaos together drastically reduces the possibility of forecasting the future dynamics of an invaded ecological network.

For the reasons discussed earlier, open adaptive systems, such as invaded ecological networks, exhibit a state of paradox. Retrospectively, we can identify and explain clear non-random features, but we cannot accurately predict a future that unfolds with surprises and novelties. Invasion science has shown exactly this feature: despite strong identifiable covariates of

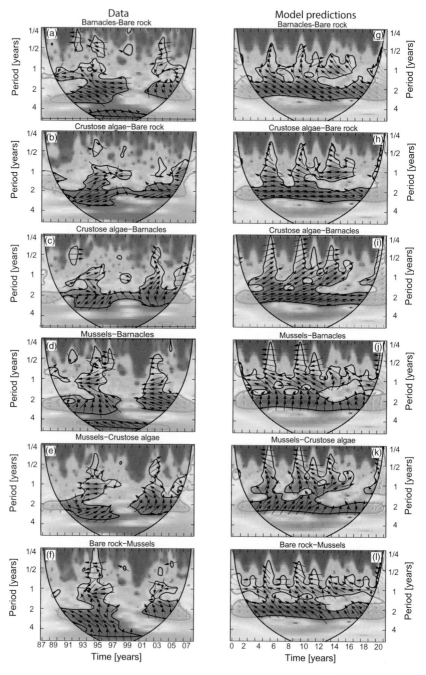

Figure 5.3 Cross-wavelet spectra of all species pairs of a rocky intertidal community in the Cape Rodney-Okakari Point Marine Reserve on the North Island of

invasion performance, prediction power is low. Invasive species often have specific features (Pyšek and Richardson 2007), while ecosystems vulnerable to invasions often exhibit diverse but unique network structures (Chapters 3 and 4). However, predictability based on these identified features is not always promising (Capinha et al. 2018). Unpredictability resulting from contingency and computational irreducibility leaves space for management that could potentially steer the system towards a preferred state. However, the complexity of socioecological systems means that conflicts between stakeholders often produce 'wicked' management problems (Rittel and Webber 1973; Evans et al. 2008; Seastedt 2014; Woodford et al. 2016). The surprises and novelties, together with management complications, often cause control or restoration actions to fail (Dana et al. 2019), as highlighted in the counterintuitive Hydra effect (Cortez and Abrams 2016). These factors have fuelled much debate on the value of prediction and future forecasting in invasion biology (Elliott-Graves 2016).

Armstrong (2009) reviewed six reasons for selecting a forecasting method: convenience, popularity, structured judgement [expert opinion], statistical criteria, relative track records and guidelines from prior research. Invasion science currently relies heavily on ecological niche modelling for forecasting (e.g., Fletcher et al. 2016; Fournier et al. 2019), probably for convenience and popularity. Tetlock (2005) noticed some interesting differences in the performance of long-term predictions: those highly specialised experts (hedgehogs) knew 'one big thing', while the investigators (foxes) knew 'many little things'. Foxes can adjust and adapt their predictions, akin to the heuristic learning strategy of 'Win-Stay, Lost-Shift' (see Chapter 2). Hedgehogs, on the other hand, have made up their minds and only collect evidence (which is often biased) to reinforce and justify their predictions, which is akin to decision making

---

←

Figure 5.3 (*cont.*) New Zealand. Cross-wavelet spectra of (a–f) the observed time series and (g–l) the model predictions. The spectra show how common periodicities in the fluctuations of two species (y axis) change over time (x axis). Colour indicates cross-wavelet power (from low power in blue to high power in red), which measures to what extent the fluctuations of the two species are related. Black contour lines enclose significant regions, with > 95% confidence that cross-wavelet power exceeds red noise. Arrows indicate phase angles between the fluctuations of the two species. Arrows pointing right represent in-phase oscillations (0°), and arrows pointing upward indicate that the first species lags the second species by a quarter period (90°). Shaded areas on both sides represent the cone of influence, where edge effects may distort the results. From Beninca et al. (2015), reprinted with permission. *A black and white version of this figure will appear in some formats. For the colour version, please refer to the plate section.*

(Kahneman 2013). This implies that better forecasting of invasion performance in an ecological network requires us to bridge many disciplines and consult diverse stakeholders, to collate many little, but useful, things. It also means that we need to shift our focus from targeting only the focal non-native species to considering many interlinked species in invaded ecosystems to achieve system-level dynamic forecasting. Following Armstrong's flowchart (see Figure 5.4), we focus on quantitative and causal models for forecasting the dynamics in invaded ecological networks. We cover four topics. First, we introduce early-warning signals for imminent regime shifts and transitions caused by biological invasions and other system-level perturbations. Second, we elucidate how network transition can be captured as the temporal turnover of its residing species and network interactions. Third, to forecast network transition we introduce the concept of a weather vane as an indicator of the transient dynamics of network turnover. Finally, we link the peculiar roles of rare species along the commonness–rarity staircase that explains the weather vane of network transition. We also discuss the potential physical meaning of network turnover and transition as drivers of entropy in ecosystems affected by biological invasions.

5.2 Early Warning Signals

If an alien species establishes positive feedbacks in a recipient ecosystem, it has the potential to trigger a regime shift (Chapter 4; Scheffer et al. 2009; Gaertner et al. 2014). For instance, before the introduction of largemouth bass, *Micropterus salmoides*, a Michigan freshwater lake was dominated by minnows (planktivorous fish). This dominance by minnows hampered the growth and recruitment of juvenile largemouth bass (Walters and Kitchell 2001; Carpenter et al. 2008, 2011), preventing the introduced juvenile bass from establishing. After the introduction of several adult largemouth bass, the freshwater ecosystem jumped to a new basin of attraction, allowing adult bass to become formidable top predators (Carpenter et al. 2008). The regime shift between minnow dominance and bass dominance can be clearly illustrated in the phase portrait, where the minnow catch declines during the study (Figure 5.5 left). Before the first addition of bass, the trajectory circles around an attractor. The trajectory during the transition period indicates a limit cycle of fluctuating population dynamics. After the last bass addition, the trajectory follows a new attractor close to the extirpation of minnows. The standardised residuals from an autoregressive model (Figure 5.5 right)

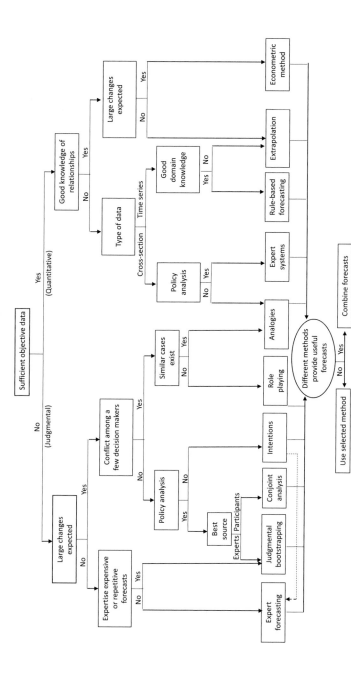

Figure 5.4 A flowchart for selecting forecasting methods. Given enough data, quantitative methods are more accurate than judgemental methods. When large changes are expected, causal methods are more accurate than naive methods. Simple methods are preferable to complex methods; they are easier to understand, less expensive and seldom less accurate. To select a judgemental method, determine whether there are large changes, frequent forecasts, conflicts among decision makers and policy considerations. To select a quantitative method, consider the level of knowledge about relationships, the amount of change involved, the type of data, the need for policy analysis and the extent of domain knowledge. When selection is difficult, combine forecasts from different methods. Redrawn from Armstrong (2009).

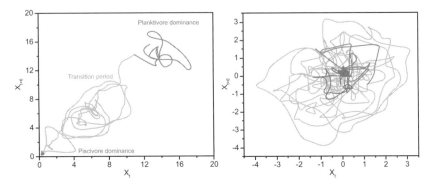

Figure 5.5 Phase portrait of minnow trap time series (left) and standardised residuals from the generalized autoregressive conditional heteroscedasticity model. The blue trajectory is the period prior to the first addition of largemouth bass in the manipulated Peter Lake in Michigan. The light grey trajectory is the transition period. The dark grey trajectory is the period after the last largemouth bass addition. The system is initially at a point attractor, but enters into a limit cycle after the first largemouth bass addition. The system has returned to a new point attractor by the time of the last largemouth bass addition. From Seekell et al. (2013), reprinted with permission.

make the circular pattern of a limit cycle more evident during the transition period (the grey trajectory). In many cases, changes in ecosystem functioning associated with regime shifts are detrimental to human well-being, and management strategies need to be deployed to prevent such regime shifts (Boettiger et al. 2013; Shackleton et al. 2018). Early warning signals (EWS) are first-line forecasting tools that often require little knowledge of the system itself (Scheffer et al. 2009, 2012).

Regime shifts require the system to offer multiple alternative stable states. This normally requires the system to operate near a bifurcation point so that minute changes in system parameters can trigger changes in the system regime (Figure 5.6). However, not all regime shifts are dramatic or catastrophic; some involve smooth transitions, especially when the system is under the influence of demographic stochasticity combined with realistic features such as limited mobility and spatial heterogeneity (Martín et al. 2015). There are seven types of bifurcations in dynamical systems, although most metrics of early warning signals are designed around a particular type – a saddle-node (fold) bifurcation, where two alternative regimes are divided by a saddle point known as the tipping point in state variables. When the system approaches the tipping point, the leading eigenvalue of the system Jacobian matrix approaches zero (Chapter 4); consequently, the system dynamics show

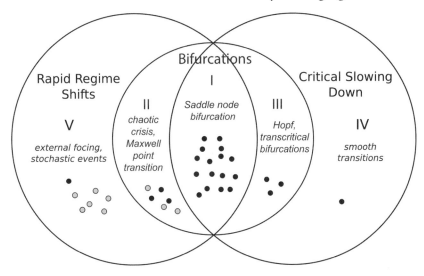

Figure 5.6 The intersecting domains of rapid regime shifts, bifurcations and critical slowing down. Labels below the Roman numerals are examples of phenomena that occur in each domain. Each dot represents a study in the domain. Dots in grey represent publications that do not explicitly test early warning signals. The centre domain (I), where all three phenomena intersect, is the most extensively researched domain in the early warning signal field. Literature outside this charted research is less extensive, but hints at how existing signals based on critical slowing down may be insufficient or misleading. For instance, bifurcations that show no signs of critical slowing down before a regime shift, known as crises, are characterised by the sudden appearance of chaotic attractors. From Boettiger et al. (2013), reprinted with permission.

critical slowing down, and the variance or autocorrelation and return time of the state variables of the system increase greatly (Carpenter and Brock 2006; Dakos et al. 2011). Critical slowing down in perturbed systems can be identified either directly after perturbation (Veraart et al. 2012; Figure 5.7a), or by comparing generic indicators to a control treatment (Drake and Griffen 2010; Figure 5.7b), although such noise-free time series are rare in practice. In contrast, Carpenter et al. (2011) found that the abundance of largemouth bass flickered between alternative states one year before the regime shift to piscivore dominance in the lake (Figure 5.7c). When flickering, indicators of such time series will exhibit much larger variation than the smooth curves of critical slowing down and only variance can serve as a good indicator of upcoming transitions (Dakos et al. 2013). Establishing the mechanisms behind a saddle-node bifurcation is, therefore, essential for assessing the feasibility

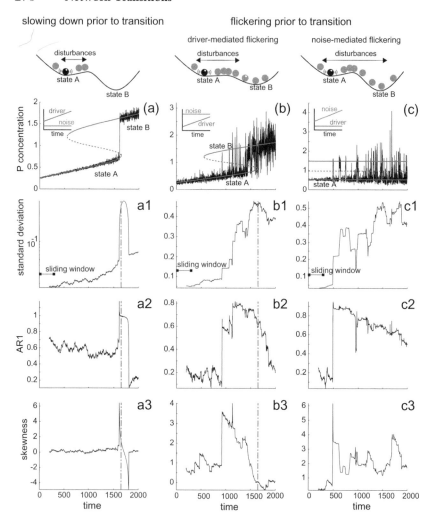

Figure 5.7 Critical slowing down versus flickering. In the critical slowing down scenario, the system state stays within the basin of attraction of the current state under weak disturbances (upper left panel), while in the flickering scenarios stronger disturbances can push the system state across the basin of attraction when the basin shrinks (driver-mediated flickering, middle panel) or when noise intensity increases (noise-mediated flickering, left panel). An example of a typical time series derived from transient simulations under critical slowing down (a), driver-mediated flickering (b) and noise-mediated flickering (c) scenarios. Generic leading indicators [variance measured as standard deviation (a1, b1, c1), autocorrelation at-lag-1 (AR1, a2, b2, c2) and skewness (a3, b3, c3)] were estimated for each scenario within sliding windows of 200 points across the time series (solid red lines denote stable equilibria, dotted red lines unstable equilibria and red dash dot lines the threshold at which the critical transition to the eutrophic state would occur in the deterministic case). From Dakos et al. (2013), reprinted with permission.

of using critical slowing down as a warning signal. Saddle-node bifurcations have been confirmed in some systems, such as lakes undergoing eutrophication (Scheffer et al. 2001), lakes with a trophic cascade (Carpenter et al. 2008) and forest/savanna transitions (Hirota et al. 2011; Staver et al. 2011). However, regime shifts from other types of bifurcations do not necessarily approach a tipping point of zero eigenvalue during transition. This means that relying solely on metrics of critical slowing down could miss important signals of imminent regime shifts (Hastings and Wysham 2010; Boettiger and Hastings 2012).

Some rapid regime shifts are not attributable to bifurcations at all. An internal stochastic event may switch a system between dynamical regimes. Large environmental shocks may change the behaviour of a system without warning. For instance, Schooler et al. (2011) found that lakes invaded by the floating fern *Salvinia molesta* and herbivorous weevils alternated between low- and high-*Salvinia* states driven by disturbances from regular external flooding events. System stochasticity could also shift systems from one state to another even when environmental conditions remain constant (Hastings and Wysham 2010). In such exogenous regime shifts with no underlying bifurcations, identifying early warning signals becomes irrelevant, and the critical slowing down is undetectable (Ditlevsen and Johnsen 2010; Wang et al. 2012). These indicators will also not respond to step changes in control variables. Moreover, using these indicators often requires making arbitrary calculations. For example, the power of lag-1 autocorrelation could be modified by changing methods of data aggregation, detrending and filtering signal bandwidth (Lenton et al. 2012). Nevertheless, strong stochasticity weakens the detectability of early warning signals (EWS) for imminent critical transitions (Perretti and Munch 2012). Such floundering of early warning signals has led many researchers to propose more sophisticated metrics, multiple early-warning indicators (Lindegren et al. 2012) and composite indices (Drake and Griffen 2010). Pettersson et al. (2020) recently designed a metric to predict a system's likely response to change, both in terms of single-species extinction and system-wide collapse.

Early warning signals and the linked concepts of regime shifts and tipping points are often used when we have little knowledge of the system's features, such as its composition and structure. EWS normally only target the time series analyses of selected state variables, such as annual population density of indicator species. Two features of EWS limit their usefulness for monitoring the dynamics of invaded ecological networks. First, EWS are a black-box tool with targeted state variables/species that do not necessarily reflect the dynamic regime of a system; as a result they can be either over- or under-sensitive. Predictive power could be drastically impeded by excluding other key species in ecosystems (e.g., those important to interaction loops) and fail quantitative bifurcation analyses from incomplete knowledge (Milkoreit et al. 2018). Second, EWS require long-term high-frequency monitoring. The throughput time series required for detecting EWS also restricts their application to limited target species and environmental variables, thus excluding many other components of ecological networks. The use of EWS often leads to the ubiquitous impression that many or most socioecological systems are operating near tipping points. Our tentative answer in Chapter 4 is that ecological networks, and perhaps also other open adaptive systems, are in constant transition, while EWS rely on the system crossing the zero eigenvalue. This does not reflect the reality of most systems undergoing persistent transition where positive leading eigenvalues are maintained and there are no obvious basins of attraction for system regimes. Consequently, discussing alternative stable regimes is problematical for open adaptive systems that operate at self-organised criticality. Forecasting the dynamics and transitions of ecological networks under invasions, therefore, requires indices and measures other than typical EWS.

5.3 Temporal Turnover of Ecological Networks

As open adaptive systems, invaded ecological networks constantly receive new arrivals and lose species, temporally or permanently. To capture this persistent transition in ecological networks subjected to biological invasions, we could consider the most direct measure of network openness – the temporal turnover of species composition and interactions. The concept of temporal turnover in ecological communities, a key component in island biogeography theory, posits a dynamic balance between the immigration of new species to an island and local extinctions. Early literature emphasised the effects of island size and isolation on the rates of colonisation and local extinction (MacArthur and Wilson 1967). At this dynamic equilibrium of the number of harboured species, island biogeography theory also posits a constant temporal turnover of residing species of the insular biota (Figure 5.8). This means that, although it is possible for an extirpated species to recolonise the island, persistence is ephemeral. Much of the early literature of island biogeography focused on natural dispersal and colonisation of regional natives. Biological invasions triggered by human-mediated dispersal breach the barrier of

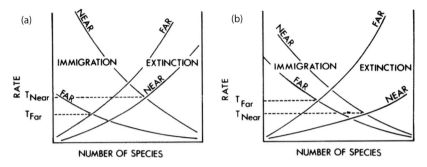

Figure 5.8 MacArthur and Wilson's (1967) theory of island biogeography.
Hypothetical extinction and immigration rates for two islands of equal area and
different degrees of isolation. The rescue effect reduces the extinction rate on the
near island. a. An increase in isolation has a greater effect on the immigration rate
than on the extinction rate. The turnover rate is predicted to be greater on the near
island. b. An increase in isolation has a greater effect on the extinction rate. The
turnover rate is predicted to be greater on the far island. From Wright (1985),
reprinted with permission.

isolation, thereby negating many of the effects of covariates related to
island size. Attention must thus shift to understanding the rate of propagule immigration and establishment. Island biogeography theory needs
to be revisited to account for the ubiquity of human-mediated dispersal
to reshape the immigration curve (Figure 5.8). Nevertheless, traditional
island biogeography theory provides a sound theoretical foundation for
considering species turnover in ecological networks driven by colonisation or invasion of alien propagules. In this section, we examine insights
from classic studies and some recent developments to explore the implications for network invasions.

Early work on island biogeography theory focused on elucidating how
immigration and extinction rates are influenced by island size and isolation, without giving due attention to temporal turnover. We, however,
selected some important pieces on temporal turnover. In particular,
Diamond (1969) reported a negative relationship (cor = −0.86)
between the percentage turnover of bird species from 1917 to 1968
and the average number of species residing on the nine Channel Islands
off the coast of California; percentages ranged from 62 per cent turnover
in Santa Barbara (smallest) with an average of 7 to 8 bird species, to
17 per cent turnover in Santa Cruz (largest) hosting on average
36 species. This means that islands with more species experience less
(and slower) temporal turnover in their biota. Diamond (1969) used a

typical turnover rate metric, $100 \times (E + F - A)/(S_1 + S_2 - A)$, where E is the number of extinct species (absent in the second survey), F the number of new species added in the second survey, and S_1 and S_2 are the number of breeding species in the two surveys. Interestingly, Diamond singled out these alien species (A), which are game birds intentionally introduced by humans during the period between the two surveys (California quail *Callipepla californica*, Gambel's quail *Callipepla gambelii*, pheasant *Phasianus colchicus* and chukar partridge *Alectoris chukar*). While the richness of the two surveys show a strong correlation (cor $= 0.90$), indicative of a dynamic equilibrium of its biota, the relationship between species extinctions and additions is negative (cor $= -0.57$). This finding could be serendipitous and attributable to the small sample size. Diamond (1969) ends with a prelude to invaded ecological networks

Ecological consequences of these insular invasions, extinctions, and variations in species diversity include striking expansions and compressions of the niche of a given species, depending upon the competing species pool it faces.

Diamond and May (1977) revisited the theory of temporal turnover, measured as

$$T(t) = \frac{1}{t}\frac{I + E}{S_1 + S_2}, \tag{5.1}$$

with I and E the numbers of immigrations and extinctions between the time elapse t, using annual avian surveys on the Farne Islands of Britain from 1946 to 1974. A Markov process model was developed based on the probability of immigration per unit time for absent species i, γ_i (introduction probability), and the probability of extinction per unit time of present species i, e_i (extinction probability), from which the expected turnover rate over period t can be estimated (Figure 5.9),

$$T(t) = \frac{1}{tS^*}\sum_{i=1}^{S_R} \frac{\gamma_i e_i \left(1 - (1 - \gamma_i - e_i)^t\right)}{(\gamma_i + e_i)^2}, \tag{5.2}$$

where $S^* = \sum_i \gamma_i/(\gamma_i + e_i)$ is the dynamic equilibrium of species richness in the island, and S_R the total number of regional species that are feasible to colonise the island. Biological invasions could drastically expand the feasible species pool (S_R) to reach an island. At the dynamic equilibrium, the expected numbers of species addition and extinction during the period t are equal, $\mathrm{Add}(t) = \mathrm{Ext}(t) = tS^* T(t)$. The observations from the Farne

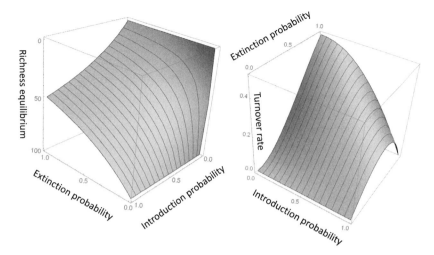

Figure 5.9 An illustration of richness equilibrium (S^*, left) and turnover rate ($T(t = 2)$, right) as a function of extinction probability e_i and introduction probability γ_i per unit time (Eq. 5.1; Diamond and May 1977). For simplicity, we ignored the cross-species differences in the extinction and introduction probabilities per unit time. Black curves are contours of introduction probability.

Islands support Diamond and May's (1977) formulation, although the emphasis was laid on the effect of the time elapse t between two surveys on the rate of temporal turnover. In Figure 5.9, instead, we focus on how the dynamic equilibrium of species richness and temporal turnover, as well as the associated extinction probability, are affected by the introduction probability. The roles of island area and isolation on the dynamic equilibrium of species richness have been intensively explored in literature (e.g., Brown and Kodric-Brown 1977; Williamson 1978; Gilbert 1980; Wright 1985). However, the extinction rate has not been formulated to date; it is driven by the immigration rate and the biotic interaction game within the insular ecological interaction network. We suggest that this merits considerable research attention in invasion science.

Early studies on island biogeography theory largely ignored the role of biotic interactions on islands in affecting the rates of successful colonisation and extinction. There has been an upsurge in research on the impacts of interaction network structures within recipient ecosystems on the rate of invasion success and extinction (Chapters 3 and 4). However, whereas the recent revival of island biogeography has seen a new appreciation of the role of embedded ecological interaction networks, temporal turnover in invaded ecological networks has not received the attention it deserves. Gravel et al. (2011) developed a trophic theory of island biogeography by considering the role of trophic interactions on the rate of immigration and extinction (Figure 5.10). While the model certainly refined the classic island biogeographic theory by assuming specific increments or decrements to the rates of introduction and extinction resulting from biotic interactions, it has limitations, as the effect of trophic interactions on these rates (and their increments or decrements) were assumed in the model, rather than emerging from the model. Consequently, it cannot assess how and why some species succeed while others fail in an invaded network, or how local extinctions emerge as a consequence from the impact of biotic interactions in the multiplayer game. It is therefore a direct extension of the meta–population models of Diamond and May (1977), Gilpin and Diamond (1982)

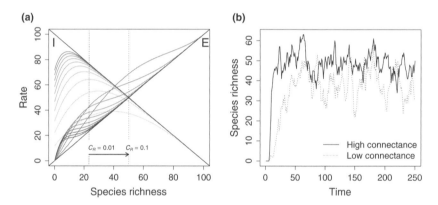

Figure 5.10 The trophic theory of island biogeography. (a) Adding a trophic constraint on species immigration and extinction affects the equilibrium species richness in a local community (note all consumers have the same diet breadth for the illustration). The classic theory of island biogeography is depicted with black lines. Species richness in a local community is found when the immigration rate equals the extinction rate (when the respective I and E curves cross, depicted by vertical dotted lines). Grey lines for the trophic theory of island biogeography are obtained following the analytical approximation in the online supporting information. The connectance in the regional species pool is varied between 0.01 (light grey for the immigration and extinction curves, respectively and 0.1 (dark grey curves, respectively). At connectance larger than 0.1, the trophic theory of island biogeography no longer differs from the classic one for these immigration and extinction rates. (b) Stochastic simulations show the assembly dynamics for local communities with low and high regional connectance. From Gravel et al. (2011), reprinted with permission.

and Hanski and Gyllenberg (1997). Nevertheless, this is a move in the right direction.

Not only species composition is under persistent turnover as a result of biological invasions; other dimensions of biodiversity such as functional diversity of traits can also undergo persistent transitions. Using non-metric multidimensional scaling ordination (Figure 5.11), Hewitt et al. (2016) explored the compositional and functional turnover in the Baltic Sea resulting from constant disturbance in the form of eutrophication and accompanying hypoxia that has created opportunities for the invasion of

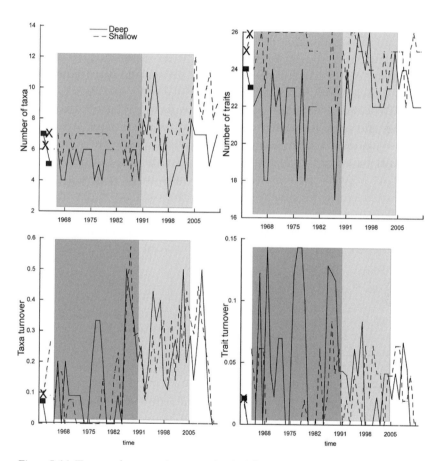

Figure 5.11 Temporal patterns in taxa and trait richness and turnover at the two sites. The solid square and cross are data at the deep and shallow sites, respectively, between 1926 and 1928. *Marenzelleria* spp. invaded the sites in 1991 and by 2005 was considered to be established. Dark and light grey shading designate the before and arrival time periods. From Hewitt et al. (2016), under CC-BY Licence.

an infaunal polychaete complex, *Marenzelleria* spp., in benthic communities. Richness was found to be greater during the phase of establishment of invaders (Figure 5.11a), while temporal turnover of species composition increased markedly just before the invasion (Figure 5.11b). However, the effects of biological invasions on functional traits and their turnover are not consistent between sites (Figure 5.11c and d). The lack of consistent responses in functional diversity to biological invasions again suggests context-dependent invasion performance and the role of community assembly in dictating network invasibility (Davis 2003; Strayer et al. 2006; Heino et al. 2015). Chapter 3 discussed how community assembly processes jointly steer the integration of incoming alien species into invaded ecological networks. A key challenge, therefore, is to determine how the traits of an alien species in relation to the trait dispersion of resident species of recipient ecosystems affect its invasiveness; we revisit this topic in Chapter 7.

An ecological network does not undergo temporal turnover of its species and their functional traits only; biotic interactions also change due to random or adaptive interaction rewiring (Chapters 2 and 3). To quantify interaction similarity or dissimilarity (turnover) of species compositions between two surveys, we can compare the species lists (composition vectors); for instance, say, the compositional vector of species ABCDE from survey 1 is 10110, while the vector from survey 2 is 01101. We can therefore quantify the number of gains and losses, and thus the temporal turnover of species between the two surveys. To compare the interaction similarity between the two surveys, following Poisot et al. (2012), we need to compare the vectorised interaction matrices ($\text{vec}(\boldsymbol{A})$) of the two surveys; vectorisation simply involves stacking the column vectors of a matrix on top of one another into a new vector; for instance,

$$\text{vec} \begin{bmatrix} a_{11} & a_{12} \\ a_{21} & a_{22} \end{bmatrix} = \langle a_{11}, a_{21}, a_{12}, a_{22} \rangle^T. \tag{5.3}$$

Interaction turnover can then be defined following the same dissimilarity beta diversity measures. Temporal turnover of network interactions can rise from two sources (CaraDonna et al. 2017): interaction rewiring and species turnover (Figure 5.12). For instance, when using Whittaker's (1960) dissimilarity index as interaction turnover, $\beta_{int} = 2(a + b + c)/(2a + b + c) - 1$, where a is the number of pairwise interactions shared between the networks, and b and c are the number of pairwise interactions unique to each of the networks,

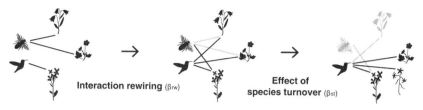

Interaction rewiring (β_{rw})   Effect of
species turnover (β_{st})

Figure 5.12 Conceptual diagram illustrating the two components of interaction turnover (β_{int}). The first is species turnover: interactions are lost or gained because of the loss or gain of species (β_{st}). The second is interaction rewiring: interactions are reassembled because of changes in who is interacting with whom; that is, the same species interact in different combinations across time (β_{rw}). Both components of interaction turnover can simultaneously occur from one transition to the next, but are shown separately for clarity. From CaraDonna et al. (2017), reprinted with permission.

respectively, we can partition interaction turnover into the contribution of species turnover (β_{st}) and interaction rewiring (β_{st}), $\beta_{int} = \beta_{st} + \beta_{rw}$ (Poisot et al. 2012; CaraDonna et al. 2017). When coupled with adaptive interaction switching models (Figure 2.8; see Sections 2.4 and 3.4.3) and higher-order zeta diversity (Hui and McGeoch 2014; McGeoch et al. 2019), we could further test whether the interaction rewiring is random or adaptive, and whether network turnover is a result of the gain and loss of specialised or generalised species or specific interactions (see Chapter 6).

As already formulated in Eq.(5.2) by Diamond and May (1977), the temporal turnover of species composition is scale dependent. The temporal turnover of network interactions and associated architectures is also scale dependent (Figure 5.13; Schwarz et al. 2020). Such scale dependency can be a result of both the scale dependency of interaction strength (Chapter 2) and the sampling effect of missing those species that are absent from both surveys but which occur temporally during the period between two surveys. For instance, Schwarz et al. (2020) discovered, besides the obvious sampling effect, strong scale dependency in the temporal turnover of network structures as a result of phenological turnover of species and links. Temporal turnover between two consecutive surveys will have a high level of shared species; the temporal turnover therefore largely reflects interaction rewiring over a short time span. In contrast, temporal turnover between two surveys with a large time lapse will normally have few shared species; the temporal turnover in this case therefore largely reflects the transition in species assemblage. Consequently, comparing

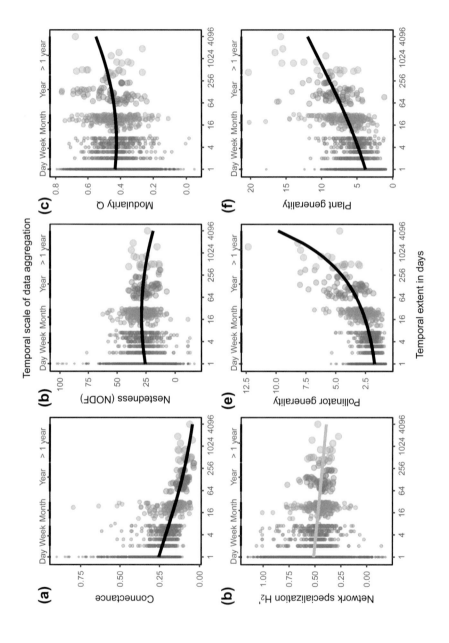

drivers of temporal turnover at different scales could help to differentiate drivers that are responsible for interaction rewiring from those that drive species turnover. The outcome and impact of biological invasions are temporally scale dependent, as reflected in the concepts of extinction debt (Tilman et al. 1994) and invasion debt (Rouget et al. 2016). Consequently, exploring network turnover at different temporal scales can help to elucidate short- to long-term responses to biological invasions and other drivers of change. Temporal turnover in ecological networks is, therefore, the most appropriate measure of network openness and transition. In the next section, we address the direction and magnitude of temporal turnover and consider network interactions and biological invasions as the main forces driving this turnover.

5.4 Weather Vane of Network Transitions

Temporal turnover is an assemblage-level summary of network transition. For conservation and invasion management, we need to know the response of all species in the recipient ecological network, both natives and aliens, to biological invasions. From an invader-centric perspective, we previously addressed issues relating to the performance and dynamics of a particular invader (Hui and Richardson 2017). In particular, we explored the velocity and dynamic behaviour of invasive species, the role of dispersal kernels, modes of invasive spread, the consequence of such non-equilibrium dynamics and the potential impacts and management strategies of targeted invaders in a community. Here, we focus instead on the future forecasting of an invaded ecological network, including all its components, not just a specific species of concern. Just as a weather vane (Figure 5.14) that actually reflects the direction of current wind can also be used for short-term wind pattern forecasting

---

Figure 5.13 Effects of the temporal scale of data aggregation on (a) connectance, (b) nestedness (NODF), (c) modularity Q, (d) network specialization, (e) pollinator generality and (f) plant generality. Trend lines are based on predictions of linear mixed models testing the effect of temporal extent as single and quadratic term. Black and grey trend lines indicate significant and no effects of temporal extent, respectively. Statistical fits for connectance, pollinator generality and plant generality are based on log-transformed data, whereas figure axes represent the scale of the original data. These partial residual plots correct for baseline differences among studies and sites (specified as random effects in linear mixed models), which occasionally leads to data points lying outside the range of the variable. From Schwarz et al. (2020), under CC-BY Licence.

Figure 5.14 A weather vane only indicates the direction of wind based on the pressure from airflow. It does not forecast wind dynamics but only responds to ambient wind at the time. However, because of the continuation of the airflow, it can be indicative of the wind direction for the short-term future. Photo from statticflickr.com under CC-BY Licence.

(because of the momentum and viscosity of air movement), the method we introduce here serves as a weather vane that is reflective and indicative of the current and short-term dynamics of system evolution and network dynamics. In Chapters 3 and 4 we detailed the changes to the assembly and structure of ecological networks resulting from invasions. Our aim here is to propose an approach for elucidating and therefore providing the means for predicting potential future dynamics in invaded ecological networks.

The dynamic behaviour of an ecological network can be depicted mathematically using dynamical systems of differential equations,

$$\dot{\boldsymbol{N}} = \boldsymbol{F}(\boldsymbol{N}), \tag{5.4}$$

where $\boldsymbol{N} = \langle N_1, N_2, \ldots \rangle^T$ is the abundance vector of resident species and $\langle \cdot \rangle^T$ its transpose; $\dot{\boldsymbol{N}}$ the time derivative of $\boldsymbol{N}$; $\boldsymbol{F}(\boldsymbol{N}) = \langle F_1(N), F_2(N), \ldots \rangle^T$ is a set of smooth, often non-linear, population growth functions. To predict the precise dynamics of an ecological network, we could simply solve $\boldsymbol{N}(t)$ in these differential equations from the current reference point of observation ($\boldsymbol{N}_0 = \boldsymbol{N}(0)$), either analytically or numerically. Although this is theoretically feasible, it does not reflect the reality facing ecologists, who often have only snapshots of species abundances and interaction strengths (or their proxies, e.g., based on the degree of trait matching) in ecological networks (Hui et al. 2016). The exact form of $\boldsymbol{F}$ is often unknown. As in many theoretical models, we could assume that $\boldsymbol{F}$ follows a specific form based on our knowledge of the study system or from first principles. The forecasted (or pre-dicted) dynamics of these systems, however, only represents the behaviour of the specific modelled ecosystem. This greatly reduces our ability to generalise results from specific forms of F to a wider range of ecosystems. As George Box (1976) argued, all such models are wrong as they cannot capture the reality of network structures and interactions, but some are useful and can provide efficient and robust forecasts of system evolution. This is exactly the intention of our proposed weather vane for the dynamics of invaded ecological net-works (Hui and Richardson 2019).

Without considering any specific forms of F in the generic network of Eq. (5.4), we can still approximate these functions around a point of observation N_0 as their tangent plane; we thus have

$$\dot{n} = F(N_0) + Jn, \tag{5.5}$$

where $n = N - N_0$ is the anticipated change of species abundances from the point of observation and J its Jacobian matrix, $J = \langle \partial F_i / \partial N_j |_{N=N_0} \rangle$. Note, N_0 does not have to be a particular equilibrium ($F(N_0) = 0$) for this linearization; however, linearization at an equilibrium point does greatly simplify the system,

$$\dot{n} = Jn. \tag{5.6}$$

As introduced in Chapter 2, the element on the i-th row and j-th column, $\partial F_i / \partial N_j |_{N=N_0}$, measures the sensitivity of species i's population growth rate to the abundance change of species j at this point of observation. In an ecological interaction network, this sensitivity at an equilibrium point is often used as the interaction strength. That is, at this ecological equilib-rium, we could consider the interaction strength matrix as the system Jacobian matrix (Figure 5.15A), $M = J$; the dynamics around this equilib-rium can thus be approximated by the linear system $\dot{n} = Mn$.

At this equilibrium point we can improve the precision of our predic-tion by considering a more exact approximation of system dynamics; for instance, considering the second-order approximation of F as

$$\dot{n} = F(N_0) + Jn + \frac{1}{2} I \otimes n^\mathsf{T} Hn, \tag{5.7}$$

where H is the vector formatted Hessian matrix of dimension $S^2 \times S$ consisting of S vertically concatenated symmetric matrices of $\partial^2 F_i / \partial N_k \partial N_l |_{N=N_0}$, depicting how the change in species k's abundance affects

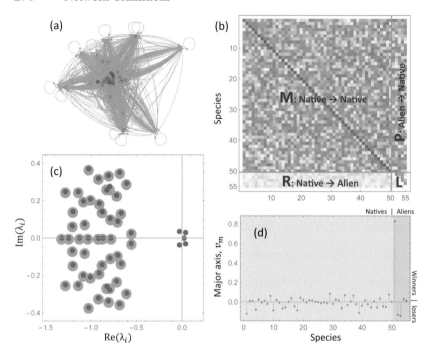

Figure 5.15 Dynamic responses of an ecological network to biological invasions. (a) The graph diagram of an ecological network of biotic interactions (50 species with interaction strengths randomly assigned from a normal distribution with zero mean and standard deviation $\sigma = 0.1$, with diagonal elements $-\mu = -1$). The network is illustrated by high-dimensional embedding using Wolfram Mathematica, with open black dots nodes and green directed lines pairwise interactions. (b) The adjacency matrix of the invaded network, with the orange lines dividing the matrix into four blocks: representing, respectively, the original native–native adjacency matrix M (top left; dimension: 50 by 50), invasion impacts on resident species (matrix P, top right; dimension: 5 by 50), biotic resistance to invasions (matrix R, bottom left, with $\sigma = 0.01$ for illustration; dimension: 50 by 5) and alien–alien interactions (matrix L, bottom right, with diagonal zeros for illustration; dimension: 5 by 5). The colours of matrix elements correspond to interaction strength; darker shading denotes stronger interactions. (c) Visualisation of all eigenvalues in the original network (50 green disks) and invaded network (55 red dots). Each eigenvalue is composed of a real part $\text{Re}(\lambda_i)$ and an imaginary part $\text{Im}(\lambda_i)$; the two parts of an eigenvalue allow us to locate it in the complex plane. Note, the real part of the lead eigenvalue for the original network (the rightmost green disk) is negative and becomes positive for the invaded network (the rightmost red dot), where the alien species define the lead eigenvalue because of its low abundance. (d) The major axis (i.e., the eigenvector of the lead eigenvalue) of the joint matrix (note, v_m in the figure is the same as v_1 in the text here), revealing how each species

the interaction strength of species l on species i. This provide a theoretical connection of higher-order interactions in ecological networks (Chapter 2). We could find no fully parameterised Hessian matrices for real ecological networks in the literature. Looking at the ecological feasibility of future forecasting using a parameterised system of Eq. (5.7), we need to estimate an excessively large number of parameters and initial conditions. In particular, we need to estimate the initial abundances of all involved species, N_0 with dimension $S \times 1$, and their initial population change rates $F(N_0)$, also with dimension $S \times 1$. Moreover, we also need to have estimates of the pairwise interaction strength as proxies for the elements of the system Jacobian J, with dimension $S \times S$, and the elements of the Hessian matrix with dimension $S^2 \times S$.

As ecological surveys are normally conducted only over short periods with a regular interval τ (e.g., annual or seasonal surveys), we can consider the equivalent Euler discrete-time form of the model by replacing $\dot{N} = (N_{t+\tau} - N_t)/\tau$ for the small time interval; that is, $\dot{N}_i = (N_{i,t+\tau} - N_{i,t})/\tau$ with $N_{i,t} = N_i(t)$. The system for $\tau = 1$ thus becomes $N_{t+1} = F(N_t) + IN_t$, with I the identity matrix. Considering demographic and environmental stochasticity during the course of one time step, we can describe the network using the following system equation,

$$N_{t+1} = \rho(F(N_t) + IN_t) + e_t, \qquad (5.8)$$

with ρ normally a Poisson process with $F(N_t) + IN_t$ as its mean to represent demographic stochasticity, and e_t environmental noise. We could therefore use generalised statistical models to estimate parameters of F and elements of its J and H. This follows a typical autoregressive statistical method for estimating interaction forms and strengths in an ecological network. Moreover, non-Poisson demographic stochasticity and coloured environmental noise that follows a specific covariance structure can also be considered. Much research is underway on this

---

Figure 5.15 (*cont.*) dynamically responds to the invasion (species 1 to 50 are the original resident species; species 51 to 55 marked with red circles are the newly introduced species). Note, introduced species 51 became highly invasive; species 52 and 53 experienced invasion failures; species 54 and 55 established but are yet not invasive. Some resident species responded positively, others negatively, and many resident species are insensitive to the invasion. From Hui and Richardson (2019), reprinted with permission. *A black and white version of this figure will appear in some formats. For the colour version, please refer to the plate section.*

topic; key focal areas are the elucidation of the effects of demographic and environmental stochasticity on the probability of invasive establishment, invasion performance and species coexistence (e.g., Cuddington and Hastings 2016; Legault et al. 2019). Our focus here is on linearized systems, but we do discuss some results from non-linear systems.

In an ecological network, the population dynamics of different residing species are co-varying, not independently from each other; this is exactly the silver line that defines interaction strengths between species. This, however, implies that in the S-dimensional vector space of $\boldsymbol{n}$, these reference axes are not orthogonal to each other. To reduce this interdependence, we need to use the orthogonal projection; this approach is related to the matrix ordination that is widely used by ecologists (Legendre and Legendre 1998). In particular, let λ_i be the i-th eigenvalue of the Jacobian, and $\boldsymbol{v}_i$ and $\boldsymbol{w}_i$ its corresponding right- and left-hand eigenvectors; that is, $\boldsymbol{J}\boldsymbol{v}_i = \boldsymbol{v}_i\lambda_i$ and $\boldsymbol{w}_i^T\boldsymbol{J} = \lambda_i\boldsymbol{w}_i^T$. We can rank these eigenvalues according to the values of their real parts: $\mathrm{Re}(\lambda_i) \geq \mathrm{Re}(\lambda_j)$ for $i \leq j$; that is, λ_1 has the largest real part among all eigenvalues. If we assign these eigenvalues as the diagonal elements of a matrix and zeros as the non-diagonal elements, we have a diagonal matrix known as the eigenvalue matrix, $\boldsymbol{\Lambda} = \mathrm{diag}(\lambda_i)$. If we simply put these right-hand eigenvectors together, we have the right-hand eigenvector matrix, $\boldsymbol{V} = \langle \boldsymbol{v}_1, \boldsymbol{v}_2, \ldots \rangle$. By definition, we have $\boldsymbol{J}\boldsymbol{V} = \boldsymbol{V}\boldsymbol{\Lambda}$ and therefore $\boldsymbol{J} = \boldsymbol{V}\boldsymbol{\Lambda}\boldsymbol{V}^{-1}$ (equivalently, $\boldsymbol{V}^{-1}\boldsymbol{J}\boldsymbol{V} = \boldsymbol{\Lambda}$), with $\boldsymbol{V}^{-1}$ simply the inverse matrix of the right-hand eigenvector matrix. If we put these left-hand eigenvectors together, we have the left-hand eigenvector matrix $\boldsymbol{W} = \langle \boldsymbol{w}_1, \boldsymbol{w}_2, \ldots \rangle^T$. By definition, we have $\boldsymbol{W}\boldsymbol{J} = \boldsymbol{\Lambda}\boldsymbol{W}$ and therefore $\boldsymbol{J} = \boldsymbol{W}^{-1}\boldsymbol{\Lambda}\boldsymbol{W}$ or $\boldsymbol{W}\boldsymbol{J}\boldsymbol{W}^{-1} = \boldsymbol{\Lambda}$.

We can now distort the abundance vector space into a new one using the eigenvector matrix. Looking at the pre-invasion linearized network, $\dot{\boldsymbol{n}} = \boldsymbol{M}\boldsymbol{n}$, the interrelatedness of the ecological network is reflected by the adjacency matrix $\boldsymbol{M}$, which is not a diagonal matrix. Let $\boldsymbol{n} = \boldsymbol{V}\boldsymbol{\varepsilon}$, and $\boldsymbol{\varepsilon}$ the new vector space. This transformation is necessary because the dynamics of each species are not independent from the others (the issue of multi-collinearity). After this transformation, each transformed node behaves independently from others, and we have $\dot{\boldsymbol{\varepsilon}} = \boldsymbol{\Lambda}\boldsymbol{\varepsilon}$ and thus $\varepsilon_i = c_i e^{\lambda_i t}$, where $c_i = \varepsilon_i(0)$ is a constant and represents the initial condition. Consequently, we have the dynamics of species i,

$$N_i = N_i^* + \sum_j v_{ji} c_j e^{\lambda_j t}, \qquad (5.9)$$

where v_{ji} is the i-th element of the eigenvector v_j. With the elapse of time t, we can see that

$$n_i \rightarrow v_{1i}c_1 e^{\lambda_1 t}, \tag{5.10}$$

where λ_1 is the leading eigenvalue (the one with the largest real part). It is evident that only if the real part of the lead eigenvalue is negative, $\text{Re}(\lambda_1) < 0$ (Figure 5.15C), do we have an asymptotically stable network.

When an ecological network is invaded by a number of alien species, we can formulate this invaded network using a joint dynamical system, $\dot{Z} = F'(Z)$, where $Z = \langle N, A \rangle^T$ is the joint abundance vector of native species N and alien species A. Assuming the pre-introduction community is settled at its ecological equilibrium $(F(N^*) = 0)$, after linearization at the point of introduction $(n = N - N^*, a = A - A^*$, with $A^* = 0$ the marginal equilibrium representing the absence of biological invasions), we have $\dot{z} = M'z$, where the joint adjacency matrix M' includes four submatrices (Figure 5.15B),

$$\begin{bmatrix} \dot{n} \\ \dot{a} \end{bmatrix} = \begin{bmatrix} M & P \\ R & L \end{bmatrix} \begin{bmatrix} n \\ a \end{bmatrix}, \tag{5.11}$$

where M represents the original native–native adjacency matrix with its element on the i-th row and j-th column representing the interaction strength of native species j on native species i; P, the alien–native matrix with its element representing the interaction strength of alien species j on native species i; R, the native–alien matrix with its element representing the interaction strength of native species j on alien species i; L, the alien–alien matrix with its element representing the interaction strength of alien species j on alien species i. To assess which alien species can invade or be repelled, and which resident species are favoured or suppressed as a result of such multispecies biological invasions, we need to know the dynamics of this invaded network near the point of invasion.

As stated in Chapter 4, invasion performance and invasibility are related to the loss of network stability, more precisely to network instability. We have assumed that the pre-invasion network is settling around its equilibrium, while biological invasions (with at least one successful invasion) can push the system away from this equilibrium. This implies that the real part of the leading eigenvalue of the joint matrix M' is positive $(\text{Re}(\lambda_1') > 0)$ at the point of introduction $(n = N - N^*, a = A - A^*$, with $A^* = 0)$. When an abundance vector is unstable, the abundance of species i will move away from the current abundance vector, representing anticipated temporal turnover from biological invasions. We have Eq. (5.9) and Eq. (5.10) which have the same form, but for the invaded ecological network (e.g., $n_i \rightarrow v'_{1i}c'_1 e^{\lambda'_1 t}$), while the alien species dynamics follow the following trends,

$$A_k = \sum_j v'_{jk}c'_j e^{\lambda'_j t}, \tag{5.12}$$

and therefore over time the invasion dynamics thus approach

$$A_k \rightarrow v'_{1k}c'_1 e^{\lambda'_1 t}. \tag{5.13}$$

Importantly, at this point of invasion the response of species i, both its direction and magnitude, to the invasion is determined solely by v'_{1i} of the eigenvector of the leading eigenvalue of the joint matrix $\boldsymbol{M}'$), noting again that $c'_1 e^{\lambda'_1 t}$ are the same for all species. Therefore, the response of each species to the invasion is revealed by the eigenvector of the leading eigenvalue of the adjacency matrix calculated for the standing abundances of resident species and the initial abundances of alien species vector.

In ecological terms, eigenvector $\boldsymbol{v}'_1$ is the first principal component (PC1) from a principal component analysis (PCA) of the joint adjacency matrix $\boldsymbol{M}'$, and its i-th element v'_{1i} is the coordinate of species i along PC1 axis, representing the projection of species i's vector on the principal component. PC1 can be readily calculated once the adjacency matrix is available; for instance, by running the default R function *prcomp* for $\boldsymbol{M}'$. If $v'_{1i} > 0$, species i's current population size is expected to increase; if $v'_{1i} < 0$, species i will converge back to its original equilibrium. if $v'_{1i} \approx 0$, species i is insensitive to biological invasions (Figure 5.15D). Therefore, the first principal component (PC1, i.e., the major axis) of the adjacency matrix for the invaded network simply reveals how each species, native or alien, responds dynamically to the biological invasion. Consequently, it is the 'weather vane' of the temporal turnover in an invaded ecological network.

If the pre-invasion network is already in transition, $\boldsymbol{F}(\boldsymbol{N}_0) \neq 0$, and the dynamics of this ecological network near this reference point of $\boldsymbol{n}_0$ is therefore,

$$\begin{bmatrix} \dot{n} \\ \dot{a} \end{bmatrix} = \boldsymbol{g}_0 + \begin{bmatrix} \boldsymbol{M} & \boldsymbol{P} \\ \boldsymbol{R} & \boldsymbol{L} \end{bmatrix} \begin{bmatrix} \boldsymbol{n} \\ \boldsymbol{a} \end{bmatrix}, \tag{5.14}$$

where $\boldsymbol{g}_0 = (\boldsymbol{F}(\boldsymbol{N}_0), \boldsymbol{g}_a)^T$. At this non-equilibrium point, we can still consider the invaded system moving away from the current trend of dynamic transitions $(F(N_0))$; g_a is the background rate of alien propagule introduction. In such a case, biological invasions can steer the community succession away from its original transient path. Following the same procedure of using the eigenvector matrix of the joint Jacobian matrix as an orthogonal operator, $z = V\varepsilon$; that is, the new state variables are linear combinations of the original abundance variables $(V^{-1}z = \varepsilon)$. In this orthogonal space, the network dynamics becomes, $\dot{\varepsilon} = b + \Lambda\varepsilon$ (that is, $\dot{\varepsilon}_i = b_i + \lambda_i\varepsilon_i$), where $b = V^{-1}g_0$ with b_i the vector's i-th element. This means that, $\varepsilon_i = c_i e^{\lambda_i t} + b_i t$ and therefore,

$$N_i(t) - N_i(0) = g_{0i}t + \sum_j v'_{ji}c'_j e^{\lambda'_j t}. \tag{5.15}$$

If there is at least one successful invader $(\mathrm{Re}(\lambda'_1) > 0)$, the population change rate will shift from following the pre-invasion and propagule pressure trend, g_0, to move gradually towards the vector direction of the weather vane. If none of the invaders can establish $(\mathrm{Re}(\lambda'_1) < 0)$, the system will eventually revert to its pre-invasion dynamic trend. Once again, the eigenvector of the leading eigenvalue points at the direction of change. The major axis, the eigenvector of the leading eigenvalue is therefore, again, the weather vane of network transitions.

A concept that is closely related to our weather vane metaphor for network transition is the notion of transient dynamics in matrix population models (Caswell 2018). Transient dynamics are non-equilibrium dynamics in complex systems that are not operating near a stable regime (Hastings 2001, 2004). It is intriguing to explain why most, or at least a large portion of, complex systems show prolonged transient dynamics, which are by definition of short duration and ephemeral and therefore not likely to be observed (Hastings et al. 2018). We should anticipate that most observed systems behave according to their asymptotic dynamics. A matrix population model for stage structured populations, can be formulated as the following for future projection,

$$n_{t+1} = An_t. \tag{5.16}$$

After m steps of iteration, the population structures can be estimated as $n_{t+m} = A^m n_t$. Ezard et al. (2010) provided a succinct summary of a number of indices that can be computed based on the leading eigenvalue and its eigenvectors of matrix A. In particular, the asymptotic population growth is the dominant eigenvalue λ_1 and is also related to the intrinsic rate of growth $(r = \ln(\lambda_1))$. The right eigenvector of λ_1 gives the stable distribution of each stage; the left eigenvector of λ_1 is the reproductive value of each stage. Sensitivity of λ_1 to a matrix element, $\partial\lambda_1/\partial a_{ij}$, measures the influence of a specific stage to the population growth. Elasticity of λ_1 to a matrix element, $\partial\log(\lambda_1)/\partial\log(a_{ij})$, measures the relative influence of a specific interaction to the long-term network growth rate. Damping ratio $\rho = \lambda_1/|\lambda_2|$ is the speed of convergence to the stable age-distribution, where λ_2 is the eigenvalue with the second largest real part. Stott et al. (2011) further reviewed a number of indices particularly designed to expose the short- and long-lasting impacts of transient dynamics on population dynamics, including reactivity, attenuation, amplification and inertia (Figure 5.16). Some of these metrics can be borrowed in exploring ecological networks. For instance, using the reactivity metric, Chen and Cohen (2001) showed that when food webs approached their stability margin, the size and duration of transient response to perturbation increases (critical slowing down). However, the value of matrix population models has been questioned by Crone et al. (2011) because forecasts of long-term asymptotic dynamics often assume that transient dynamics can be extended indefinitely. Further developments in matrix projection models, such as incorporating non-linear perturbation analysis, have been proposed (Stott et al. 2012; Stott 2016). Transient dynamics analyses in matrix projection models are great assets for understanding and predicting invasion dynamics during the early phases of naturalisation and establishment, which are by definition in transient mode (Chapter 1; also Hui and Richardson 2017).

There has been an overwhelming focus on the leading eigenvalue and its role in mediating network stability in the literature, and little attention has been paid to the major axis of the network adjacency matrix. More work is needed to elucidate what is and what determines the major axis (v_m) of the network adjacency matrix. The eigenvector of the leading eigenvalue has two layers of meaning. First, it indicates the direction of the steepest decline in the effective potential of a dynamical system (Martín et al. 2015), analogous to the path of water flow over uneven terrain. Second, in terms of signal processing, the network adjacency matrix can be considered the cross-covariance matrix that converts the signal of relative abundances into the signal of population change rates; the eigenvector v_m is simply a signal that will not be distorted from the converter but will only be amplified by a factor of λ_m. Along this principal component the covariance between the abundance and the population change rate of a species becomes the greatest. In fact, we could potentially estimate the adjacency matrix of interaction strength as

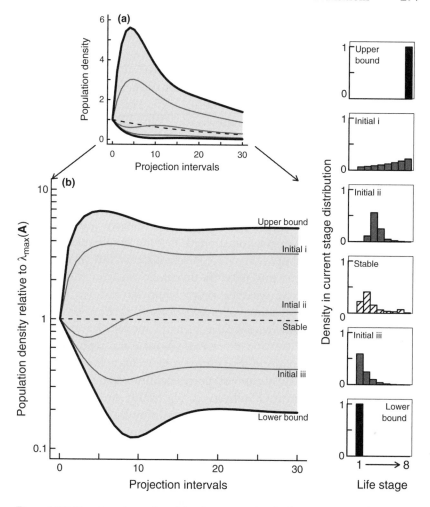

Figure 5.16 Transient dynamics of the desert tortoise *Gopherus agassizii* with medium fecundity (Doak et al. 1994). (a) Absolute population dynamics, including both transient and asymptotic influences; (b) standardised transient dynamics, excluding the influence of asymptotic growth. All demographic distributions are scaled, so that overall initial population density equals 1. Bold lines and black barplots indicate transient bounds (note that sometimes, unlike in this case, different stage-biased projections define the amplification envelope at different times in the projection); thin lines and grey barplots indicate case-specific initial demographic distributions; dashed lines and hatched barplots indicate the stable demographic distribution. Areas shaded in light grey indicate the transient envelope, which is the range of values in which all case-specific projections will lie. From Stott et al. (2011), reprinted with permission.

the cross-covariance matrix of species abundances and species change rates, with the element on the i-th row and j-th column being the covariance between species i's change rate and species j's abundance, $cov(r_i, n_j)$. Indeed, the diagonal element of the matrix, $cov(r_i, n_i)$, is essentially the growth-density covariance that has been shown to greatly affect species persistence (Chesson et al. 2005; Hui et al. 2017). Our proposal to use the major axis as the weather vane for invaded networks highlights that biological invasions are driven by the demographic dynamics of abundance and population growth (Gurevitch et al. 2011). This systematically breaks away from invasion assessment based solely on trait matching (Carboni et al. 2016; Van Kleunen et al. 2010). We are certainly not suggesting that traits are irrelevant, but rather that they should be assessed in terms of their impacts on density and population growth, so as to assess interaction strength beyond simple trait matching (Lurgi et al. 2014; Cadotte et al. 2015; Romanuk et al. 2017). We discuss how to incorporate traits into network invasibility analyses in Chapter 7.

We anticipate precise weather vanes being devised for ecological networks with specific forms of interactions (i.e., $\boldsymbol{F}$). For instance, Rossberg (2013) considered a Lotka–Volterra model of a competitive community; let $\mu = E(a_{ij})$ and $\sigma^2 = Var(a_{ij})$, for $i \neq j$, representing the mean and variance of interspecific competition strength. Rossberg (2013) showed the dynamic equilibrium of species richness can be approximated as $S^* = (1 - \mu^2)/(2\sigma^2)$. O'Sullivan et al. (2019) further considered meta-community structures to allow species to be linked with constant flows of propagules between local networks, deriving a similar result. Following the same line, Rossberg and Barabás (2019) criticised our proposal of a weather vane for network transitions from biological invasions. They argued that the generic model in Eq. (5.11) could be too crude, and that a specific, more constrained, network model, $\dot{\boldsymbol{N}} = \boldsymbol{N}f(\boldsymbol{N})$, might be more appropriate. They considered a network with the first S_1 species natives and the next S_2 species aliens ($S = S_1 + S_2$). At the point of invasion, the network of S_1 native species are set at a positive stable equilibrium, and the aliens at the marginal equilibrium, $(\boldsymbol{N}^*, 0)$. They provided four propositions regarding the shape of the Jacobian matrix and its eigenvalues and eigenvectors. In particular, the eigenvalues of the post-invasion network are the same as the eigenvalues of the pre-invasion network, plus those per-capita growth rates of aliens, while the eigenvectors of the post-invasion network are the S_1 pre-invasion eigenvectors extended with zeros for their last S_2 elements, and S_2 vectors of all zeros but with one non-zero in the last S_2 elements. This, they suggest, makes the weather vane trivial and implies that the initial per-capita growth rate of aliens is the driver of the leading eigenvalue of the invaded network. Total abundance will remain the same in an invaded network but abundance varies randomly among species (Rossberg and Barabás 2019),

$$\text{Var}(\boldsymbol{N} - \boldsymbol{N}^*) = \frac{N_A^2 \sigma^2}{(1 - \mu^2) - S\sigma^2}. \tag{5.17}$$

Note that when $S\sigma^2$ approaches $(1 - \mu^2)$, the system behaves erratically, signalling the onset of structural instability. This is true for ecological networks following $\dot{\boldsymbol{N}} = \boldsymbol{N}f(\boldsymbol{N})$ and is congruent with our proposal of network transition at marginal instability. However, our view, in responding to Rossberg and Barabás (2019), is that this constrained form does not accommodate the openness of an invaded ecological network, as evidently $\dot{\boldsymbol{N}} = 0$ when $\boldsymbol{N} = 0$ (Hui and Richardson 2019b). Consequently, the propositions do not apply to the generic case of Eq. (5.14). In particular, the rate of introduction was not formulated in this specific network, $\dot{\boldsymbol{N}} = \boldsymbol{N}f(\boldsymbol{N})$. We will revisit this issue when we discuss meta-networks in Chapter 6, and we provide a tentative solution for an open adaptive network in Chapter 7. Similar to the pros and cons of matrix population modelling and its associated transient dynamics analysis, we think the weather vane of network transition has merit to be both crude but sensitive in capturing the network transitions and turnover. If the eigenvector of the leading eigenvalue can be considered as the weather vane of an invaded ecological network, the Philosopher's Stone for network resilience and dynamics is the leading eigenvalue. We now turn to capture this Philosopher's Stone of an invaded ecological network.

5.5 Peculiarity of Rarity

It is evident that drivers controlling the leading eigenvalue, and therefore its associated eigenvectors, mediate not only network resilience but also the direction and magnitude of network transition and turnover. Several clues point to rare species, especially those that lack self-regulation, determining the leading eigenvalue λ_m. For instance, consider an ecological network depicted by a generic Lotka–Volterra model, $\dot{N}_i = N_i\left(r_i + \sum_j a_{ij}N_j\right)$. At an equilibrium $\boldsymbol{N}^*$ the Jacobian matrix is simply $\boldsymbol{J} = \boldsymbol{AD}$, where $\boldsymbol{A} = \{a_{ij}\}$ and $\boldsymbol{D} = \text{diag}(\boldsymbol{N}^*)$. This Jacobian matrix can be further partitioned into a diagonal matrix $\boldsymbol{B} = \text{diag}(a_{ii}N_i^*)$ and a zero-diagonal matrix

$C = \{a_{ij}N_i^*\}_{i\neq j}$ as $J = B + C$. According to the perturbative expansion (Kato 1995), $\lambda_i(J) \approx \lambda_i(B) + v_i^T C v_i$, where v_i is the eigenvector of B and has only one non-zero element and thus $v_i^T C v_i = C_{ii} = 0$. Therefore, $\lambda_i(J) \approx a_{ii}N_i^*$. If all species have a standard level of self-regulation ($a_{ii} = -1$), the leading eigenvalue is thus $\lambda_m \approx -\min_i(N_i^*)$ (Suweis et al. 2013; Stone 2018; Nnakenyi et al. 2021); that is, the rarest species determines the leading eigenvalue and thus the network resilience ($-\mathrm{Re}(\lambda_m)$). More specifically, the leading eigenvalue of an ecological network is controlled by rare species (small N_i^*), especially those also lacking density-dependent self-regulation (negative a_{ii} but close to zero). This peculiar role of rare species in an ecological network highlights their important functional roles (Lyons et al. 2005). This view echoes the findings from an increasing number of research that found that perturbations to rare species can have the most profound effect on system stability, besides the obvious vulnerability of rare species to extinction that also attracts attention of conservationists (Johnson 1998; Matthies et al. 2004; Enquist et al. 2019; Säterberg et al. 2019). However, although endangered species are by definition rare, not all rare species are close to extinction (Mace and Lande 1991). Introduced species are often initially rare, while many successful invaders do seem to lack density regulation, or disturbance and the novel environment has loosened the self-regulation mechanism. Due to such peculiar roles of rare species to stability, biological invasions of rare alien propagules can easily flip an invaded ecological network from stable to unstable (Figure 5.15C). Understanding how rare species are regulated in a community will, therefore, explain how introduced species fare in such networks.

Besides their role in dictating system stability and resilience, rare species are also important in maintaining ecosystem functionality. Flather and Sieg (2007) summarised three prevailing views on the functional roles in ecosystems (Figure 5.17): (i) the complementarity hypothesis argues that niche differentiation between species can lead to a reduction of ecosystem functions from any species loss (Chapin et al. 1998; Loreau et al. 2001; van Ruijven et al. 2003); (ii) the redundancy hypothesis suggests that species' functions are substitutable or often nested (Schwartz et al. 2000; Hector et al. 2001); and (iii) the facilitative hypothesis implies an acceleration of ecosystem functionality with biodiversity because of a faster increase in the ways of biotic interactions anticipated from the increase in the number of species (Cardinale et al. 2002). It is, however, difficult to assess the role of rare versus common

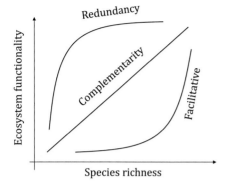

Figure 5.17 Three alternative perspectives on the relationship between biodiversity (species richness) and ecosystem functionality. Redrawn from Flather and Sieg (2007).

species in these hypotheses, as the sequence of species addition or removal is often unspecified in tests. As He and Hubbell (2011) demonstrated, it is more difficult to extirpate a species from habitat loss than to discover a species from adding an equal amount of lost habitat; consequently, estimating species gain or loss from habitat loss should follow the endemic area relationship rather than the typical species accumulation curve. Similarly, Leitão et al. (2016) found that losing rare species could lead to the more severe reduction of functional specialisation, richness and community originality than was the case with random species loss (Figure 5.18).

There is a gradient, a staircase, of commonness and rarity in any ecological community. Not only do we need to understand the mechanisms that generate this gradient but also the drivers of waxing and waning of species on this staircase. While an alien species can climb this gradient to become common and invasive, an abundant species may step down the same gradient to become rare and extinct. Rarity is the door to both entering and exiting an ecological community (Figure 1.11; Colautti et al. 2017). If the leading eigenvalue is the door, then its eigenvector – the weather vane – is the staircase of species commonness-rarity in an ecological network. The peculiarity of rarity, therefore, also means that rare species often lack persistence, and the gain and loss of rare species contributes excessively to the observed network transition and turnover. Indeed, Schoener and Spiller (1987) showed that the probability of persistence drops faster in rarer species, suggesting not only shorter periods of each stay but also more frequent flickering

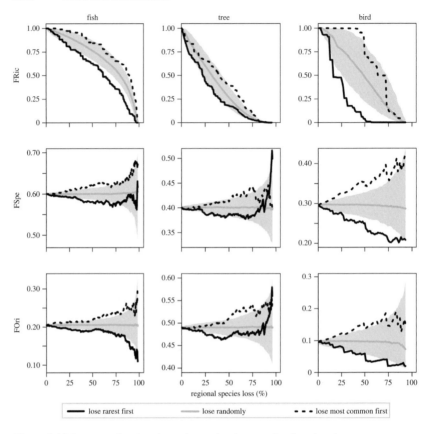

Figure 5.18 Impact of regional species extinction on the functional structure (functional richness, FRic; specialization, FSpe; and originality, FOri) of three tropical assemblages: stream fish from the Brazilian Amazon, rainforest trees from French Guiana and birds from the Australian Wet Tropics. Rarest species sequential loss (black solid line) is compared with the opposite scenario where most common species are lost first (black dashed line) and with a null scenario simulating a random sequential extinction (grey line indicates the median of this scenario among 1,000 replicates and the 95% confidence interval is represented as the shaded area). As a result of computation constraints, the maximum species removals from the regional pools of fish, trees and birds were, respectively, 99%, 96% and 93%. From Leitão et al. (2016), reprinted with permission.

(Figure 5.19). Recent studies have confirmed that species with small population sizes are responsible for the persistent temporal turnover in ecological networks (Figure 5.20; McCollin 2017). This association between rarity and temporal turnover prompted Hanski (1982) to conjecture the core–satellite hypothesis that dichotomises resident species,

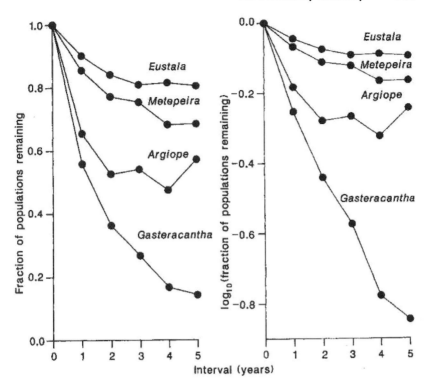

Figure 5.19 Population persistence curves for the four commonest orb-spider species on 108 islands of the central Bahamas. Curves give the fraction of populations remaining after *n* years on an arithmetic (left) and semilogarithmic (right) scale. From Schoener and Spiller (1987), reprinted with permission.

where core species are abundant and dominant species that persist through perturbation and invasions, from satellite species that are dynamically unstable. Similarly, in an ecological network there is also a persistent core–periphery structure, with periphery species (having low linkage level in a network) also experience stronger variations in persistence (Figure 5.21; Olesen et al. 2011). However, this core–periphery structure does not necessarily align perfectly with the commonness–rarity gradient. Miele et al. (2020) revealed a persistent core–periphery structure in plant–pollinator networks; although they found the structural position of a species highly dynamic (Figure 5.22), many species remained peripheral, while some temporally became core in the network. This means that the weather vane is not constant but highly dynamic, in this case reflecting the seasonal fluctuation of temporal niches.

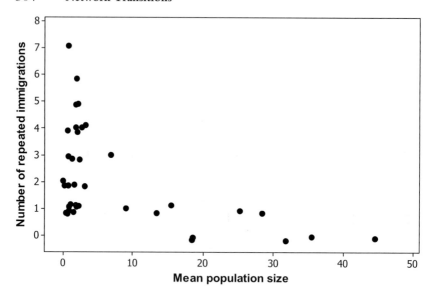

Figure 5.20 Number of repeated immigrations per species versus mean population size. Most species involved in turnovers have small population sizes. Data have been jittered to display multiple species at the same points. From McCollin (2017), under CC-BY Licence.

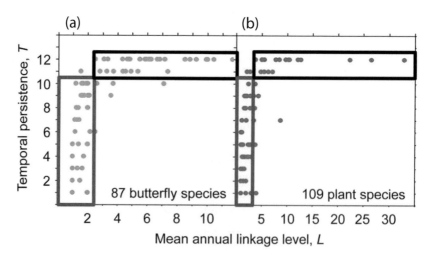

Figure 5.21 Relationship between temporal persistence of species and their linkage level. Relationship between temporal persistence T (number of years observed) of butterfly species (a) and plant species (b) and their mean annual linkage level L. Two non-overlapping regions with respect to T and L were distinguished: The grey frame areas are regions of specialists, varying a lot in their T, and the black ones are regions of stable species, varying a lot in their L. From Olesen et al. (2011), under CC-BY Licence.

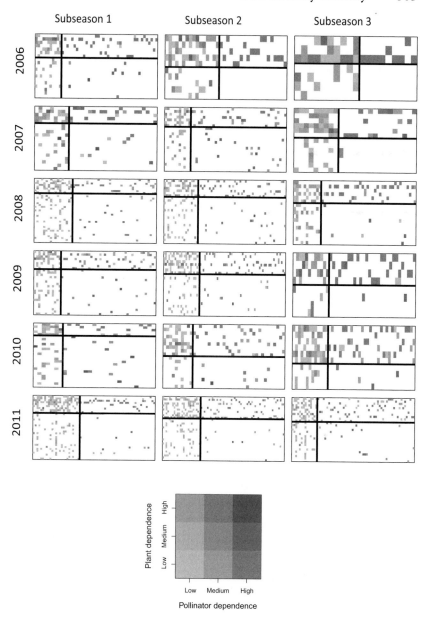

Figure 5.22 Temporal dynamics of Villavicencio plant–pollinator network for three sub-seasons of the flowering season for six study years (2006–2011). For each matrix, cells represent the plant and pollinator dependence values between a plant (rows) and pollinator (columns) species, with the grey level computed as a mixture of the two dependence values according to the legend. Dark lines separating each matrix delineate the group boundaries (core/peripheral group of plants above/below the horizontal line; core/peripheral group of pollinators on the left/right of the vertical line). From Miele et al. (2020), under CC-BY Licence.

If rare species hold the key to network instability and invaded networks undergo rapid temporal turnover, are there any functional traits associated with rarity? For instance, a typical rare species has a slow life pace (e.g., low growth rate, small litter size, long generation time, few reproductive episodes; McKinney 1997; Pilgrim et al. 2004). High specialisation, high trophic level, large habitat requirement, low vagility and large body size have also been suggested as other associates of rarity (McKinney 1997; Purvis et al. 2000; Flather and Sieg 2007). As rare species that rapidly increase their population sizes capture the key feature of invasiveness, these rarity traits are largely opposite to those invasive traits (Van Kluenen and Richardson 2007). Other non-functional traits, such as taxon age and size, niche and range size, are also strong covariates of both rarity and invasiveness. For instance, diverse taxa of plants and insects often have more rare species than would be the case if species were arranged along a range size/abundance continuum (rare to abundant and widespread) randomly (Schwartz and Simberloff 2001; Ulrich 2005), while the opposite is true for birds and mammals (Purvis et al. 2000). Younger lineages (e.g., *Acacia* species) have a larger portion of highly invasive species than the evolution-arily older eucalypts (*Eucalyptus* and closely related genera) (Miller et al. 2017), while large range size is positively correlated with local commonness and invasiveness (Hui et al. 2014). Although these non-functional traits are good predictors at the macroecological scale, they are, however, not good predictors within an ecological network. Linking interaction pressure received by and imposed from resident species probably explains more the commonness–rarity gradient within a local ecological network. Of course, resource and disturbance levels in the environment certainly also mediate species rarity within a local community (Harper 1977; Boughton and Malvadkar 2002; Pärtel et al. 2005).

Despite the peculiar role of rare species in dictating network invasibility and transition and profiling invasive traits, both common and rare species are essential to the functioning of ecosystems (Whittaker 1965; Smith and Knapp 2003; Gaston 2010; McGeoch and Latombe 2016; Chapman et al. 2018; Arnoldi et al. 2019). The weather vane (the eigenvector of the leading eigenvalue), according to the transient dynamics of the matrix projection model, provides the stable species abundance distribution in an ecological network. It is a staircase or gradient of commonness and rarity. Considering the role of abundance (and its change) to both v_m and λ_m, we need to revisit Rabinowitz's (1981) classification of species rarity in three dimensions: geographical range, local density and niche/habitat specificity (also see a recent extension that quantifies species commonness and spread potential; McGeoch and Latombe 2016). Violle et al. (2017) further highlighted the necessity to consider both species rarity (scarcity) and functional distinctiveness (functional rarity). We discuss the role of novel and rare traits in invaded ecological networks in Chapter 7. We highlight the need to explore commonness–rarity gradients in ecological networks and their covariates. In particular, we need to clarify the relationships among demography, traits and abundance in local networks, and the role of environmental heterogeneity on these relationships in meta-networks. Moreover, because interaction specificity can greatly interfere with the cascading effect of species removal and invasion in ecological networks (Bosc et al. 2018, 2019), we also need to consider the third dimension of rarity – interaction specificity (and related interaction promiscuity) – when exploring these potential covariates with and effects on the rarity staircase.

5.6 Advice from a Caterpillar

Turnover, instability, abundance (rarity) and interaction networks are intertwined features of an open ecological network. In each community, there is a gradient of commonness and rarity: the commonness of rarity and the rarity of commonness. That is, many or most species are actually rare, while only a few are common and abundant. The relative abundance distributions are therefore largely highly skewed, following a lognormal-like curve. This gradient can be measured in three dimensions – abundance, range and niche width; this means that there are seven forms of rarity (Rabinowitz 1981). In her classic paper, Rabinowitz (1981) proposed the classification of rare species and contemplated the causal relationship between species' competitiveness and their population size. She concluded that competitiveness is not a regulator of population size, but mainly a strategy to offset the disadvantage of being locally rare. Looking at the Jacobian at the beginning of Section 5.5, $J = B + C$, we see that each diagonal element (say, $a_{ii}N_i^*$) in the diagonal matrix B represents the strength of intraspecific interaction. The corresponding row sum and column sum of elements in C represent, respectively, the pressure of interspecific interactions experienced and imposed by species i in the network. These interspecific pressures need to be balanced with the strength of intraspecific self-regulation. Abundance is therefore a badge of demographic success in an ecological network. Consequently, rare species hold the key to network instability and invasibility, while the commonness–rarity gradient, captured by the weather vane, gives us the direction and magnitude of temporal turnover. These are features of an open adaptive system in transition, learning and coping with its standing context. An ecological network under biological invasions experiences persistent transitions and turnover at marginal instability. At this criticality, the open adaptive system has tentatively maximised its entropy production (Martyushev and Seleznev 2006; Chakrabarti and Ghosh 2009, 2010; Andrae et al. 2010; Frank 2017; Muñoz 2018), and consequently such persistent transition represents a physical attractor of system evolution.

Until now we have argued that relationships between network topology and invasion performance could arise from constraints on network function and assembly (such as the stability criterion and its many variates), and that the consequence of invasion in an ecological network is related to its instability when the standing resident species are confronting initially rare alien incursions. Each species within the ecological network responds dynamically to invasions, while a weather vane exists to indicate how each species behaves in such an invaded ecological network and can be statistically estimated as the major axis of the network adjacency matrix. When the invaded network functions as a linear system, the major axis is equivalent to the eigenvector v_m of the leading eigenvalue λ_m of the Jacobian matrix, and thus could indicate both the short-term transient and long-term asymptotic dynamics. However, when the invaded network is non-linear, the major axis could indicate only short-term transient dynamics; forecasting medium- to long-term dynamics in non-linear systems also requires information of at least the Hessian matrix. For short-term invasion prediction and risk assessment, the major axis can serve as a weather vane of the invaded network. As the invaded network is unstable at the point of successful invasion, the major axis of the adjacency matrix indicates how the dynamical system will spontaneously and instantaneously deviate from the current standing point when facing invasions, thereby revealing the dynamic trajectory of each species (native or alien) involved in the network (Figure 5.15D). We have returned to the concept of eigenvalues and eigenvectors to describe complex dynamic systems. This tool is applied in almost all fields of modern science, following its inaugural use in celestial mechanics when Cauchy (1829) worked on the equations to determine the secular inequalities in the movement of planets. Sylvester (1852) and Poincaré (1894) brought this tool to the calculation of matrix determinants that quickly transformed modern science. Hilbert (1904) connected the tool to a list of integral equations that is related to modelling many invasion dynamics (Hui and Richardson 2017). Hilbert used *eigenwerte* and *eigenschaften* to refer to eigenvalues; *eigen* can be understood from its literal meaning of self, proper and characteristic of a complex system.

We are on an interesting journey, exploring how open adaptive systems transform through persistent transitions and turnover in response to biological invasions and co-evolution. At this stage you, the reader, may be feeling a little giddy and uncomfortable, like Alice in Wonderland. '[B]eing so many different sizes in a day is very confusing', she complains. 'It isn't', replies the Caterpillar of the absurd world (or the master of system transformation – metamorphosis; Figure 5.23). Alice,

Figure 5.23 The Caterpillar. One of John Tenniel's illustrations from the first edition of Alice's Adventures in Wonderland (1865). The illustration is noted for its ambiguous central figure, whose head can be viewed as being a human male's face with pointed nose and protruding lower lip or being the head end of an actual caterpillar, with the right three 'true' legs visible. ('And do you see its long nose and chin? At least, they look exactly like a nose and chin, don't they? But they really are two of its legs. You know a Caterpillar has got quantities of legs: you can see more of them, further down.' [Carroll, Lewis. *The Nursery 'Alice'*. Dover Publications (1966), p. 27]). In Public Domain.

not one to give up easily, puts her case. 'Well, [...] when you have to turn into a chrysalis [...] and then after that into a butterfly, I should think you'll feel it a little queer, won't you?' 'Not a bit', retorts the Caterpillar. Indeed, dealing with open system transformation is a crucial new frontier in ecology. The advice from the Caterpillar to Alice is, 'Keep your temper', 'You'll get used to it in time', and 'One side [of the mushroom] will make you grow taller, and the other side will make you grow shorter'. Eigenvalues and eigenvectors are the 'temper' and 'mushroom' of network metamorphosis.

References

Andrae B, et al. (2010) Entropy production of cyclic population dynamics. *Physical Review Letters* 104, 218102.

Armstrong JS (2009) *Principles of Forecasting*. Boston: Springer.

Arnoldi J, Loreau M, Haegeman B (2019) The inherent multidimensionality of temporal variability: How common and rare species shape stability patterns. *Ecology Letters* 22, 1557–1567.

Beckage B, Gross LJ, Kauffman S (2011) The limits to prediction in ecological systems. *Ecosphere* 2, 1–12.

Benincà E, et al. (2015) Species fluctuations sustained by a cyclic succession at the edge of chaos. *Proceedings of the National Academy of Sciences USA* 112, 6389–6394.

Boettiger C, Hastings A (2012) Quantifying limits to detection of early warning for critical transitions. *Journal of the Royal Society Interface* 9, 2527–2539.

Boettiger C, Ross N, Hastings A (2013) Early warning signals: The charted and uncharted territories. *Theoretical Ecology* 6, 255–264.

Bosc C, et al. (2018) Interactions among predators and plant specificity protect herbivores from top predators. *Ecology* 99, 1602–1609.

Bosc C, et al. (2019) Importance of biotic niches versus drift in a plant-inhabiting arthropod community depends on rarity and trophic group. *Ecography* 42, 1926–1935.

Boughton D, Malvadkar U (2002) Extinction risk in successional landscapes subject to catastrophic disturbances. *Ecology and Society* 6, 2.

Box GEP (1976) Science and statistics. *Journal of the American Statistical Association* 71, 791–799.

Brown JH, Kodric-Brown A (1977) Turnover rates in insular biogeography: Effect of immigration on extinction. *Ecology* 58, 445–449.

Cadotte MW, et al. (2015) Predicting communities from functional traits. *Trends in Ecology & Evolution* 30, 510–511.

Capinha C, et al. (2018) Models of alien species richness show moderate predictive accuracy and poor transferability. *NeoBiota* 38, 77–96.

CaraDonna PJ, et al. (2017) Interaction rewiring and the rapid turnover of plant–pollinator networks. *Ecology Letters* 20, 385–394.

Carboni M, et al. (2016) What it takes to invade grassland ecosystems: Traits, introduction history and filtering processes. *Ecology Letters* 19, 219–229.

Cardinale BJ, Palmer MA, Collins SL (2002) Species diversity enhances ecosystem functioning through interspecific facilitation. *Nature* 415, 426–429.

Carpenter SR, Brock WA (2006) Rising variance: A leading indicator of ecological transition. *Ecology Letters* 9, 311–318.

Carpenter SR, et al. (2011) Early warnings of regime shifts: A whole-ecosystem experiment. *Science* 332, 1079–1082.

Carpenter SR, et al. (2008) Leading indicators of trophic cascades. *Ecology Letters* 11, 128–138.

Carroll L (1865) *Alice's Adventures in Wonderland*. London: Macmillan.

Caswell H, Vindenes Y (2018) Demographic variance in heterogeneous populations: Matrix models and sensitivity analysis. *Oikos* 127, 648–663.

Cauchy A (1829) Sur la resolution des équivalences don't lest modules se réduisent à des nombres premiers. *Exercices de Mathematiques* 4, 174–195.

Chakrabarti CG, Ghosh K (2009) Non-equilibrium thermodynamics of ecosystems: Entropic analysis of stability and diversity. *Ecological Modelling* 220, 1950–1956.

Chakrabarti CG, Ghosh K (2010) Maximum-entropy principle: Ecological organization and evolution. *Journal of Biological Physics* 36, 175–183.

Chapin FS, et al. (1998) Ecosystem consequences of changing biodiversity. *BioScience* 48, 45–52.

Chapman ASA, Tunnicliffe V, Bates AE (2018) Both rare and common species make unique contributions to functional diversity in an ecosystem unaffected by human activities. *Diversity and Distributions* 24, 568–578.

Chen X, Cohen JE (2001) Global stability, local stability and permanence in model food webs. *Journal of Theoretical Biology* 212, 223–235.

Chesson P, et al. (2005) Scale transition theory with special reference to species coexistence in a variable environment. In Holyoak M, Leibold MA, Holt RD (eds.), *Metacommunities: Spatial Dynamics and Ecological Communities*, pp. 279–306. Chicago: The University of Chicago Press.

Colautti RI, et al. (2017) Invasions and extinctions through the looking glass of evolutionary ecology. *Philosophical transactions of the Royal Society B: Biological Sciences* 372, 20160031.

Cortez MH, Abrams PA (2016) Hydra effects in stable communities and their implications for system dynamics. *Ecology* 97, 1135–1145.

Crone EE, et al. (2011) How do plant ecologists use matrix population models? *Ecology Letters* 14, 1–8.

Cuddington K, Hastings A (2016) Autocorrelated environmental variation and the establishment of invasive species. *Oikos* 125, 1027–1034.

Dakos V, et al. (2011) Slowing down in spatially patterned ecosystems at the brink of collapse. *The American Naturalist* 177, E153–E166.

Dakos V, van Nes EH, Scheffer M (2013) Flickering as an early warning signal. *Theoretical Ecology* 6, 309–317.

Dana ED, et al. (2019) Common deficiencies of actions for managing invasive alien species: A decision-support checklist. *NeoBiota* 48, 97–112.

Davis MA (2003) Biotic globalization: Does competition from introduced species threaten biodiversity? *BioScience* 53, 481–489.

Diamond JM (1969) Avifaunal equilibria and species turnover rates on the channel islands of California. *Proceedings of the National Academy of Sciences USA* 64, 57–63.

Diamond JM, May RM (1977) Species turnover rates on islands: Dependence on census interval. *Science* 197, 266–270.

Ditlevsen PD, Johnsen SJ (2010) Tipping points: Early warning and wishful thinking. *Geophysical Research Letters* 37, L19703.

Doak D, Kareiva P, Kleptka B (1994) Modeling population viability for the desert tortoise in the Western Mojave Desert. *Ecological Applications* 4, 446–460.

Dornelas M, Madin JS (2020) Novel communities are a risky business. *Science* 370, 164–165.

Drake JM, Griffen BD (2010) Early warning signals of extinction in deteriorating environments. *Nature* 467, 456–459.

Elliot-Graves A (2016) The problem of prediction in invasion biology. *Biology & Philosophy* 31, 373–393.

Enquist BJ, et al. (2019) The commonness of rarity: Global and future distribution of rarity across land plants. *Science Advances* 5, eaaz0414.

Evans JM, Wilkie AC, Burkhardt J (2008) Adaptive management of nonnative species: Moving beyond the "either-or" through experimental pluralism. *Journal of Agricultural and Environmental Ethics* 21, 521–539.

Ezard THG, et al. (2010) Matrix models for a changeable world: The importance of transient dynamics in population management. *Journal of Applied Ecology* 47, 515–523.

Flather CH, Sieg CH (2007) Species rarity: Definition, causes, and classification. In Raphael MG, Molina R (eds.), *Conservation of Rare or Little-Known Species: Biological, Social, and Economic Considerations*, pp. 40–66. Washington: Island Press.

Fletcher DH, et al. (2016) Predicting global invasion risks: A management tool to prevent future introductions. *Scientific Reports* 6, 26316.

Fournier A, et al. (2019) Predicting future invaders and future invasions. *Proceedings of the National Academy of Sciences USA* 116, 7905–7910.

Frank SA (2017) Universal expressions of population change by the Price equation: Natural selection, information, and maximum entropy production. *Ecology and Evolution* 7, 3381–3396.

Gaertner M, et al. (2014) Invasive plants as drivers of regime shifts: Identifying high-priority invaders that alter feedback relationships. *Diversity and Distributions* 20, 733–744.

Gaston KJ (2010) Valuing common species. *Science* 327, 154–155.

Gilbert FS (1980) The equilibrium theory of island biogeography: Fact or fiction? *Journal of Biogeography* 7, 209–235.

Gilpin ME, Diamond JM (1982) Factors contributing to non-randomness in species co-occurrences on islands. *Oecologia* 52, 75–84.

Gould SJ (1989) *Wonderful Life: The Burgess Shale and the Nature of History*. New York: Norton & Company.

Gravel D, et al. (2011) Trophic theory of island biogeography. *Ecology Letters* 14, 1010–1016.

Green KC, Armstrong JS (2015) Simple versus complex forecasting: The evidence. *Journal of Business Research* 68, 1678–1685.

Gurevitch J, et al. (2011) Emergent insights from the synthesis of conceptual frameworks for biological invasions. *Ecology Letters* 14, 407–418.

Hanski I (1982) Dynamics of regional distribution: The core and satellite species hypothesis. *Oikos* 38, 210–221.

Hanski I, Gyllenberg M (1997) Uniting two general patterns in the distribution of species. *Science* 275, 397–400.

Harper JL (1977) *Population Biology of Plants*. London: Academic Press.

Hastings A (2001) Transient dynamics and persistence of ecological systems. *Ecology Letters* 4, 215–220.

Hastings A (2004) Transients: The key to long-term ecological understanding? *Trends in Ecology & Evolution* 19, 39–45.

Hastings A, et al. (2018) Transient phenomena in ecology. *Science* 361, eaat6412.

Hastings A, Wysham DB (2010) Regime shifts in ecological systems can occur with no warning. *Ecology Letters* 13, 464–472.

He F, Hubbell SP (2011) Species–area relationships always overestimate extinction rates from habitat loss. *Nature* 473, 368–371.

Hector A, et al. (2001) Conservation implications of the link between biodiversity and ecosystem functioning. *Oecologia* 129, 624–628.

Heino J, Melo AS, Bini LM (2015) Reconceptualising the beta diversity-environmental heterogeneity relationship in running water systems. *Freshwater Biology* 60, 223–235.

Hewitt JE, et al. (2016) Species and functional trait turnover in response to broad-scale change and an invasive species. *Ecosphere* 7, ee01289.

Hilbert D (1904) Über eine Anwendung der Integralgleichungen auf eine Problem der Funktionentheorie. In Krazer A (ed.), *Verhandlungen des 3. Internationalen Mathematiker-Kongresses: in Heidelberg vom 8. bis 13*, pp. 233–240. Leipzig: Teubner.

Hirota M, et al. (2011) Global resilience of tropical forest and savanna to critical transitions. *Science* 334, 232–235.

Hirsch H, Richardson DM, Le Roux JJ (2017). Tree invasions: Towards a better understanding of their complex evolutionary dynamics. *AoB Plants* 9, plx014.

Hui C, McGeoch MA (2014) Zeta diversity as a concept and metric that unifies incidence-based biodiversity patterns. *The American Naturalist* 184, 684–694.

Hui C, Richardson DM (2017) *Invasion Dynamics*. Oxford: Oxford University Press.

Hui C, Richardson DM (2019a) How to invade an ecological network. *Trends in Ecology & Evolution* 34, 121–131.

Hui C, Richardson DM (2019b) Network invasion as an open dynamical system: Response to Rossberg and Barabás. *Trends in Ecology & Evolution* 34, 386–387.

Hui C, Fox GA, Gurevitch J (2017) Scale-dependent portfolio effects explain growth inflation and volatility reduction in landscape demography. *Proceedings of the National Academy of Sciences USA* 114, 12507–12511.

Hui C, et al. (2014) Macroecology meets invasion ecology: Performance of Australian acacias and eucalypts around the world foretold by features of their native ranges. *Biological Invasions* 16, 565–576.

Hui C, et al. (2016) Defining invasiveness and invasibility in ecological networks. *Biological Invasions* 18, 971–983.

Johnson CN (1998) Species extinction and the relationship between distribution and abundance. *Nature* 394, 272–274.

Kahneman D (2013) *Thinking, Fast and Slow.* New York: Farrar, Straus and Giroux.

Kato M (1995) Numerical analysis of the Nernst-Planck-Poisson system. *Journal of Theoretical Biology* 177, 299–304.

Legault G, et al. (2019) Demographic stochasticity alters expected outcomes in experimental and simulated non-neutral communities. *Oikos* 128, 1704–1715.

Legendre P, Legendre L (1998) *Numerical Ecology*, 2nd edition. Amsterdam: Elsevier.

Leitão RP, et al. (2016) Rare species contribute disproportionately to the functional structure of species assemblages. *Proceedings of the Royal Society B: Biological Sciences* 283, 20160084.

Lenton TM, et al. (2012) Early warning of climate tipping points from critical slowing down: Comparing methods to improve robustness. *Philosophical transactions of the Royal Society A: Mathematical, Physical and Engineering Sciences* 370, 1185–1204.

Lindegren M, et al. (2012) Early detection of ecosystem regime shifts: A multiple method evaluation for management application. *PLoS ONE* 7, e38410.

Loreau M, et al. (2001) Biodiversity and ecosystem functioning: Current knowledge and future challenges. *Science* 294, 804–808.

Low-Décarie E, Chivers C, Granados M (2014) Rising complexity and falling explanatory power in ecology. *Frontiers in Ecology and the Environment* 12, 412–418.

Lurgi M, et al. (2014) Network complexity and species traits mediate the effects of biological invasions on dynamic food webs. *Frontiers in Ecology and Evolution* 2, 36.

Lyons KG, et al. (2005) Rare species and ecosystem functioning. *Conservation Biology* 19, 1019–1024.

MacArthur RH, Wilson EO (1967) *The Theory of Island Biogeography.* Princeton: Princeton University Press.

Mace GM, Lande R (1991) Assessing extinction threats: Toward a re-evaluation of IUCN threatened species categories. *Conservation Biology* 5, 148–157.

Martín PV, et al. (2015) Eluding catastrophic shifts. *Proceedings of the National Academy of Sciences USA* 112, E1828–E1836.

Martyushev LM, Seleznev VD (2006) Maximum entropy production principle in physics, chemistry and biology. *Physics Reports* 426, 1–45.

Matthies D, et al. (2004) Population size and the risk of local extinction: Empirical evidence from rare plants. *Oikos* 105, 481–488.

May RM (1976) Simple mathematical models with very complicated dynamics. *Nature* 261, 459–467.

McCollin D (2017) Turnover dynamics of breeding land birds on islands: Is island biogeographic theory 'true but trivial' over decadal time-scales? *Diversity* 9, 3.

McGeoch MA, Latombe G (2016) Characterizing common and range expanding species. *Journal of Biogeography* 43, 217–228.

McGeoch MA, et al. (2019) Measuring continuous compositional change using decline and decay in zeta diversity. *Ecology* 100, ee02832.

McKinney ML (1997) Extinction vulnerability and selectivity: Combining ecological and paleontological views. *Annual Review of Ecology and Systematics* 28, 495–516.

Miele V, Ramos-Jiliberto R, Vázquez DP (2020) Core–periphery dynamics in a plant–pollinator network. *Journal of Animal Ecology* 89, 1670–1677.

Milkoreit M, et al. (2018) Defining tipping points for social-ecological systems scholarship: An interdisciplinary literature review. *Environmental Research Letters* 13, 033005.

Miller JT, et al. (2017) Is invasion success of Australian trees mediated by their native biogeography phylogenetic history or both? *AoB Plants* 9, plw080.

Muñoz MA (2018) Criticality and dynamical scaling in living systems. *Reviews of Modern Physics* 90, 031001.

Nnakenyi CA, Hui C, Minoarivelo HO, Dieckmann U (2021) Leading eigenvalue of block-structured random matrices. In prep.

O'Sullivan JD, Knell RJ, Rossberg AG (2019) Metacommunity-scale biodiversity regulation and the self-organised emergence of macroecological patterns. *Ecology Letters* 22, 1428–1438.

Olesen JM, Stefanescu C, Traveset A (2011) Strong, long-term temporal dynamics of an ecological network. *PLoS ONE* 6, e26455.

Pärtel M, et al. (2005) Grouping and prioritization of vascular plant species for conservation: combining natural rarity and management need. *Biological Conservation* 123, 271–278.

Perretti CT, Munch SB (2012) Regime shift indicators fail under noise levels commonly observed in ecological systems. *Ecological Applications* 22, 1772–1779.

Pettersson S, Savage VM, Jacobi MN (2020) Predicting collapse of complex ecological systems: Quantifying the stability–complexity continuum. *Journal of the Royal Society Interface* 17, 20190391.

Pilgrim ES, Crawley MJ, Dolphin K (2004) Patterns of rarity in the native British flora. *Biological Conservation* 120, 161–170.

Poincaré H (1894) Sur la théorie cinétique des gaz. *Revue générale des sciences pures et appliquées* 5, 513–521.

Poisot T, et al. (2012) The dissimilarity of species interaction networks. *Ecology Letters* 15, 1353–1361.

Purvis A, et al. (2000) Nonrandom extinction and the loss of evolutionary history. *Science* 288, 328–330.

Pyšek P, Richardson DM (2007). Traits associated with invasiveness in alien plants: Where do we stand? In Nentwig W (ed.), *Biological Invasions*, pp. 97–125. Berlin: Springer.

Rabinowitz D (1981) Seven forms of rarity In Synge H (ed.), *The Biological Aspects of Rare Plant Conservation*, pp. 205–217. New York: John Wiley & Sons Ltd.

Rittel HW, Webber MM (1973) Dilemmas in a general theory of planning. *Policy Sciences* 4, 155–169.

Romanuk TN, et al. (2017) Robustness trade-offs in model food webs: Invasion probability decreases while invasion consequences increase with connectance. *Advances in Ecological Research* 56, 263–291.

Rossberg AG (2013) *Food Webs and Biodiversity: Foundations, Models, Data*. Chichester: John Wiley & Sons Ltd.

Rossberg AG, Barabás G (2019) How carefully executed network theory informs invasion ecology. *Trends in Ecology & Evolution* 34, 385–386.

Rouget M, et al. (2016) Invasion debt: Quantifying future biological invasions. *Diversity and Distributions* 22, 445–456.

Sabin P (2013) *The Bet: Paul Ehrlich, Julian Simon, and Our Gamble over Earth's Future.* New Haven: Yale University Press.

Sakai AK, et al. (2001) The population biology of invasive species. *Annual Review of Ecology and Systematics* 32, 305–332.

Säterberg T, et al. (2019) A potential role for rare species in ecosystem dynamics. *Scientific Reports* 9, 11107.

Scheffer M, et al. (2001) Catastrophic shifts in ecosystems. *Nature* 413, 591–596.

Scheffer M, et al. (2009) Early-warning signals for critical transitions. *Nature* 461, 53–59.

Scheffer M, et al. (2012) Anticipating critical transitions. *Science* 338, 344–348.

Schoener T, Spiller D (1987) High population persistence in a system with high turnover. *Nature* 330, 474–477.

Schooler SS, et al. (2011) Alternative stable states explain unpredictable biological control of *Salvinia molesta* in Kakadu. *Nature* 470, 86–89.

Schwartz MW, Simberloff D (2008) Taxon size predicts rates of rarity in vascular plants. *Ecology Letters* 4, 464–469.

Schwartz MW, et al. (2000) Linking biodiversity to ecosystem function: Implications for conservation ecology. *Oecologia* 122, 297–305.

Schwarz B, et al. (2020) Temporal scale-dependence of plant-pollinator networks. *Oikos* 129, 1289–1302.

Seastedt TR (2014) Biological control of invasive plant species: A reassessment for the Anthropocene. *New Phytologist* 205, 490–502.

Seekell DA, et al. (2013) Evidence of alternate attractors from a whole-ecosystem regime shift experiment. *Theoretical Ecology* 6, 385–394.

Shackleton RT, et al. (2018) Social-ecological drivers and impacts of invasion-related regime shifts: Consequences for ecosystem services and human well-being. *Environmental Science and Policy* 89, 300–314.

Smith MD, Knapp AK (2003) Dominant species maintain ecosystem function with non-random species loss. *Ecology Letters* 6, 509–517.

Staver AC, Archibald S, Levin SA (2011) The global extent and determinants of savanna and forest as alternative biome states. *Science* 334, 230–232.

Stokstad E (2009) On the origin of ecological structure. *Science* 326, 33–35.

Stone L (2018) The feasibility and stability of large complex biological networks: a random matrix approach. *Scientific Reports* 8, 8246.

Stott I (2016) Perturbation analysis of transient population dynamics using matrix projection models. *Methods in Ecology and Evolution* 7, 666–678.

Stott I, Hodgson DJ, Townley S (2012) Beyond sensitivity: Nonlinear perturbation analysis of transient dynamics. *Methods in Ecology and Evolution* 3, 673–684.

Stott I, Townley S, Hodgson DJ (2011) A framework for studying transient dynamics of population projection matrix models. *Ecology Letters* 14, 959–970.

Strayer DL, et al. (2006) Understanding the long-term effects of species invasions. *Trends in Ecology & Evolution* 21, 645–651.

Suweis S, et al. (2013) Emergence of structural and dynamical properties of ecological mutualistic networks. *Nature* 500, 449–452.

Sylvester JJ (1852) A demonstration of the theorem that every homogeneous quadratic polynomial is reducible by real orthogonal substitutions to the form of a sum of positive and negative squares. *Philosophical Magazine* IV, 138–142.

Tetlock PE (2005) *Political Judgment: How Good Is It? How Can We Know?* Princeton: Princeton University Press.

Tilman D, et al. (1994) Habitat destruction and the extinction debt. *Nature* 371, 65–66.

Ulrich W (2005) Regional species richness of families and the distribution of abundance and rarity in a local community of forest Hymenoptera. *Acta Oecologia* 28, 71–76.

Van Kleunen M, Richardson DM (2007) Invasion biology and conservation biology: Time to join forces to explore the links between species traits and extinction risk and invasiveness. *Progress in Physical Geography* 3, 447–450.

Van Kleunen M, Weber E, Fischer M (2010) A meta-analysis of trait differences between invasive and non-invasive plant species. *Ecology Letters* 13, 235–345.

Van Ruijven J, Berendse F (2003) Positive effects of plant species diversity on productivity in the absence of legumes. *Ecology Letters* 6, 170–175.

Veraart AJ, et al. (2012) Recovery rates reflect distance to a tipping point in a living system. *Nature* 481, 357–359.

Violle C, et al. (2017) Functional rarity: The ecology of outliers. *Trends in Ecology & Evolution* 32, 356–367.

Walters C, Kitchell JF (2001) Cultivation/depensation effects on juvenile survival and recruitment: Implications for the theory of fishing. *Canadian Journal of Fisheries and Aquatic Sciences* 58, 39–50.

Wang R, et al. (2012) Flickering gives early warning signals of a critical transition to a eutrophic lake state. *Nature* 492, 419–422.

Whittaker RH (1960) Vegetation of the Siskiyou Mountains, Oregon and California. *Ecological Monographs* 30, 279–338.

Whittaker RH (1965) Dominance and diversity in land plant communities. *Science* 147, 250–260.

Williamson GB (1978) A comment on equilibirium turnover rates for islands. *The American Naturalist* 112, 241–243.

Woodford DJ, et al. (2016) Confronting the wicked problem of managing biological invasions. *NeoBiota* 31, 63–86.

Wright SJ (1985) How isolation affects rates of turnover of species on islands. *Oikos* 44, 331–340.

Zhang F, Dieckmann U, Metz JAJ, Hui C (2021) Condition for evolutionary branching and trapping in multidimensional adaptive dynamics. In prep.

6 · *Network Scaling*

The essence of spatial ecology is that the spatial structure of ecological interactions affects populations as much as do average birth and death rates, competition and predation.

Hanski (1998)

6.1 The Rise of Alien Biomes

Maps of global biomes or ecoregions show geographical clusters – unique assemblages of plants and animals that are spatially tied with associated geomorphologic and climatic features. Biomes are typically defined on the basis of broad vegetation types and the biophysical features that impose fundamental controls on the distribution of plants (Cox and Moore 2000). The concept of biomes has a deep history in ecology and has experienced waves of knowledge synthesis, reaching a recent consensus of seven points (Mucina 2019), one of which caught our attention: 'A biome incorporates a complex of fine-scale biotic communities; it has its characteristic flora and fauna and it is home to characteristic vegetation types and animal communities.' Macro-scale biodiversity patterns, therefore, reflect the overarching geophysical structures of the globe such as the well-known latitudinal gradients of biodiversity (Willig et al. 2003) and associated ecosystem functioning (e.g., litter decomposition in streams via detritivores; Boyero et al. 2015). Nevertheless, within constantly changing environments, the species composition and geographical boundaries of biomes (called ecotones) are not fixed, but are fluid over evolutionary timescales (Haywood et al. 2019). This biodiversity–environment coupling has been disrupted by agriculture and urbanisation, and the appetite of humans for resources and raw materials and their carelessness in handling waste. Humans are steadily altering land cover and modifying ecological processes across the globe, creating a new ecological order of anthropogenic biomes (anthromes;

sensu Ellis and Ramankutty 2008). Natural biomes are facing unprecedented pressures to change, shift, dissolve, merge and emerge, at a pace on par with the most tumultuous periods of the biosphere's history.

Biological invasions add a new layer of biological reshuffling to natural biomes and anthromes, with alien species acting both as drivers and passengers of global change. Humans have been moving organisms around the world intentionally or accidentally for centuries (Elton 1958; Seebens et al. 2017), and many species have established self-sustaining populations outside their native ranges defined by biogeographic barriers (van Kleunen et al. 2010; Richardson 2011). Some alien species become invasive (i.e., spread widely) and some of these have profound effects in recipient ecosystems (e.g., Gaertner et al. 2009; Vilà et al. 2011). As discussed in Chapter 4, this grand reshuffling of species has transformed, and to some extent homogenised, regional landscapes in many parts of the world. The spatial patterns and dynamics of these non-native species ['alien biomes' (Rouget et al. 2015) or 'alien [plant] species assemblage zones' (Richardson et al. 2020)], and their interplay with natural biomes and anthromes over regional scales, are receiving increasing attention (Pyšek et al. 2005; Hui et al. 2013; Richardson et al. 2020). The interactions among native species, non-native ones and humans in natural biomes, anthromes and alien species assemblages can alter these spatial clusters of species and reshape our perceptions of the structure and functioning of these novel ecosystems (Hobbs et al. 2009; Hui 2021).

Most invasion hypotheses, however, aim to explain the invasion success of single species in novel environments. An increasing number of studies are highlighting the need to consider alien species as a labelled group in invaded ecosystems and to gradually and meticulously determine how these labelled species interact, function and become assimilated in recipient ecosystems (Roura-Pascual et al. 2016; Vizentin-Bugoni et al. 2021). Based on key drivers of biological invasions synthesised from such studies (Catford et al. 2009), Rouget et al. (2015) proposed six biogeographical hypotheses pertaining to alien biomes (Table 6.1), with human activities and propagule pressure included as drivers of global change and introduction dynamics. For instance, strong biotic interactions between alien and native species during the invasion process should generate a clear signal in large-scale spatial patterns. The absence of any correspondence between alien and native community patterns would suggest that other processes such as introduction history are at play (Something In the Way You Move, Table 6.1), or that a mix of many different processes specific to each particular species are influential

Table 6.1 *Hypotheses regarding alien biomes (definitions and discussions follow Rouget et al. 2015; Hui et al. 2020).*

Hypothesis	Definition
The weed-shaped hole (Mack et al. 2000; Buckley et al. 2007; Rouget et al. 2015)	Invasions are facilitated by certain levels of disturbance; these may be anthropogenic or natural (and so an inherent function of the invasibility of an ecosystem).
The biome decides (Alpert et al., 2000; Rouget et al. 2015)	Native communities differ inherently in invasibility, such that there is selectivity regarding which invasive species can and cannot invade.
Goldilocks (Rouget et al. 2015)	Native communities and invasive species share broadly similar abiotic requirements, such that they occupy the same niches.
A new world order (regime shift) (Simberloff and Von Holle 1999; Sax et al. 2007; Rouget et al. 2015)	A new set of boundaries is formed by suites of interacting native and alien species. These new associations lead to novel ecosystems that need not be similar in nature to those in native communities.
Something in the way you move (Wilson et al. 2009; Donaldson et al. 2014; Rouget et al. 2015)	Distribution patterns in invasions are dominated by drivers associated with introduction histories of alien species.
Random tessellation (Daleo et al. 2009; Rouget et al. 2015)	The distributions of alien species is inherently idiosyncratic. When many species are considered, it appears that the relative clustering of distributions is simply constrained by geometry.
Decreasing stochasticity (Hui et al. 2013)	Stochasticity is especially important at the early stages of invasions, and older assemblages (comprising species that arrived in the environment a long time ago) should display more deterministic composition than new ones.

(Random Tessellation, Table 6.1). Testing these hypotheses is largely based on comparisons of biodiversity patterns: species richness of native and alien species, compositional and functional turnover in natural and alien biomes. Inferring processes from patterns is, nonetheless, challenging, as different processes can produce the same patterns, and clear signals in observed patterns may be generated by either biotic (the Biome Decides, Table 6.1) or abiotic selection processes (Goldilocks, Table 6.1). It is therefore important to account for the effects of both abiotic and biotic processes when interpreting the spatial organisation of biodiversity.

Although anthropogenic drivers clearly influence the establishment and spread of alien species (Pyšek and Richardson 2006; Pyšek et al. 2010), the broad-scale distributions of some alien biomes can largely be explained by the same environmental factors that delineate biomes based on native vegetation. These broad environmental factors serve as ecological barriers, dividing the environmental space into distinct clusters with characteristic native and alien species confined within them (Cowling and Pressey 2001; Rouget et al. 2015), as is evident in Figure 6.1 for the alien biomes defined by naturalised plants in South Africa (Figure 6.1). In contrast, Stotz et al. (2020) found that the same species show different patterns of association depending on where they are – whether they are in their native or non-native ranges. That is, novel ecosystems are assembled by origin-dependent associations, where alien-rich patches exist within a mosaic of native-dominated communities. Nonetheless, what is clear is that human-mediated introductions are reshuffling species at regional and global scales, thereby altering both regional biotas and their interaction networks (Figure 6.2; Fricke and Svenning 2020), leading to a New World Order (Table 6.1). Biological invasions break down the natural geographic bounds that confine spatial clusters of species, and establish new, often supergeographic clusters of species (Figure 6.3; Capinha et al. 2015), shifting the dominant drivers from distance and climate to also include anthropogenic factors especially relating to trade volumes of selected commodities.

As established in Chapter 3, biological invasions not only reshuffle species composition but also rejig associated network structures and ecosystem services. To cope with the new environments and contexts in invaded ecological networks, successful invaders are often generalists that can utilise a broad base of resources. They often also have high interaction fidelity and promiscuity. For instance, in plant-pathogen networks, biological invasions can increase linkage density as successful alien pathogens are typically generalists (Figure 6.4; Andreazzi et al. 2017; Bufford et al. 2020). Consequently, the level of nestedness is also expected to increase, while the level of modularity will decline as a result of the negative nestedness–modularity relationship. However, it is important to note that, in most of such comparative studies, multiple ecological networks can be described within one local community. For instance, invaded networks considering both alien and native species, pre-invasion networks (different from invaded networks with alien species removed), networks of alien species, and network interactions between only alien species, between alien and native species and between

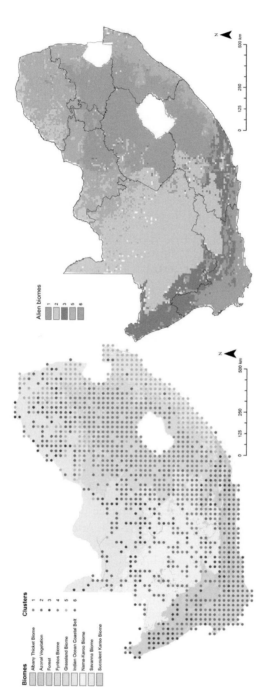

Figure 6.1 (Left) Clusters of invasive alien plant species in relation to biomes defined on the basis of natural vegetation (shading). Clusters were derived based on presence/absence of invasive alien species at the resolution of 15-minute grid cells (shown as circles). (Right) The potential distribution of alien biomes. Environmental determinants of the distribution of invasive alien species clusters were identified from a classification tree and were mapped over the full range of environmental conditions in South Africa. From Rouget et al. (2015), reprinted with permission. *A black and white version of this figure will appear in some formats. For the colour version, please refer to the plate section.*

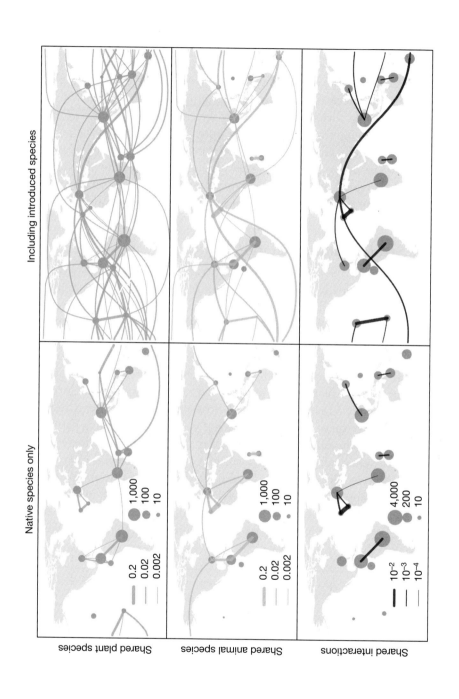

Native species only   Including introduced species

Shared plant species

1,000
100
10

0.2
0.02
0.002

Shared animal species

1,000
100
10

0.2
0.02
0.002

Shared interactions

4,000
200
10

10^{-2}
10^{-3}
10^{-4}

only native species; each such sub-network can be structured differently. As we will see, biological invasions not only transform ecological networks of local communities but also affect spatial structures of biodiversity in recipient regional ecosystems. We look at the details of the patterns, processes, dynamics and stability of meta-networks. We start by exploring the possible effects of biological invasions on the scaling patterns of spatial meta-networks that are interlinked by the exchange and spread of propagules. We then introduce the stability criterion and weather vane of meta-networks, emphasising the roles of both biotic interaction and dispersal. To connect the dots of diverse network scaling patterns and to elucidate the impacts of biological invasions to structural emergence in spatial networks, we end by proposing a physical model of network percolation at criticality.

6.2 Spatial Scaling Patterns

Complex systems, such as alien biomes and associated meta-networks, often exhibit spatial scaling patterns. In other words, the structure and function of an ecological network change with the spatial scales (grain and extent) at which they are quantified. Such scale dependence in structure and function creates both problems and opportunities when making predictions and inferences cross spatial scales. Let us explore the spatial scaling patterns of different biodiversity currencies and network structures; how such scaling patterns can tentatively reveal the propensity of alien species to invade; and how recipient ecosystems respond and rearrange themselves in space. First, we clarify the scaling pattern of the distribution range of a single species and its demographic performance. We then address the association, co-occurrence and interactions between two species. Next, we discuss the scaling patterns of multi-species compositional turnover and dissimilarity that result in the typical species–area curve and the scale dependence of different biodiversity currencies (e.g., alpha and beta) and the number of differently labelled

←——————————————————————————

Figure 6.2 Shared species and biotic interactions across regions in the presence and absence of anthropogenic species introductions. Points represent the number of species or interactions observed in each bioregion or oceanic island system, with the point's coordinates at the centroid of the study locations for each region. The thickness of lines, plotted along circles, shows the proportion of shared species or interactions between bioregions or oceanic island systems. Note the long arcs in the right column connecting Europe to New Zealand to the east and the Azores to New Zealand to the west. From Fricke and Svenning (2020), reprinted with permission.

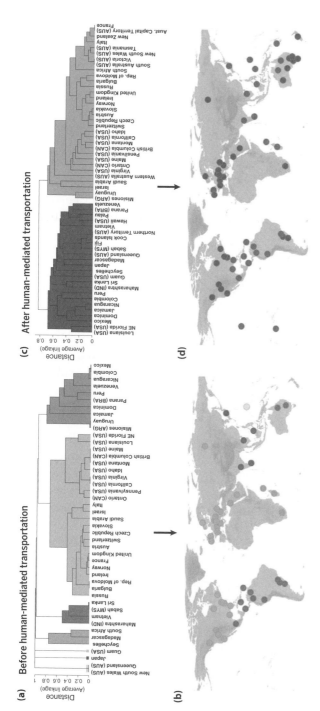

Figure 6.3 Dendrogram and map of compositional similarities among lists of alien terrestrial gastropods. (a and b) Before dispersal by humans. (c and d) After dispersal by humans. Compositional dissimilarity was measured by the β_{sim} index. Clusters were built through the minimization of the average compositional dissimilarity of one location to the others [i.e., UPGMA (unweighted pair group method with arithmetic mean) grouping]. Grey levels indicate main clusters identified by the dendrogram and their corresponding locations in the world map. From Capinha et al. (2015), reprinted with permission.

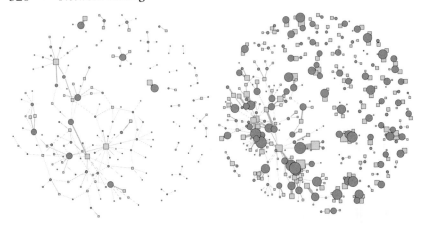

Figure 6.4 Bipartite subnetworks of native New Zealand plant species and their alien (left) and native (right) pathogen species. Each node represents a species; plants are shown as dark grey circles, pathogens as light grey squares and lines connecting pathogens to plants they have been recorded on. The size of each node is proportional to the square root of the number of records for that species, the width of the link is proportional to the number of records of that association and the position of nodes is determined by the Fruchterman-Reingold algorithm. In the alien subnetwork, the pathogens connected to the most hosts were *Phytophthora cinnamomi* (20 hosts), *Fusarium solani* (8) and *Claviceps purpurea* (8), while the most connected pathogens in the native subnetwork were *Armillaria novae-zelandiae* (15) and *Ilyonectria coprosmae* (14). From Bufford et al. (2020), reprinted with permission.

species (e.g., total, endemic, core, satellite). Finally, we discuss the scaling patterns of network structures (e.g., link density, connectance, nestedness and modularity). Our treatment does not seek to provide a comprehensive review of the field. Rather, we have selected topics that are most important to understand as background to themes addressed in later sections.

Species distributions are spatially aggregated because of species' limited dispersal capacity and because environmental heterogeneity is spatially autocorrelated. When exploring species distributions over multiple spatial scales, such spatial aggregation can give rise to the scaling pattern of occupancy summarised in the Modifiable Areal Unit Problem (Figure 6.5; Openshaw 1984). This scaling pattern can be captured in two ways. First, if we overlay a grid of square cells of size *a* over spatially scattered individuals of a species, or randomly deploy a number of sample plots of size *a*, some grid cells or plots will be empty whereas others have a number of individuals present. If these individuals are independent

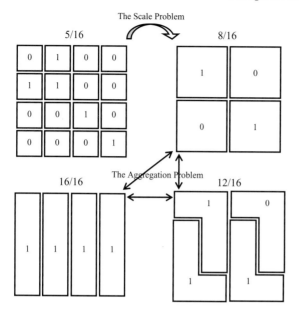

Figure 6.5 A simplified illustration of the scale and aggregation problems in the Modifiable Areal Unit Problem (MAUP). Distributions of a hypothetic species surveyed under four different grain sizes and shapes. 1: presence; 0: absence. The occupancy (proportion of occupied cells; above each plot) is sensitive to the size and shape of the grain. Redrawn from Hui (2009).

from each other, the probability of having n individuals in a sample plot follows a Poisson probability distribution, $p_a(n) = (\mu_a)^n \exp(-\mu_a)/n!$, where μ_a is the mean abundance in a sample plot and equals the density times the plot size ($\mu_a = d \cdot a$). Therefore, the scaling pattern of occupancy under this setting is, Occupancy $= A(1 - \exp(-\text{Abundance} \cdot a/A))$, where A is the spatial extent. Species occupancy increases when examined from finer (smaller a) to coarser (larger a) scales. This scaling pattern of occupancy also allows us to estimate species occupancy from density (abundance) and vice versa (Wright 1991; He and Gaston 2000) and connect geographic expansion to abundance rise, $\Delta p_a^+ = \left(1 - p_a^+\right)\Delta\mu_a$ (Hui et al. 2012a). Other probability distributions can also be used to reflect different assumptions and yield congruent scaling patterns of occupancy with better precision (see review, Hui et al. 2009; Barwell et al. 2014; Kunin et al. 2018). Another way to picture the formation of occupancy scaling is through coalescing adjacent grid cells into bigger ones and recounting species presence and absence (Hui et al. 2006; Hui

2009). Besides the overall scale dependence, typical insights from such spatial coalescing method also include: (i) the scaling pattern of occupancy is steeper (and thus more scale dependent) when the spatial correlation between adjacent occupied cells is weak; (ii) sample plots with more irregular shapes (higher edge-to-size ratio) overestimate occupancy but underestimate density (Figure 6.5).

It is widely established that the spatial and temporal scales of ecological processes are correlated (e.g., Figure 3.1). In other words, over its correlated spatiotemporal scale, an ecological process can be responsible for both the spatial structure of species distributions and the temporal fluctuations of population dynamics. This also means that the scaling pattern of occupancy carries information on a species' temporal dynamics and demographic performance (Wilson et al. 2004; Borregaard and Rahbek 2006; Hui 2011). As revealed for the major tree genera *Acacia* and *Eucalyptus*, occupancy patterns convey important insights on potential invasiveness (Figure 6.6; Hui et al. 2011, 2014). For instance, comparing the occupancy scaling of acacias and eucalypts in their native ranges suggests that (i) invasive acacias have a shallower occupancy scaling than average acacia species, whereas invasive eucalypts have a steeper occupancy scaling than average eucalypt species; (ii) shallower occupancy

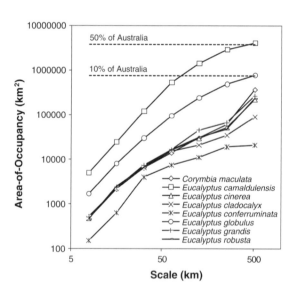

Figure 6.6 Scaling patterns of the native range size, measured by the area-of-occupancy (AOO, km2), for the eight invasive Australian *Eucalyptus* species. From Hui et al. (2014), reprinted with permission.

scaling is indicative of faster range expansion and greater invasiveness (i.e., acacias), whereas steeper occupancy scaling curves tend to be associated with retracting ranges and lower invasiveness (i.e., eucalypts). Over an even longer temporal scale, the evolutionarily younger *Acacia* lineage could be inherently well equipped for rapid colonisation of new ranges, whereas the evolutionarily older *Eucalyptus* will likely colonise new suitable ranges more slowly. Not only is an invader's native geographic range an important predictor of its potential inherent invasiveness, but the scaling pattern of its geographic range provides further clues regarding its invasion potential.

A species' distribution is realised by the incidences of births and deaths, and the vital rates of population demography are also scale dependent. The incidence of births and deaths within a sampling plot over a fixed window of time, and consequently the birth and death rates, also declines with the decrease of spatial grain (plot size). The population growth rate over a landscape, a crucial component of invasion performance and spread, is also scale dependent and typically increases with the enlargement of spatial extent (Figure 6.7). Such augmented performance over larger spatial extent is caused by three factors: (i) spatial covariance between the growth rates of local populations (landscape demography; Gurevitch et al. 2016; Hui et al. 2017), (ii) covariance between the growth rate and population size of local populations (Chesson 2012) and (iii) the rescue effect of meta-populations (Hanski and Gilpin 1997). Such intertwined demographic features give rise to the so-called demography accumulation curve (Hui et al. 2017). Positive spatial covariance between population dynamics (coincident changes in the abundance or other characteristics of geographically disjunct populations that vary over large space apart) is known as spatial synchrony (Liebhold et al. 2004). Spatial synchrony can be generated from the interplay of multiple processes (see Section 6.3), while elevated demographic performance can explain the success of many invasive species such as gypsy moths in northeastern North America (Liebhold et al. 2004; Hui et al. 2017) and the failure of local-scale eradications when handling regional-scale invasions because of persistent recurrent invasions such as the case of controlling oriental fruit fly invasions in California (Zhao et al. 2019). Importantly, the demography accumulation curve highlights the different role of local populations, depending on their locations and population size, to the overall landscape-scale invasiveness. By strategically allocating and positioning its propagules, an alien species can optimise and boost its chances of invading and its performance as an invader. This

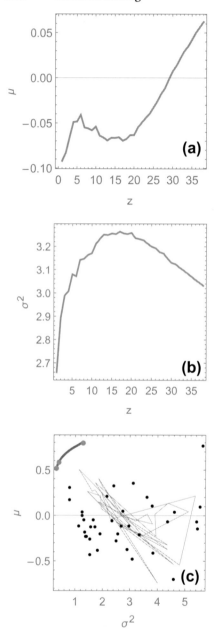

Figure 6.7 Landscape demography and demography accumulation curves of 38 populations of gypsy moth (*Lymantria dispar*). Accumulation curves for expectation (a) and variance (b) of ensemble relative growth rates as a function of the

implies that spatial context can result in diverse outcomes (in terms of demographic performance) for the same alien species in the same invaded landscape. Identifying these strategic positions in the regional communities is a priority for effective invasion management (see Section 6.6).

The spatial association and co-occurrence of two species are also scale dependent. This is directly related to ecological network analyses as the strength and incidence of biotic interactions between species are often inferred by their spatial association and co-occurrence patterns (e.g., Diamond 1975; Bosc et al. 2018). This can become evident when we drape grid cells on a landscape. There are four scenarios for a randomly chosen grid cell (Hui 2009): species A and B co-occur, p_a^{AB} (called the joint occupancy); only species A occurs, $p_a^{A\bar{B}}$; only species B occurs, $p_a^{\bar{A}B}$; neither occurs, $p_a^{\bar{A}\bar{B}}$. A positive association between species A and B can be defined as $p_a^{AB} > p_a^A p_a^B$, or simply define an index $I_a^{AB} = p_a^{AB}/p_a^A p_a^B$ and in this case $I_a^{AB} > 1$, representing a higher chance of co-occurrence than expected from spatially random encounters. A negative association can be defined as $p_a^{AB} < p_a^A p_a^B$ ($I_a^{AB} < 1$), and species independence as $p_a^{AB} = p_a^A p_a^B$ ($I_a^{AB} = 1$). By coalescing adjacent cells into bigger ones, we can formulate the scaling pattern of species association and co-occurrence (Hui 2009). Importantly, we see that species independence is scale-free, regardless of the spatial aggregation structure of each species. In general, the joint occupancy of species A and B, p^{AB}, changes in a non-linear fashion when scaling up (Figure 6.8). Even if two species do not co-occur ($p_a^{AB} = 0$), their joint occupancy at coarser scales can still become positive ($p_{4a}^{AB} > 0$). Nevertheless, the type of species association (positive or negative) remains the same across scales, although its intensity declines towards random co-occurrence when estimated at increasingly coarser scales (Hui 2009). This explains the scale dependence when inferring the strength of biotic interactions from co-occurrence (Araújo and Rozenfeld 2014). That is, significant biotic interactions are more easily detected at finer spatial scales, while at coarse scales biotic interactions weaken and become indiscernible from random or neutral forces (Rouget and Richardson 2003). This does not mean that biotic

---

Figure 6.7 (cont.) number of included local populations (z). Grey lines indicate averages of 1,000 rarefaction curves. (c) Black dots show expectation and variance of the relative growth rate for specific populations. Thin lines show the demographic trajectory for the ensemble. The thick curve denotes the efficient frontier (defined as portfolios for the ensemble of populations that have the lowest volatility but the highest growth) for the ensemble. From Hui et al. (2017), reprinted with permission.

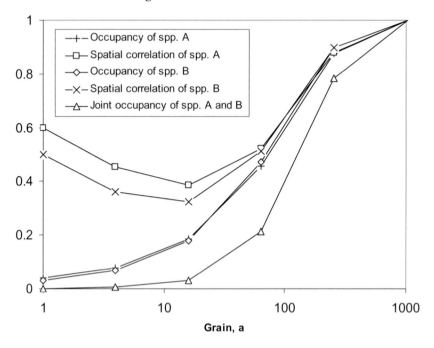

Figure 6.8 An example of the scaling patterns of species occupancy and joint occupancy as a function of spatial grain. From Hui (2009), reprinted with permission.

interactions are not important drivers of large-scale patterns, just that they cannot be discerned statistically, and are masked by randomness.

The scale dependence of co-occurrence can rather elegantly explain the so-called invasion paradox. According to niche-based theories, species- and functionally rich communities are less susceptible to invasion because establishment opportunities for invaders are rare (Tilman 2004; Renne et al. 2006). This notion has received strong support from fine-scale (plot-level) experiments (e.g., Naeem et al. 2000; Levine 2001). In contrast, a positive correlation between the numbers of native and alien species, known as the 'the rich get richer' hypothesis (Palmer and Maurer 1997; Stohlgren et al. 2003), has been mainly observed and supported at landscape and regional scales (e.g., Richardson et al. 2005). Fridley et al. (2007) identified many potential mechanisms that could flip the correlation of native and alien richness when crossing spatial scales (Figure 6.9): statistical artefact; Eltonian biotic resistance; invasional meltdown; neutral process plus spatial variance in immigration rates or in disturbance rates, spatial environmental heterogeneity; biotic acceptance plus non-

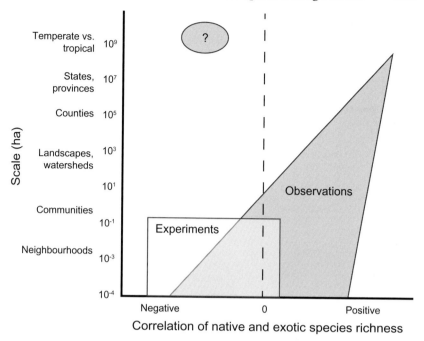

Figure 6.9 The invasion paradox, illustrated as the contrasting signs of correlation between native and alien species richness from experimental and observational studies, could be potentially explained by the scale of study. From Fridley et al. (2007), reprinted with permission.

equilibrium conditions; and facilitation. As discussed previously, all these factors are scale dependent, but perhaps the most obvious reason is that the key component in the invasion paradox – the co-occurrence pattern itself – is scale dependent. It is simply more likely that co-occurrence of two species will be observed at broader spatial scales. Taken together, at fine spatial scales, density dependence and resource competition within sites place constraints on both native and alien richness (Levine 2000), thereby generating the checkerboard pattern of co-occurrence and a negative correlation. At broad spatial scales, resource heterogeneity between sites (Shea and Chesson 2002) and within sites (Davies et al. 2005) allows more species to be accommodated, irrespective of whether they are native or alien. Environmental heterogeneity could increase invasion success but also reduce the impact of invasion by promoting coexistence that is otherwise impossible in homogenous environments (Figure 6.10; Melbourne et al. 2007). One aspect of the alien–native richness correlation that remains to be tested is whether the correlation is

Large spatial scales

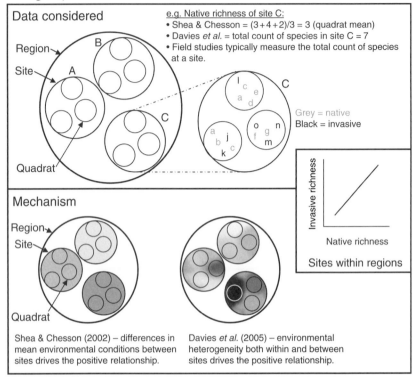

Data considered

e.g. Native richness of site C:
- Shea & Chesson = (3+4+2)/3 = 3 (quadrat mean)
- Davies *et al.* = total count of species in site C = 7
- Field studies typically measure the total count of species at a site.

Grey = native
Black = invasive

Mechanism

Invasive richness

Native richness

Sites within regions

Shea & Chesson (2002) – differences in mean environmental conditions between sites drives the positive relationship.

Davies *et al.* (2005) – environmental heterogeneity both within and between sites drives the positive relationship.

Small spatial scales

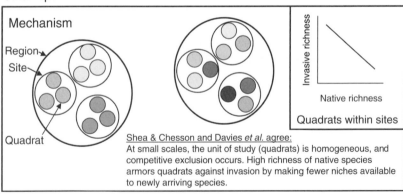

Mechanism

Invasive richness

Native richness

Quadrats within sites

Shea & Chesson and Davies *et al.* agree:
At small scales, the unit of study (quadrats) is homogeneous, and competitive exclusion occurs. High richness of native species armors quadrats against invasion by making fewer niches available to newly arriving species.

Figure 6.10 Differences between the models of Shea and Chesson (2002) and Davies et al. (2005) regarding the diversity–invasibility paradox. Small spatial scales are those at which individuals interact (e.g., experience inter- and intraspecific competition). Large spatial scales are those greater than the scale of individual interactions. Shading represents variation in an environmental (exogenous) factor. Scales of species richness: alpha diversity is the mean diversity of quadrats within a site; beta–diversity is the difference in species composition between quadrats within a site; gamma–diversity is the diversity of a site (i.e., total count of species). From Melbourne et al. (2007), reprinted with permission.

spatially explicit and contingent to the location, quality and the level of isolation of sites (see Section 6.6).

Because co-occurrence is scale-dependent, compositional similarity and turnover are also scale dependent. Using zeta diversity metrics we can derive different biodiversity components; in particular, the total number of species across m sites (S_m); the number of species that occupy i sites out of the total m sites surveyed ($F_{i,m}$); and the number of *locally* endemic species that only occur in the selected i sites ($E_{i,m}$) (Hui and McGeoch 2014). By pooling samples into bigger ones and by quantifying numbers of shared and newly detected species in spatially arranged plots, we can further elucidate the scale dependence of compositional similarity and the distance decay of similarity (Figure 6.11). Importantly, the scaling patterns differ for common versus rare species (McGeoch et al. 2019), implying that alien species at different stages of the introduction-naturalization-invasion continuum (which is often related to their level of commonness and rarity) can operate under influence of different demographic processes (Richardson and Pyšek 2012). Regression on spatially explicit compositional similarity and turnover allow us to dissect assembly

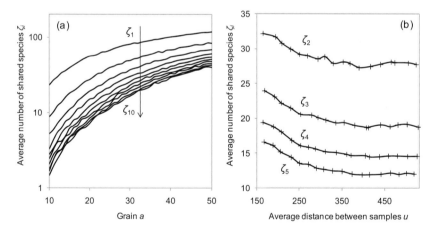

Figure 6.11 a, The scaling relationship (ζ-diversity as a function of sampling grain a [in m]). b, The distance decay of ζ diversity (average distance between samples u [in m] resulting from expanding sampling extent) for the 20 by 20-m quadrate samples of 307 tree species in the 50-ha plot on Barro Colorado Island (Condit 1998; Hubbell et al. 1999, 2005). Lines from top to bottom in c, indicated by the arrow, are for ζ_1 to ζ_{10}; those in d are for ζ_2 to ζ_5 at a grain of 20 m by 20 m. From Hui and McGeoch (2014), reprinted with permission.

processes operating at different scales and for species at different levels of rarity (see Section 6.3).

As different types of species and their interaction strengths are both scale dependent, network structures are inevitably scale dependent, and consequently, the roles of species in affecting network features (e.g., specialisation, centrality) are also scale dependent (e.g., Brose et al. 2004; Poisot et al. 2014). In particular, Brose et al. (2004) depicted the unified spatial scaling of species and their trophic interactions, where two established scaling patterns were combined: the power-law species area relationship, $S = cA^z$, and the power-law scaling of feeding links between species, $L = bS^u$, with c, z, b and u constants. Two specific hypotheses address how the number of links scales up in a network: the link-species scaling law (where $u = 1$; Cohen and Briand 1984) and the constant connectance hypothesis (where $u = 2$; Martinez 1992). The number of feasible links in a well-mixed local network is $n = S(S - 1) \sim S^2$, and thus the connectance of the network is $C = L/n = bS^{u-2}$; this means that $C = b$ for the constant connectance hypothesis. Under the link-species scaling law, we have $C = b/S = b/c \cdot A^{-z}$; that is, network connectance declines with increasing species richness or geographical area. The reality lies between these two extremes, $1 \leq u \leq 2$ (e.g., Bersier and Sugihara 1997; Schmid-Araya et al. 2002). Through the overriding effects of network size (species richness) and connectance on other network structures (e.g., nestedness and modularity; Figure 3.8), the unified scaling in spatial networks also produce anticipated changes in network architectures. Galiana et al. (2018) made three universal predictions regarding network area relationships (Figure 3.9): (i) network degree distribution preserves its skewness across spatial scales, but specialism increases with area; (ii) species–area relationships vary across trophic levels, while the proportion of omnivorous links increases with area promoting an increase of food chain length; (iii) network modularity is constant across spatial scales in homogeneous landscapes.

Many studies have explored network structures across multiple scales by pooling spatially explicit local interactions collected from different sites to come up with regional-scale network representations (Figure 6.12; Traveset et al. 2013; Vizentin-Bugoni et al. 2019). Caution is needed when examining the structure of such pooled ecological networks for at least two reasons. First, pooling biotic interactions is not strictly generating a scaling pattern as a result of the spatial variation in sampling effort for collecting interaction data. Areas outside local

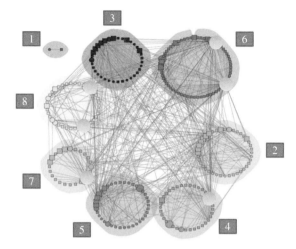

Figure 6.12 Modules in a network of 60 plants and their 220 pollinators in the Galapagos Islands. The size of a node (species) depicts the different network roles, from peripherals (smallest) to network hubs (largest, indicated in grey circles). Plant species are represented by circles and animals by squares. Links of alien species are indicated in red, whereas those of the remaining species are in black (native, endemic or of unknown origin). Alien links represent 34% of all links among modules. Numbers in squares refer to the module number. From Traveset et al. (2013), reprinted with permission.

sampling sites but within the larger study extent therefore likely harbour many species and interactions that are not accounted for when pooling data. For this reason, pooling site-specific interaction data only allows us to explore network accumulation curves, not network area curves or network scaling. Second, assuming that the individuals of species A and B in local communities have had opportunities to encounter each other over the duration of sampling (the scale dependence of encounter rate is important), we can explain the presence of an interaction between these two species by their trait complementarity, and the absence of an interaction by their trait mismatching. However, in metacommunities with interlinked local communities via dispersal of individuals, or when exploring a community over an expanding spatial extent, the presence/ absence of interactions is also mediated by the probability of individuals encountering each other, which is constrained by both dispersal limitation and the physical distance between sites. Pooling local interactions into so-called regional networks thus confounds the assumption behind the presence or absence of an interaction. For such pooled networks, comparisons at different scales are needed, such as between local versus pooled regional networks, to single out the change of encounter probability becuase of physical distance and dispersal limitation. For instance, the absence of an interaction can be the result of trait mismatching, infeasibility of encounters or a short observation window. All three factors and their interactions will affect the observed network structures and their interpretations.

6.3 Dissecting Spatial Communities

In a trait-mediated local network, the functional roles of species are fully reflected by their topological positions in the ecological network (based on the interaction strength matrix). This allows us to define core and peripheral species in terms of network functioning. In contrast, the physical isolation and the dispersal-related connectivity between local communities also form a spatial affinity network, which can define the core and peripheral sites in a regional spatial community. How the role of core and peripheral species with regard to network functioning varies over the core and peripheral sites in the spatial affinity network warrants future attention. More work is needed to unravel the underlying drivers of spatial variation in network interactions and structures. At the local community scale, biological invasions can penetrate the recipient ecological network and drive it to undergo persistent transition and temporal turnover at marginal instability (Chapter 3, 4 and 5). At regional to global scales, we first resort to pattern analyses to infer underlying drivers of changes and then assess how these alien biomes have broken down natural bounds.

The scaling patterns of spatial communities and their embedded metanetworks are driven by several processes each of which operates at a specific range of spatial and temporal scales (Figure 3.1). Over regional scales, drivers of the spatial variation of species co-occurrence and interactions can be grouped into multiple blocks (Figure 6.13): distance-limited processes (such as dispersal limitation) that are responsible for the pattern of isolation-by-distance (Wright, 1943); spatial covariance in environmental forcing (known as the Moran effect; Hansen et al. 2020) that is responsible for the pattern of isolation-by-resistance (Meynard et al. 2013); as well as biotic interactions within and/or between functional guilds. These three blocks of drivers, together, form the pattern of distance decay of similarity in spatial communities (Soininen et al. 2007; Morlon et al. 2008). Biological invasions can interrupt and alter such

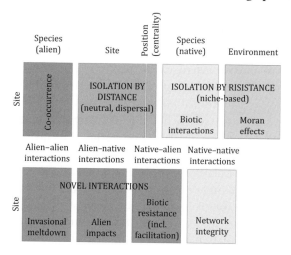

Figure 6.13 Standard data frames for exploring spatial variations of species co-occurrence (top) and network interactions (bottom) across sites. To dissect invaded spatial networks, we can follow two types of analytic framework. First, the co-occurrence structure of a focal group of species (here, alien species) is captured as a site-by-species matrix and considered as the response variable. Second, interactions between native and aliens (e.g., directed interactions) are captured by site-by-interaction matrices and considered as response variables.

universal patterns, for example when successful alien species increase network connectance, thereby breaking down the distance decay of similarity (Figure 6.3; Capinha et al. 2015).

Indeed, within a local network, biotic interactions function as a magnet to impose viscosity in the co-occurrence of interacting species. The feedback loops from intransitive interactions and between species and their habitat can modify the geophysical drivers of species distributions, creating bistability at the geographic boundaries of a species' range (e.g., Mohammed et al. 2018) and sharp geographic transitions of the entire assemblage of a functional guild even at the global scale (as evident in the latitudinal gradient of dominant root-associated microbial symbionts; Steidinger et al. 2019). Such neat construction and modification of biotic interactions decouples the relationship between abiotic niches and species distributions (Han and Hui 2014), inevitably resulting in poor model transferability when using species distribution models to forecast the distribution of species in novel environments (Liu et al. 2020). Consequently, it is necessary to include biotic interactions when exploring species–environment relationships. To be more specific, when exploring variations in species co-occurrence and network interactions we need to consider three sets of processes (Figure 6.13): neutral processes (e.g., dispersal and drift), niche processes (habitat and environmental heterogeneity) and biotic interactions. The first type of analyses (Figure 6.13 top) allows us to explain co-occurrence patterns of the focal functional guild as a function of isolation by distance as a result of neutral processes such as dispersal limitation, cross-guild biotic interactions and isolation by resistance from niche differentiation and the Moran effect of spatial structures in environmental gradients. The second type of analyses (Figure 6.13 bottom) sheds light on specific types of novel interactions as explained by other types of interactions and any site-specific features. Using examples, we demonstrate the possible procedures and interpretations of dissecting an invaded spatial community. Clearly, as the response variables are matrices in Figure 6.13, rather than the typical single column vector, we can use analytic methods such as correspondence analysis (Legendre and Legendre 1998) and dissimilarity modelling (Ferrier et al. 2007; Latombe et al. 2017). As examples of network interactions and turnover from studies in enough sites to permit robust statistical analyses are lacking (Traveset et al. 2013; Vizentin-Bugoni et al. 2021), we confine our discussion here to a few examples of co-occurrence networks.

Just as two tuning forks with the same pitch resonate, so do ecological patterns and processes that operate at the same scale. This phenomenon of scale resonance also prevails in the spatial variations of species distributions and network compositions. In particular, species distributions are regulated by a variety of abiotic and biotic processes working in concert, but at different scales (McGill 2010; Peterson et al. 2011), whereas identified key processes (for example, those using multivariate statistics) resonate with deployed spatial grain and extent of the survey. That is, information that is detected is filtered by the scale and resolution of measurement, and reflects only a snapshot of the intrinsic cross-scale mechanism. This brings into question many regional management planning practices that are based on the upscaling extrapolation of local-scale studies (Guisan et al. 2013). For instance, the distribution of *Acacia deal-bata* in the Iberian Peninsula is driven by multiple factors. The factors identified as being significant predictors of distribution and abundance depend on the spatial resolution: landscape complexity trumps climatic variables in importance as a driver of invasion as the spatial scale of interest moves from coarse to fine (Figure 6.14; Vicente et al. 2019). Strategies in invasion management at different spatial scales should target

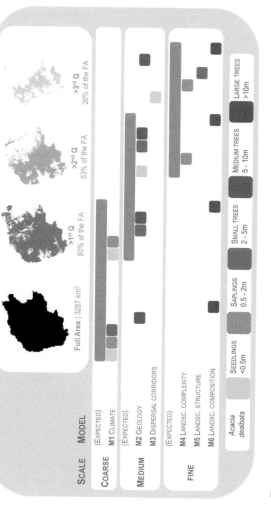

Figure 6.14 Scales of spatial structure/influence (coarse-, medium- and fine-scale) and associated models (M1–M6; competing models representing environmental factors) selected by multi-model inference for each *Acacia dealbata* height class (seedlings, saplings, small trees, medium trees and large trees) for each nested area/extent (full area, area above the first quartile, area above the second quartile and area above the third quartile). Horizontal grey bars represent the expected patterns based on the research hypothesis and on previous research. From Vicente et al. (2019), reprinted with permission.

their corresponding factors. In particular, optimising resources between regional-scale policies on climate change mitigation and local-scale landscape management practices would be a sensible recommendation, while regional-scale landscape management with local climate change mitigation will go against the scale-dependent nature of processes at work. Further research is needed to determine how management effort can be balanced for optimal solutions across scales, in addition to the spatial optimisation targeting core or peripheral populations under feasible constraints (Van Wilgen et al. 2011; Hui and Richardson 2017).

In a regional-scale study of the co-occurrence of 2,054 vascular plant species in 302 nature reserves in the Czech Republic, network compartmentalisation analyses revealed congruent spatial and functional modules for three assemblages of species categorised according to their residence time (Hui et al. 2013). Natives species that have been present in the region since the last glaciation, archaeophytes are historical immigrants introduced to Europe between the initiation of agricultural activities during the Neolithic period (ca. 4000 BC) and the European exploration of the Americas (ca. 1500 AD), and neophytes are modern invaders introduced to Europe after 1500 AD (Pyšek et al. 2012). The modularity increasingly deviates from the null model expectation as we move from young to mature assemblages, representing the transition over millennia, from neophytes through archaeophytes to natives, supporting the settling-down hypothesis of diminishing effect of stochasticity with increasing residence time. Moreover, species within a module are functionally distinctive from those in other modules in terms of taxonomy and phylogeny (e.g., Figure 6.15; Hui et al. 2013), supporting the hypothesis that modules within assemblages become more distinctive with increasing residence time. Furthermore, using zeta diversity to differentiate the role of common versus rare species in community assemblage patterns (Hui and McGeoch 2014; McGeoch et al. 2019), Latombe et al. (2018) showed that, besides a list of key abiotic drivers, considering dissimilarity in native species composition improved our ability to explain turnover of both archaeophytes and neophytes, especially for low orders of zeta, i.e., mostly for rare species (Figure 6.16). However, to explicitly explore the role of biotic interactions, we need to consider functional diversity, for instance, using functional similarity in trait space as the proxy of interaction strength (Divíšek et al. 2018). We look into this in Chapter 7.

At a global scale, biotic interactions are often difficult to detect, for instance in the co-occurrence networks of native and alien ant communities

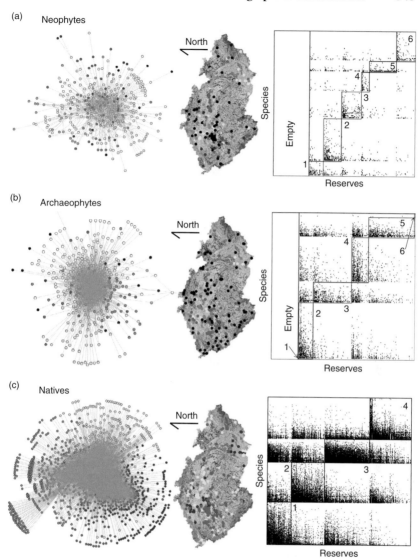

Figure 6.15 Network structures of vascular plants in the Czech Republic. Network expression, geographical location of reserves and species-by-site matrix of modules identified for (a) neophytes, (b) archaeophytes and (c) natives. In the network expression, open circles represent reserves. Points with different levels of grey in the network expression and geographical maps indicate different modules identified in each of the three assemblages. Modules in the matrices are marked by the serial numbers and a rectangle, with points indicating the presence of a species (a row) occurring in a reserve (a column) and the rectangles of 'Empty' in neophytes and archaeophytes indicating reserves where these two species assemblages do not occur. From Hui et al. (2013), under CC-BY Licence.

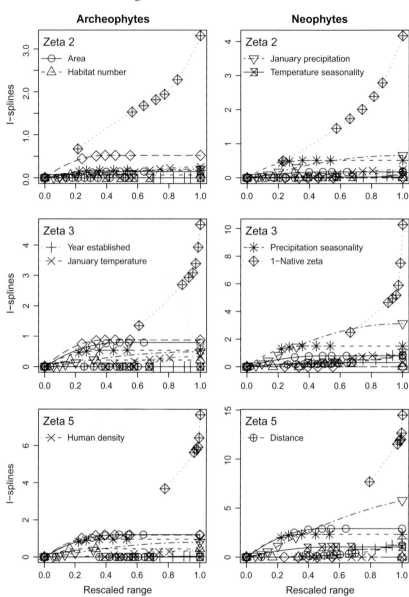

Figure 6.16 I-splines generated by the multi-site generalised dissimilarity model of zeta diversity of Archeophytes and Neophytes, considering static and seasonality environmental variables, distance and the Sorensen zeta diversity of native species as predictors. For distance and native zeta, the y-axis should be interpreted as one minus zeta similarity. From Latombe et al. (2018), reprinted with permission.

on small islands. Although clear and distinct modules can be identified for both native and alien ants on islands (Figure 6.17; Roura-Pascual et al. 2016), the richness and turnover of alien and native ant communities were found to be correlated in the Pacific islands, with the same

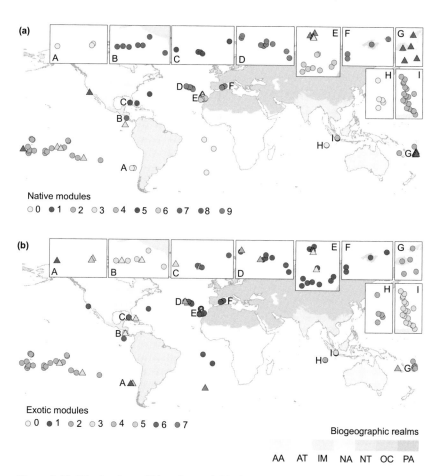

Figure 6.17 Distribution of (a) native and (b) alien modules for islands (i.e., distinct clusters of islands with a certain assemblage of species) derived from a modularity analysis. Colours indicate the module assigned by the modularity analysis, while the symbols show whether the conditional inference tree grouped that particular island in a node where the majority of islands have the same module (o) or a different one (Δ). Module 0 means that the island does not have ant species. The colours of the land masses show the biogeographical realms (AA, Australasia; AT, Afrotropical; IM, Indo-Malay; NA, Nearctic; NT, Neotropical; OC, Oceania, PA, Palaearctic; Olson et al., 2001). The insets represent an expanded view of areas where symbols overlap. From Roura-Pascual et al. (2016), reprinted with permission. *A black and white version of this figure will appear in some formats. For the colour version, please refer to the plate section.*

abiotic variables, suggesting that abiotic, rather than biotic, processes are at play in structuring alien community assembly (Figure 6.18; Latombe et al. 2019). The geographical distribution of modules suggests that the native modules reflect the long-term outcome of barriers that limit

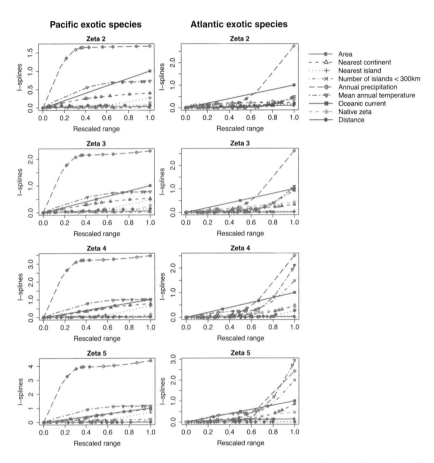

Figure 6.18 I-splines computed with the multi-site generalised dissimilarity model of zeta diversity for alien species of the Pacific and Atlantic island groups, from ζ_2 to ζ_5 for Simpson zeta diversity. The horizontal axes represent the original variables, rescaled between 0 and 1 for comparison (environmental variables are in red, geographical ones are in blue and the biotic native zeta variable is in green). The relative amplitude of each spline in a given panel therefore represents the relative importance of the corresponding variable to explain zeta diversity for a specific order. For each variable, the symbols are located at the percentiles, providing information on the distribution of values. From Latombe et al. (2019), under CC-BY Licence. *A black and white version of this figure will appear in some formats. For the colour version, please refer to the plate section.*

dispersal, whereas the alien assemblages appear to be slightly less spatially structured than the native ones. This apparent homogenization in the distribution of modules for alien species is driven by the presence of a few species with large global geographical ranges. Alien ants represent only a quarter of the total pool of ant species, but a few of these species are present in a large number of islands (the most prominent one is *Paratrechina longicornis*; Roura-Pascual et al. 2016). Biotic interactions, at least those detectable at the island scale, do not seem to influence the modular structure of these insular ant assemblages. Other variables related to environmental heterogeneity (through island area and climate), distance to the nearest land masses (such as geographical coordinates and nearest continents/islands) and human alteration (represented only by the human population size variable) do not seem to influence the compartmentalization of ant assemblages. The lack of influence of these variables could be a result of the inability of these variables (as measured) to capture the spatially nested structure of ant assemblages at the global scale. Taken together, to detect the role of biotic interactions in shaping species co-occurrence and network interactions, we suggest using sample plots of different sizes and considering three groups of variables (niche, neutral and biotic interactions) for species with different labels (e.g., natives versus aliens, functional guilds).

6.4 Meta-network Dynamics

Until now we have discussed only the spatial patterns of ecological networks affected by biological invasions. Essentially, we have assumed that observed spatial patterns of species co-occurrence result from assembly processes, including especially biotic interactions and biological invasions. However, such pattern analyses are not well suited for making inferences of causality, as processes and impacts are normally only inferred from correlative statistics. Such pattern-to-process inference faces obvious limitations because of multiple processes that can generate similar patterns. To explore spatial turnover and transition in spatial networks we therefore need to consider the spatial dynamics and the stability of meta-networks. To demonstrate the dynamics of meta-networks, the pivotal task is to demonstrate how biotic interactions affect species co-occurrence; that is, process to pattern inference, also known as spatial pattern formation or emergence. In contrast, the stability criteria of meta-networks elucidate how the interplay of multiple forces, including especially biotic interactions and dispersal, regulates species coexistence in regional communities. To this end, and to complete this inference loop, we first demonstrate the dynamics of meta-networks and, in the next section, the stability criteria governing meta-network dynamics.

The multi-site patch occupancy model has been used to study meta-population persistence and spatial dynamics (e.g., Levins 1969; Tilman 1994; Hanski 1998). The key elements of the dynamics of a meta-population are colonisation and local extinction in a landscape with a large number of suitable habitat patches: $\dot{p}_i = m_i p_i (1 - p_i) - e_i p_i$, where p_i is the occupancy of species i in the habitat-patch network (in particular, it is the proportion of occupied patches) and $\dot{p}_i$ its temporal derivative (i.e., the change rate of occupancy); m_i and e_i are the rates of colonisation and local extinction of species i, respectively. The occupancy eventually settles at a stable equilibrium, $\hat{p}_i = 1 - e_i/m_i$, implying the necessity of a greater colonisation rate than the local extinction rate ($m_i > e_i$) to ensure meta-population persistence ($\hat{p}_i > 0$).

To be able to consider the effect of biotic interactions on patterns of species co-occurrence and turnover, we need to consider the joint occupancy of species i and j, $p_{i \cap j}$, defined as the proportion of patches occupied simultaneously by both species. Clearly, only a subset of occupied patches of species i will also be occupied by species j, $p_{i \cap j} \leq p_i$. If the two species are independent from each other, we would expect the joint occupancy to be equal to the product of two species' occupancies, $p_{i \cap j} = p_i p_j$, and therefore we could use $I_{ij} = p_{i \cap j}/p_i p_j$ as an index for species co-occurrence (Hui 2009), with $I_{ij} > 1$ for species association, $I_{ij} = 1$ for species independence and $I_{ij} < 1$ species dissociation [spatial segregation]. We can partition the dynamics of joint occupancy into four events during a unit time step: (i) a patch currently only occupied by species i becomes colonised by species j at the end of the time unit; (ii) a patch only occupied by species j becomes colonised by species i; (iii) the local population of species i goes extinct in a patch occupied by both species; and (iv) the local population of species j goes extinct in a patch jointly occupied by both species. Note that the likelihood of both species simultaneously occupying an empty patch or becoming extinct from a jointly occupied patch is small and can be ignored without affecting the dynamics. This then leads to the following dynamics of joint occupancy,

$$\dot{p}_{i \cap j} = m'_j p_j (p_i - p_{i \cap j}) + m'_i p_i (p_j - p_{i \cap j}) - (e'_i + e'_j) p_{i \cap j}, \qquad (6.1)$$

where m'_i and e'_i are the colonisation and extinction rates, respectively, of species i under the influence of species j; m'_j and e'_j can be defined in parallel. Specifically, if $m_i' < m_i$ and $e_i' > e_i$, the presence of species j is antagonising species i; if $m_i' > m_i$ and $e_i' < e_i$, the presence of species j is facilitating species i; if $m_i' = m_i$ and $e_i' = e_i$, the presence of species j has no impact on species i. Obviously, there could also be the case $m_i' < m_i$ and $e_i' < e_i$ (the presence of species j reduces the colonisation rate of species i but reduces its extinction rate), or $m_i' > m_i$ and $e_i' > e_i$, although the two cases are not common in reality. With biotic interactions, we need to modify the meta-population dynamics as follows,

$$\dot{p}_i = m_i p_i\left(1 - p_i - (p_j - p_{i\cap j})\right) + m_i' p_i(p_j - p_{i\cap j}) - e_i(p_i - p_{i\cap j}) - e_i' p_{i\cap j}.$$

(6.2)

Note that the colonisation events of species i (the first two terms on the right-hand side) are divided into whether the events occurred in patches without the two species, $p_{\bar{i}\cap\bar{j}} = 1 - p_i - (p_j - p_{i\cap j})$, or in patches with species j but without species i, $p_{\bar{i}\cap j} = p_j - p_{i\cap j}$; the extinction events of species i (the last two terms) are divided into whether the events occurred in patches with species i but without species j, $p_{i\cap\bar{j}} = p_i - p_{i\cap j}$, or in patches with both species i and j, $p_{i\cap j}$.

Dynamic equilibrium can be reached with the following level of spatial association between the two species,

$$I_{ij} = \frac{m_i' + m_j'}{e_i' + e_j' + m_i'\hat{p}_i + m_j'\hat{p}_j}.$$

(6.3)

For instance, when the presence of species j affects species i negatively (e.g., interference competition, with $m_i' \leq m_i$ and $e_i' \geq e_i$), species i's occupancy will decline to between $1 - e_i'/m_i' \leq \hat{p}_i \leq 1 - e_i/m_i$ and therefore we have $1/(1 + e_i'\rho_i + e_j'\rho_j) \leq I_{ij} \leq 1$, where $1/(1 - \rho_i) = (e_i'/e_i)(m_i/m_i')$. Evidently, parameter ρ_i represents the magnitude of interaction impact experienced by species i, with $\rho_i = 0$ corresponding to the absence of interaction. Overall, we see both occupancy and spatial association decline as a result of negative interactions, while they increase as a result of positive interactions (Figure 6.19). From the altered occupancy and joint occupancy from biotic interactions, we can estimate zeta diversity metrics (Hui and McGeoch 2014; McGeoch et al. 2019). For instance, the expected number of species in a randomly selected site, i.e., alpha diversity, is the sum of normalised occupancies and its variance, $E(\zeta_1) = \sum_i^S p_i$ and $Var(\zeta_1) = \sum_i^S \sum_j^S (p_{i\cap j} - p_i p_j)$. For the number of shared species between

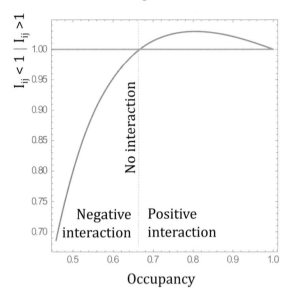

Figure 6.19 An illustration of spatial association (I_{ij} in Eq. (6.3)) between two species as a function of the interaction strength, defined as $\rho = m_i - m_i' = e_i' - e_i$ (let $m_i = m_j = 0.3$ and $e_i = e_j = 0.1$ in Eq. (6.1) and Eq. (6.2), for simplicity. The blue curve depicts the dynamic equilibrium of the meta-network for interaction strength ρ ranging from -0.1 (left end, negative interaction) to 0.1 (right end, positive interaction); the dashed line represents the absence of interaction between the two species. Positive interactions ($\rho > 0$) lead to positive spatial association ($I_{ij} > 1$), while negative interactions ($\rho < 0$) lead to negative spatial association ($I_{ij} < 1$). Spatial association represents independent distributions of the two species ($I_{ij} < 1$) when there is no interaction ($\rho = 0$).

two sites, we have, $E(\zeta_2) = \sum_i^S p_i^2$ (evidently, this is the Simpson index of evenness, H_{Sim}) and $\mathrm{Var}(\zeta_2) = \sum_i^S \sum_j^S (p_{i \cap j}^2 - p_i^2 p_j^2)$.

We have emphasised the role of biotic interactions in affecting species co-occurrence and compositional similarity. We now turn to explore the role of dispersal. The widely observed spatial synchrony in population dynamics is mainly caused by three factors: (i) spatial autocorrelation inherent in the environment (Moran 1953; Vasseur and Fox 2009); (ii) interspecific density regulation through predation and parasitism (Ims and Andreassen 2000; Gonzalez-Olivares and Ramos-Jiliberto 2003; see previously); and (iii) density-independent dispersal (Jansen 2001). For a predator–prey system, in particular, even a low level of dispersal can lead to synchrony if the environment is homogeneous (Jansen 2001).

In contrast, density-dependent dispersal (e.g., predator pursuit and prey evasion) can desynchronise spatial patterns in predator–prey systems (Figure 6.20; Li et al., 2005; Ramanantoanina et al. 2011). Moreover, density-dependent dispersal from a higher trophic level (predator pursuit) has a less desynchronizing effect than density-dependent dispersal from a lower trophic level (prey evasion). For instance, prey evasion is mainly driven by fear of predators and thus takes the minimal-effort evasion strategy (Oshanin et al. 2009). Prey can move to 'refuge' patches with low predator densities and thus grow in abundance before being detected by predators. This allows the growth of prey in 'refuge' patches while the remaining prey in the original patch are being depleted by predators, inducing a spatial asynchrony between refuge and original patches.

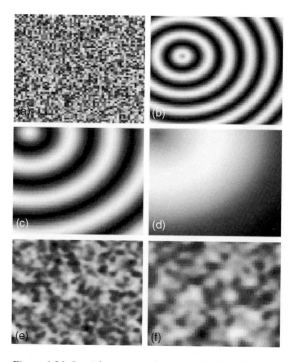

Figure 6.20 Spatial patterns of two species (predator and its prey) in a two-dimensional landscape (80 × 80). The grey-level intensity indicates the density of predator. The demonstrations have the parameters of carrying capacity, functional response and growth rate fixed, with only the dispersal rates (a and g) and the intensity of predator pursuit and prey evasion (w and v) changing. (a) $a = g = 0.005$, $w = v = 0$; (b) $a = g = 0.01$, $w = v = 0$; (c) $a = g = 0.005$, $w = v = 0$; (d) $a = g = 0.15$, $w = v = 0$; (e) $a = g = 0.01$, $w = v = 0.2$; (f) $a = g = 0.05$, $w = v = 0.2$. From Li et al. (2005), reprinted with permission.

The decline of synchrony caused by the density-dependent dispersals is more severe in predator–prey networks than in single-guild networks (Munkmuller and Johst 2008). Together, adaptive individuals under mutation and selection, spatial local interactions and dispersal can explain realistic macroecological patterns of species distributions and traits (Figure 6.21; Hui and McGeoch 2006).

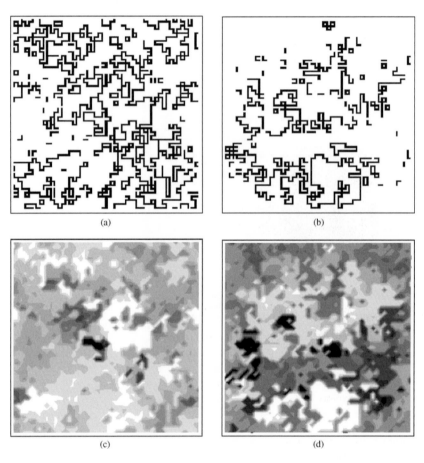

Figure 6.21 Spatial structure of species border and body size in two-dimensional habitat after 2,000 generations. Individual is represented by a binary genome. (a) and (b) are species borders of prey and predator, respectively. Black curves indicate that the two adjacent populations are reproductively isolated. (c) and (d) are spatial structure of body sizes for prey and predator, respectively. From Hui and McGeoch (2006), reprinted with permission.

As articulated in the Moran effect, habitat structures can also affect the emerged spatial networks. In particular, landscape variance has the effect of enlarging a landscape experienced by species, while landscape auto-correlation has the effect of shrinking a landscape. Both spatial hetero-geneity and productivity can dictate the diversity and structure of a spatial network. Wickman et al. (2020) investigated how the variance and autocorrelation length of primary production affect properties of evolved food webs consisting of one autotroph and several heterotrophs. Several findings emerged from their study (Figure 6.22): (1) diversity increases with landscape variance but peaks at moderate autocorrelation length; (2) trophic level increases with landscape variance but peaks at moderate autocorrelation length; (3) landscape variance promotes, while auto-correlation length reduces, the dissimilarity between the spatial distribu-tion of heterotrophs and that of the autotroph; (4) disruptive selection

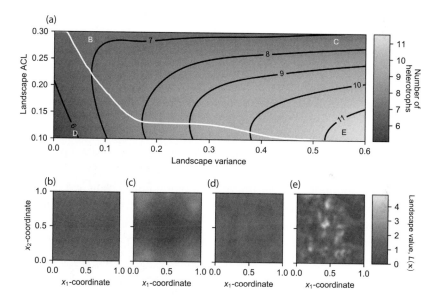

Figure 6.22 Diversity increases with landscape variance and is unimodal in autocorrelation length (ACL). (a) Average number of evolved heterotrophs in the food web plotted against landscape variance and ACL. The white line indicates the ACL that maximizes average diversity for each given landscape variance. The letters b–e in the panel cross reference examples of resource landscapes L(x) for different degrees of landscape variance and ACL, with (b) low variance and high ACL, (c) high variance and high ACL, (d) low variance and low ACL and (e) high variance and low ACL. Note that all landscapes are periodic in both spatial dimensions. From Wickman et al. (2020), reprinted with permission.

experienced by ancestral species dictates properties of the final evolved communities (priority effect). Taken together, biotic interactions, dispersal and the Moran effect of spatially autocorrelated environmental forcing are jointly responsible for the dynamics and co-occurrence patterns of meta-networks. We now turn to look at how biological invasions could interfere with such emerged spatial patterns and network stability.

6.5 Stability Criteria of Meta-networks

One key aspect of ecosystem functioning in meta-networks is the coexistence and persistence of its component species, as a result of system stability. For instance, let us first apply May's stability criterion with the universal ecological network theory (Brose et al. 2004). Interestingly, if we invoke May's stability criterion, $SC < 1/\sigma^2$, the power laws in the network theory will result in $bS^{u-1} < 1/\sigma^2$ or equivalently $S < (b\sigma^2)^{-1/(u-1)}$ or $A < c^{-1/z}(b\sigma^2)^{-1/(z(u-1))}$. Under the link-species scaling law with $u = 1$, an ecological network is stable if the number of links to a single species is less than the reciprocal of the variance of interspecific interaction strengths, $b < 1/\sigma^2$. Under the constant connectance hypothesis, a network is stable if the average number of links to a residing species is less than the reciprocal of the variance of interspecific interaction strengths, $Sb < 1/\sigma^2$. In general, $L/S < 1/\sigma^2$. That is, increasing the study area will likely violate the stability criterion, thus making a spatial network more invasible. In small areas, stability can be assured and thus the system can resist invasions. However, such a simple deduction does not consider the roles of biotic interactions and dispersal in meta-network stability. Let us now look at a dynamic model.

The dynamics of a meta-network with S species and n local networks can be formulated by adding dispersal to the standard Lotka–Volterra model (Gravel et al. 2016),

$$\dot{N}_{ix} = N_{ix}\left(r_{ix} + \sum_{j=1}^{S} a_{ijx}N_{jx}\right) + \sum_{y=1}^{n} d_{ixy}\left(N_{iy} - N_{ix}\right). \qquad (6.4)$$

where N_{ix} and r_{ix} represent the population size and intrinsic growth rate of species i in local network x; a_{ijx} is the per capita effect of species j on species i in local network x, and can be randomly drawn, with probability C, from a distribution with a zero mean and standard deviation σ; and d_{ixy} is the dispersal rate of species i from network y to x. Note, the Jacobian matrix of the meta-network can be partitioned into three

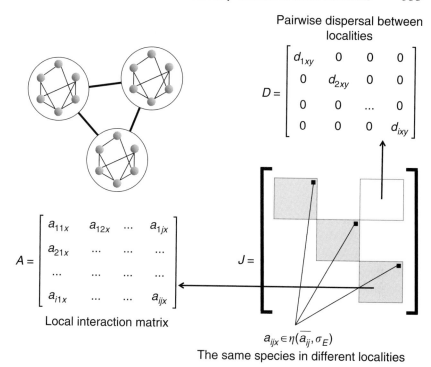

Figure 6.23 Conceptual illustration of the Jacobian matrix in random meta-ecosystems. The model represents the dynamics of a meta-ecosystem, pictured as a spatial network of interaction networks (top left). The Jacobian matrix representing all interactions among pairs of species and locations is highly structured. It is made of the sub-matrices A, M and D. The spatial heterogeneity among locations is implemented by varying interaction coefficients in space (different entries in each interaction sub-matrix) and the landscape is implemented by varying the spatial structure of the model (entries in each pairwise dispersal matrix). From Gravel et al. (2016), under CC-BY Licence.

components (Figure 6.23): $\mathbf{J} = \mathbf{M} + \mathbf{D} + \mathbf{A}$, where $\mathbf{M}$ is a diagonal matrix of intraspecific density-dependence of $-m_{ix}$; $\mathbf{D}$ is the dispersal matrix of d_{ixy}; $\mathbf{A}$ is the collection of local interaction matrices of a_{ijx} with each network a block. Note, C is not the connectance of matrix $\mathbf{A}$ but only the proportion of non-zero elements for these blocks. The effect of density dependence can further be applied and tested on Eq. (6.4) (Stone 2018), which changes the Jacobian to $\mathbf{J} = \mathbf{N}(\mathbf{A} + \mathbf{M}) + \mathbf{D}$, where $\mathbf{N}$ is a square matrix representing the population equilibrium in each local community with the non-diagonal elements zero. As discussed in

Chapter 4, the stability of meta-networks crucially depends on the leading eigenvalue of the Jacobian matrix.

With the same dispersal rate and intrinsic growth rate for all species and interaction strengths between any two species a_{ijx} correlated across networks x with ρ the correlation coefficient ($\rho = 0$ for high heterogeneity of interactions across networks; $\rho = 1$ for identical interaction strengths across networks), a simplified stability criterion can be derived (Gravel et al. 2016),

$$\sigma(C(S-1)/n_e)^{1/2} < m, \qquad (6.5)$$

with $n_e = n/(1 + (n-1)\rho)$ the effective number of ecologically independent local networks. Note, this is equivalent to May's stability criterion if $\rho = 1$: the maximal admissible complexity, $\sigma^2 C(S-1)$, is the same as one local network. If $\rho = 0$, meaning the interaction strength of two species is completely context-dependent, the meta-network can allow the maximal admissible complexity to be n times that of a local network. However, when the dispersal rate is low (small d) in a large meta-network (large S and n), the stability criterion becomes,

$$\sigma(C(S-1))^{1/2} < m + d. \qquad (6.6)$$

This means that weak dispersal can stabilise meta-networks (Figure 6.24). Overall, a meta-network becomes more stable when interaction strengths are network-specific, or when the dispersal rate is low in large meta-networks. Further analyses could also reveal the role of adaptive interaction-switching in local networks, enhancing stability by reducing the standard deviation of strengths of any realised interactions. Moreover, interaction strength a_{ijx} can be trait-mediated, i.e., a function of the trait values of species i and j in network x, which will further allow us to examine trait (co)evolution in meta-networks. Next, we argue that biological invasions can violate this stability criterion of meta-networks, thereby pushing the regional spatial network to undergo transition.

6.6 Percolation Transition

Network ecology has been expanding rapidly in two directions: multi-layer networks composed of multiple functional guilds and interaction types (e.g., Pocock et al. 2012), and meta-networks composed of multiple interdependent local networks connected via exchange of propagules (e.g., Shekhtman et al. 2018). As discussed previously, weak

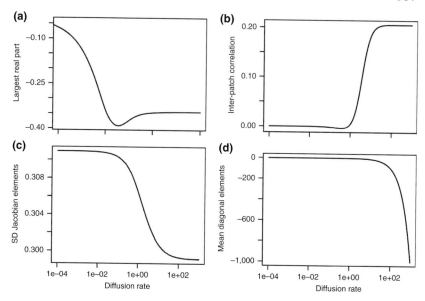

Figure 6.24 Effect of diffusion rate on Jacobian matrices and their stability. (a) The real part of the largest eigenvalue, (b) the s.d. of non-null elements of the Jacobian matrix, (c) inter-patch correlation and (d) mean of the diagonal elements. Meta-ecosystems were generated randomly with the constraint that all populations have positive equilibrium densities. Each line represents the average of 100 replicated random meta-ecosystems with increasing diffusion rate. From Gravel et al. (2016), under CC-BY Licence.

dispersal can enhance intraspecific interactions and promote the rescue effect and thus stability (Hanski and Ovaskainen 2000; Gravel et al. 2016). While density-independent dispersal promotes spatial synchrony, density-dependent dispersal breaks it down (Jansen 2001; Nguyen-Huu et al. 2006; Ramanantoanina et al. 2011). Biological invasions are challenging such spatial organisation and functioning of meta-networks. Spatial variability is a stabilising force in meta-networks, whereas biological invasions are a destabilising force (Figure 6.25; Donohue et al. 2013). Although evidence on how spatial networks and meta-networks function is accumulating (e.g., Vizentin-Bugoni et al. 2021), there is not enough understanding to draw conclusions on how biological invasions can bring about transformation in invaded meta-networks. Consequently, we end this chapter by proposing a few themes that warrant further attention.

First, dispersal is a key process that allows species to capture resources and evade harm. To this end, the spatial connectivity via dispersal

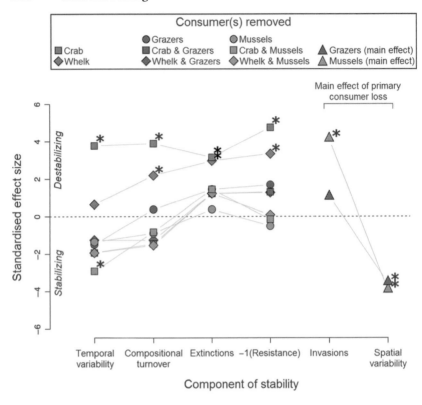

Figure 6.25 Effects of the loss of different consumer species on individual components of ecological stability. Standardised effect sizes based on the difference between experimental treatments and the corresponding treatment with no species removals. Treatments that differed significantly (P < 0.05) from the latter are highlighted with asterisks. The inverse of resistance is shown here so that all positive effect sizes correspond to reductions in stability. As the number of invasions and the spatial variability of algal cover were affected only by the loss of primary consumer species, effect sizes for these components of stability are shown only for the main effect of primary consumer loss. From Donohue et al. (2013), reprinted with permission.

between local populations should be settled near the threshold for critical percolation by natural selection. A set of clues support this proposition. In particular, Fronhofer et al. (2012) posed an interesting question: why are meta-populations so rare? Simulations in their study suggest that spatially structured populations will either be extirpated or become ubiquitous in a landscape, while the typical meta-population structure of small local populations rescuing each other only persists under narrow parameter ranges. That is, meta-population structures represent the borderline case between two spatial structure attractors (extinction versus dominance). A tentative answer to this puzzling observation has emerged in subsequent work: the eco-evolutionary dynamics of dispersal can explain such pervasive meta-population structures of spatial modules (Kubisch et al. 2014). Indeed, dispersal can be considered as an effective adaptive trait that can respond to a diverse array of selection pressures in meta-populations such as enhanced demographic performance of many island colonizers (Baker's law on the Good Coloniser Syndrome; Rodger et al. 2018). Over ecological timescales, such flexibility of dispersal strategy reflects the disperser's heuristic learning of heterogeneous environment (e.g., the *good-stay, bad-disperse* dispersal strategy; Hui et al. 2012b). In other words, dispersal, like other biotic interactions, can simply be a heuristic learning strategy that a species deploys to learn and contextualise its biotic and abiotic environment. The consequence of such dispersal evolution is that each residing species in the meta-network fine-tunes its spatially explicit dispersal rate to approach the percolation threshold (Figure 6.26; Gilarranz 2020). If this threshold is exceeded, the meta-population of the species becomes well connected and therefore faces overwhelming selection force against it (e.g., attacks from natural enemies can be felt over the entire landscape as a result of spatial connectivity). Below the threshold the meta-population degenerates into a few scattered local populations and cannot fully explore available niches in the landscape. Only when near the percolation threshold, can a species reach maximal flexibility and handle both the benefit and cost of spatial connectivity with only minor changes in its dispersal rate. Consequently, species that cannot fine-tune their dispersal rates to near the percolation threshold often face negative selection force and suffer the consequence of extirpation/extinction or failed attempts to invade and establish.

Second, as a direct consequence of this fine-tuning of dispersal rates at percolation threshold, spatial networks are extremely vulnerable and prone to collapse (Bashan et al. 2013; Shekhtman et al. 2014). Interdependency, measured as the spatial covariance between local networks, can greatly affect the threshold at which the system undergoes critical percolation transition. The stability of meta-networks, as introduced in the previous section, depends critically on the leading eigenvalue of the Jacobian matrix that consists of elements representing both biotic interactions and dispersal (Gravel et al. 2016). The capacity of an alien species to violate this stability criterion can be considered as the proxy for its capacity to invade this meta-network. In other words, alien species that lack strong self-

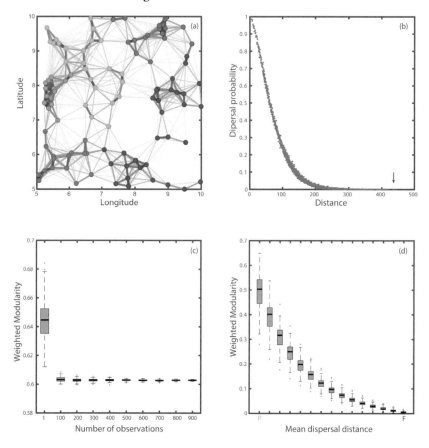

Figure 6.26 Dispersal kernel instead of dispersal threshold. (a) nodes with their respective coordinates linked by a probabilistic dispersal network. Link width is proportional to the frequency at which two nodes are connected when drawing random dispersal events following a Weibull distribution (methods). Node colours represent the module partition that maximizes modularity in weighted networks. (b) Posterior dispersal kernel obtained from panel a. It shows the dispersal probability as a function of distance between nodes. The arrow points the maximum dispersal distance recorded (c) Convergence of the value of modularity as a function of the number of observations. Each observation is a binary network where a link is established as a function of a probability dispersal kernel. The number of observations represent how many of these networks we add together to create the weighted network over which we calculate modularity. The distributions correspond to combinations of 100 weighted networks. (d) It shows how modularity decreases as a function of mean of the Weibull distribution used to determine whether a link is established for 100 random landscapes. From Gilarranz (2020), under CC-BY Licence.

regulation, possess large niche breadth, impose strong interspecific inter-actions and experience weak density-independent dispersal are likely to succeed. Certainly, more realistic stability criteria can be developed to more explicitly consider the types of biotic interactions and rates of introduction, as well as the spatial distributions of initial propagules. For a single species, a similar persistence criterion – the carrying capacity of meta-population – has been developed which is dependent also on the leading eigenvalue of the system dynamics (Hanski and Ovaskainen 2000). In addition, meta-networks are spatial multiplexes that are prone to collapses from localised and clustered invasions (Vaknin et al. 2017; Barthelemy 2018). We need to be able to quantify the degree of openness and embeddedness. A closed meta-community (such as the meta-network of the Galapagos Islands which is isolated from the continental ecosystem by a large expanse of ocean) is different from reserve networks in terrestrial ecosystems (e.g., of the Czech Republic; Divíšek et al. 2018) that are embedded in an even larger network of protected areas. Strict stability applies only to isolated meta-networks. For an embedded meta-network, instability can be presumed. In general, it is clear that these networks have unique vulnerability properties that contrast significantly with those of non-spatial networks. One big step towards a better understanding of spatially explicit invasibility is the elucidation of how the spatial network structures affect their vulnerabilities to spatially explicit invasions. Predictions of invasion dynamics at the regional scale, therefore, require the estimation of (i) local network interactions, (ii) species-specific dispersal rate between local networks and (iii) propagule introduction of both native and alien species into the regional meta-network, to allow us to parameterise its Jacobian matrix and assess its leading eigenvalue.

Finally, with the Jacobian matrix we can ask not only how each species in the meta-network waxes and wanes but also where. As it has been established that each local network undergoes persistent transition and turnover due to the effects of biological invasions at marginal instability (Chapter 5), the entire meta-network is also expected to experience transition and turnover. Eigenvectors of the meta-network Jacobian matrix can reveal the direction and magnitude of such transient dynamics. We think that the same weather vane – the coordinates along the principle component of the meta-network Jacobian – also operates to indicate the short-term demographic trend of a species' population in a particular local network. In addition, this weather vane allows us to assess how the initial propagules of an alien species should be distributed spatially to deliver maximal or minimal invasiveness. The crucial locations of a spatial network can be further assessed from changes in stability from removing particular local communities in the meta-network (Demšar et al. 2008). In particular, as the Jacobian matrix contains entries representing both biotic interactions and dispersal, the transient dynamics captured by its eigenvectors will allow us to investigate the spatial gradient of core versus peripheral communities, and the functional gradient of hub (core) and satellite (peripheral) species. For instance, whether alien species are more likely to establish in core or peripheral communities (an issue of spatial invasibility); and whether alien species need to have similar traits and distributions to those of core or peripheral resident species (functional and spatial invasiveness). This, therefore, is only a prelude to Invasion Science 2.0.

References

Alpert P, Bone E, Holzapfel C (2000) Invasiveness, invasibility and the role of environmental stress in the spread of non-native plants. *Perspectives in Plant Ecology, Evolution and Systematics* 3, 52–66.

Andreazzi CS, Thompson JN, Guimarães Jr PR (2017) Network structure and selection asymmetry drive coevolution in species-rich antagonistic interactions. *The American Naturalist* 190, 99–115.

Araújo MB, Rozenfeld A (2014) The geographic scaling of biotic interactions. *Ecography* 37, 406–415.

Barthelemy M (2018) Transitions in spatial networks. *Comptes Rendus Physique* 19, 205–232.

Barwell LJ, et al. (2014) Can coarse-grain patterns in insect atlas data predict local occupancy? *Diversity and Distributions* 20, 895–907.

Bashan A, et al. (2013) The extreme vulnerability of interdependent spatially embedded networks. *Nature Physics* 9, 667–672.

Bersier L, Sugihara G (1997) Scaling regions for food web properties. *Proceedings of the National Academy of Sciences USA* 94, 1247–1251.

Borregaard MK, Rahbek C (2006) Prevalence of intraspecific relationships between range size and abundance in Danish birds. *Diversity and Distributions* 12, 417–422.

Bosc C, Roets F, Hui C, Pauw A (2018) Interactions among predators and plant specificity protect herbivores from top predators. *Ecology* 99, 1602–1609.

Boyero L, et al. (2015) Leaf-litter breakdown in tropical streams: Is variability the norm? *Freshwater Science* 34, 759–769.

Brose U, et al. (2004) Unified spatial scaling of species and their trophic interactions. *Nature* 428, 167–171.

Buckley YM, Bolker BM, Rees M (2007) Disturbance, invasion and re-invasion: Managing the weed-shaped hole in disturbed ecosystems. *Ecology Letters* 10, 809–817.

Bufford JL, et al. (2020) Novel interactions between alien pathogens and native plants increase plant–pathogen network connectance and decrease specialization. *Journal of Ecology* 108, 750–760.

Capinha C, et al. (2015) The dispersal of alien species redefines biogeography in the Anthropocene. *Science* 348, 1248–1251.

Catford JA, Jansson R, Nilsson C (2009) Reducing redundancy in invasion ecology by integrating hypotheses into a single theoretical framework. *Diversity and Distributions* 15, 22–40.

Chesson P (2012) Scale transition theory: Its aims, motivations and predictions. *Ecological Complexity* 10, 52–68.

Cohen JE, Briand F (1984) Trophic links of community food webs. *Proceedings of the National Academy of Sciences USA* 81, 4105–4109.

Condit R (1998) *Tropical Forest Census Plots*. Berlin: Springer.

Cowling RM, Pressey RL (2001) Rapid plant diversification: Planning for an evolutionary future. *Proceedings of the National Academy of Sciences USA* 98, 5452–5457.

Cox CB, Moore PD (2000) *Biogeography: An Ecological and Evolutionary Approach*. Oxford: Blackwell.

Daleo P, Alberti J, Iribarne O (2009) Biological invasions and the neutral theory. *Diversity and Distributions* 15, 547–553.

Davies KF, et al. (2005) Spatial heterogeneity explains the scale dependence of the native-exotic diversity relationship. *Ecology* 86, 1602–1610.

Demšar U, Špatenková O, Virrantaus K (2008) Identifying critical locations in a spatial network with Graph Theory. *Transactions in GIS* 12, 61–82.

Diamond JM (1975) The island dilemma: Lessons of modern biogeography studies for the design of natural reserves. *Biological Conservation* 7, 129–146.

Divišek J, et al. (2018) Similarity of introduced plant species to native ones facilitates naturalization, but differences enhance invasion success. *Nature Communications* 9, 4631.

Donaldson JE, et al. (2014) Invasion trajectory of alien trees: The role of introduction pathway and planting history. *Global Change Biology* 20, 1527–1537.

Donohue O, et al. (2013) On the dimensionality of ecological stability. *Ecology Letters* 16, 421–429.

Ellis EC, Ramankutty N (2008) Putting people in the map: Anthropogenic biomes of the world. *Frontiers in Ecology and the Environment* 6, 439–447.

Elton CS (1958) *The Ecology of Invasions by Animals and Plants*. London: Methuen.

Ferrier S, et al. (2007) Using generalized dissimilarity modelling to analyse and predict patterns of beta diversity in regional biodiversity assessment. *Diversity and Distributions* 13, 252–264.

Fricke EC, Svenning JC (2020) Accelerating homogenization of the global plant–frugivore meta-network. *Nature* 585, 74–78.

Fridley JD, et al. (2007) The invasion paradox: Reconciling pattern and process in species invasions. *Ecology* 88, 3–17.

Fronhofer EA, et al. (2012) Why are metapopulations so rare? *Ecology* 93, 1967–1978.

Gaertner M, et al. (2009) Impacts of alien plant invasions on species richness in Mediterranean-type ecosystems: A meta-analysis. *Progress in Physical Geography* 33, 319–338.

Galiana N, et al. (2018) The spatial scaling of species interaction networks. *Nature Ecology and Evolution* 2, 782–790.

Gilarranz LJ (2020) Generic emergence of modularity in spatial networks. *Scientific Reports* 10, 8708.

González-Olivares E, Ramos-Jiliberto R (2003) Dynamic consequences of prey refuges in a simple model system: More prey, fewer predators and enhanced stability. *Ecological Modelling* 166, 135–146.

Gravel D, Massol F, Leibold MA (2016) Stability and complexity in model meta-ecosystems. *Nature Communications* 7, 12457.

Guisan A, et al. (2013) Predicting species distributions for conservation decisions. *Ecology Letters* 16, 1424–1435.

Gurevitch J, et al. (2016) Landscape demography: Population change and its drivers across spatial scales. *Quarterly Review of Biology* 91, 451–485.

Han XZ, Hui C (2014) Niche construction on environmental gradients: The formation of fitness valley and stratified genotypic distributions. *PLoS ONE* 9, e99775.

Hansen BB, et al. (2020) The Moran effect revisited: Spatial population synchrony under global warming. *Ecography* 43, 1591–1602.

Hanski I (1998) Metapopulation dynamics. *Nature* 396, 41–49.

Hanski I, Gilpin ME (1997) *Metapopulation Biology: Ecology, Genetics, and Evolution.* San Diego: Academic Press.

Hanski I, Ovaskainen O (2000) The metapopulation capacity of a fragmented landscape. *Nature* 404, 755–758.

Haywood AM, et al. (2019) What can palaeoclimate modelling do for you? *Earth Systems and Environment* 3, 1–18.

He F, Gaston KJ (2000) Estimating species abundance from occurrence. *American Naturalist* 156, 553–559.

Hobbs RJ, Higgs E, Harris JA (2009) Novel ecosystems: Implications for conservation and restoration. *Trends in Ecology & Evolution* 24, 599–605.

Hubbell SP, et al. (1999) Light-gap disturbances, recruitment limitation, and tree diversity in a Neotropical forest. *Science* 283, 554–557.

Hubbell SP, et al. (2005) Barro Colorado Forest Census Plot Data. Center for Tropical Forest Science. https://ctfs.arnarb.harvard.edu/webatlas/datasets/bci.

Hui C (2009) On the scaling pattern of species spatial distribution and association. *Journal of Theoretical Biology* 261, 481–487.

Hui C (2011) Forecasting population trend from the scaling pattern of occupancy. *Ecological Modelling* 222, 442–446.

Hui C (2021) Introduced species shape insular mutualistic networks. *Proceedings of the National Academy of Sciences USA* 118, e2026396118.

Hui C, McGeoch MA (2006) Evolution of body size, range size, and food composition in a predator–prey metapopulation. *Ecological Complexity* 3, 148–159.

Hui C, McGeoch MA (2014) Zeta diversity as a concept and metric that unifies incidence-based biodiversity patterns. *The American Naturalist* 184, 684–694.

Hui C, Richardson DM (2017) *Invasion Dynamics.* Oxford: Oxford University Press.

Hui C, et al. (2006) A spatially explicit approach to estimating species occupancy and spatial correlation. *Journal of Animal Ecology* 75, 140–147.

Hui C, et al. (2009) Extrapolating population size from the occupancy-abundance relationship and the scaling pattern of occupancy. *Ecological Applications* 19, 2038–2048.

Hui C, et al. (2011) Macroecology meets invasion ecology: Linking the native distributions of Australian acacias to invasiveness. *Diversity and Distributions* 17, 872–883.

Hui C, et al. (2012a) Estimating changes in species abundance from occupancy and aggregation. *Basic and Applied Ecology* 13, 169–177.

Hui C, et al. (2012b) Flexible dispersal strategies in native and non-native ranges: Environmental quality and the 'good-stay, bad-disperse' rule. *Ecography* 35, 1024–1032.

Hui C, et al. (2013) A cross-scale approach for abundance estimation of invasive alien plants in a large protected area. In Foxcroft LC, Pyšek P, Richardson DM, Genovesi P (eds.), *Plant Invasions in Protected Areas*, pp. 73–88. Dordrecht: Springer.

Hui C, et al. (2013) Increasing functional modularity with residence time in the co-distribution of native and introduced vascular plants. *Nature Communications* 4, 2454.

Hui C, et al. (2014) Macroecology meets invasion ecology: Performance of Australian acacias and eucalypts around the world revealed by features of their native ranges. *Biological Invasions* 16, 565–576.

Hui C, et al. (2017) Scale-dependent portfolio effects explain growth inflation and volatility reduction in landscape demography. *Proceedings of the National Academy of Sciences USA* 114, 12507–12511.

Hui C, et al. (2020) The role of biotic interactions in invasion ecology: Theories and hypotheses. In Traveset A, Richardson DM (eds.), *Plant Invasions: The Role of Biotic Interactions*, pp. 26–44. Wallingford: CAB International.

Ims RA, Andreassen HP (2000) Spatial synchronization of vole population dynamics by predatory birds. *Nature* 408, 194–196.

Jansen VAA (2001) The dynamics of two diffusively coupled predator–prey populations. *Theoretical Population Biology* 59, 119–131.

Kubisch A, et al. (2014) Where am I and why? Synthesizing range biology and the eco-evolutionary dynamics of dispersal. *Oikos* 123, 5–22.

Kunin WE, et al. (2018) Upscaling biodiversity: estimating the species–area relationship from small samples. *Ecological Monographs* 88, 170–187.

Latombe G, et al. (2018) Drivers of species turnover vary with species commonness for native and alien plants with different residence times. *Ecology* 99, 2763–2775.

Latombe G, Hui C, McGeoch MA (2017) Multi-site generalised dissimilarity modelling: using zeta diversity to differentiate drivers of turnover in rare and widespread species. *Methods in Ecology and Evolution* 8, 431–442.

Latombe G, Roura-Pascual N, Hui C (2019) Similar compositional turnover but distinct insular environmental and geographical drivers of native and exotic ants in two oceans. *Journal of Biogeography* 46, 2299–2310.

Legendre P, Legendre L (1998) *Numerical Ecology*, 2nd Edition. Amsterdam: Elsevier.

Levine JM (2000) Species diversity and biological invasions: Relating local process to community pattern. *Science* 288, 852–854.

Levine JM (2003) Local interactions, dispersal, and native and exotic plant diversity along a California stream. *Oikos* 95, 397–408.

Levins R (1969) Some demographic and genetic consequences of environmental heterogeneity for biological control. *Bulletin of the Entomological Society of America* 15, 237–240.

Li Z, et al. (2005) Impact of predator pursuit and prey evasion on synchrony and spatial patterns in metapopulation. *Ecological Modelling* 185, 245–254.

Liebhold A, Koenig WD, Bjørnstad ON (2004) Spatial synchrony in population dynamics. *Annual Review of Ecology, Evolution, and Systematics* 35, 467–490.

Lui C, et al. (2020) Species distribution models have limited spatial transferability for invasive species. *Ecology Letters* 23, 1682–1692.

Mack RN, et al. (2000) Biotic invasions: Causes, epidemiology, global consequences and control. *Ecological Applications* 10, 689–710.

Martinez ND (1992) Constant connectance in community food webs. *American Naturalist* 139, 1208–1218.

McGeoch MA, et al. (2019) Measuring continuous compositional change using decline and decay in zeta diversity. *Ecology* 100, ee02832.

McGill BJ (2010) Matters of scale. *Science* 328, 575–576.

Melbourne BA, et al. (2007) Invasion in a heterogeneous world: Resistance, coexistence or hostile takeover? *Ecology Letters* 10, 77–94.

Meynard CN, et al. (2013) Disentangling the drivers of metacommunity structure across spatial scales. *Journal of Biogeography* 40, 1560–1571.

Mohammed MMA, et al. (2018) Frugivory and seed dispersal: Extended bi-stable persistence and reduced clustering of plants. *Ecological Modelling* 380, 31–39.

Moran PAP (1953) The statistical analysis of the Canadian Lynx cycle. *Australian Journal of Zoology* 1, 291–298.

Morlon H, et al. (2008) A general framework for the distance-decay of similarity in ecological communities. *Ecology Letters* 11, 904–917.

Mucina L (2019) Biome: Evolution of a crucial ecological and biogeographical concept. *New Phytologist* 222, 97–114.

Munkmuller T, Johst K (2008) Spatial synchrony through density independent versus density-dependent dispersal. *Journal of Biological Dynamics* 2, 31–39.

Naeem S, et al. (2000) Plant diversity increases resistance to invasion in the absence of covarying extrinsic factors. *Oikos* 91, 97–108.

Nguyen-Huu T, et al. (2006) Spatial synchrony in host–parasitoid models using aggregation of variables. *Mathematical Biosciences* 203, 204–221.

Olson DM, et al. (2001) Terrestrial ecoregions of the world: A new map of life on Earth. *Bioscience* 51, 933–938.

Openshaw S (1984) *Modifiable Areal Unit Problem*. Norwich: Geo Books.

Oshanin G, et al. (2009) Survival of an evasive prey. *Proceedings of the National Academy of Sciences USA* 106, 13696–13701.

Palmer MW, Maurer TA (1997) Does diversity beget diversity? A case study of crops and weeds. *Journal of Vegetation Science* 8, 235–240.

Peterson AT, Soberón J (2012) Species distribution modeling and ecological niche modeling: Getting the concepts right. *Natureza & Conservação* 10, 102–107.

Peterson AT, et al. (2011) *Ecological Niches and Geographic Distributions*. Princeton: Princeton University Press.

Pocock MJO, et al. (2012) The robustness and restoration of a network of ecological networks. *Science* 335, 973–977.

Poisot T, Stouffer DB, Gravel D (2014) Beyond species: Why ecological interaction networks vary through space and time. *Oikos* 124, 243–251.

Pyšek P, et al. (2005) Alien plants in temperate weed communities: Prehistoric and recent invaders occupy different habitats. *Ecology* 86, 772–785.

Pyšek P, et al. (2010) Disentangling the role of environmental and human pressures on biological invasions across Europe. *Proceedings of the National Academy of Sciences USA* 107, 12157–12162.

Pyšek P, et al. (2012) Catalogue of alien plants of the Czech Republic (2nd edn): Checklist update, taxonomic diversity and invasion patterns. *Preslia* 84, 155–255.

Pyšek P, Richardson DM (2006) The biogeography of naturalization in alien plants. *Journal of Biogeography* 33, 2040–2050.

Ramanantoanina A, et al. (2011) Effects of density-dependent dispersal behaviours on the speed and spatial patterns of range expansion in predator–prey metapopulations. *Ecological Modelling* 222, 3524–3530.

Renne IJ, Tracy BF, Colonna IA (2006) Shifts in grassland invasibility: Effects of soil resources disturbance, composition, and invader size. *Ecology* 87, 2264–2277.

Richardson DM (2011) Invasion science: The roads travelled and the roads ahead. In Richardson DM (ed.), *Fifty Years of Invasion Ecology: The Legacy of Charles Elton*, pp. 397–407. Oxford: Wiley-Blackwell.

Richardson DM, Pyšek P (2012) Naturalization of introduced plants: Ecological drivers of biogeographic patterns. *New Phytologist* 196, 383–396.

Richardson DM, et al. (2005) Species richness of alien plants in South Africa: Environmental correlates and the relationship with indigenous plant species richness. *EcoScience* 12, 391–402.

Richardson DM, et al. (2020). The biogeography of South African terrestrial plant invasions. In Van Wilgen BW, Measey J, Richardson DM, et al. (eds.), *Biological Invasions in South Africa*, pp. 67–96. Berlin: Springer.

Rodger JG, et al. (2018) Heterogeneity in local density allows a positive evolutionary relationship between self-fertilisation and dispersal. *Evolution* 72, 1784–1800.

Rouget M, Richardson DM (2003) Understanding patterns of plant invasion at different spatial scales: Quantifying the roles of environment and propagule pressure. In Child L, et al. (eds.), *Plant Invasions: Ecological Threats and Management Solutions*, pp. 3–15. Leiden: Backhuys Publishers.

Rouget M, et al. (2015) Plant invasions as a biogeographical assay: Vegetation biomes constrain the distribution of invasive alien species assemblages. *South African Journal of Botany* 101, 24–31.

Roura-Pascual N, Sanders NJ, Hui C (2016) The distribution and diversity of insular ants: Do exotic species play by different rules? *Global Ecology and Biogeography* 25, 642–654.

Sax DF, et al. (2007) Ecological and evolutionary insights from species invasions. *Trends in Ecology & Evolution* 22, 465–471.

Schmid-Araya JM, et al. (2002) Connectance in stream food webs. *Journal of Animal Ecology* 71, 1056–1062.

Seebens H, et al. (2017) No saturation in the accumulation of alien species world-wide. *Nature Communications* 8, 14435.

Shea K, Chesson P (2002) Community ecology theory as a framework for biological invasions. *Trends in Ecology & Evolution* 17, 170–176.

Shekhtman L, et al. (2018) Robustness of spatial networks and networks of networks. *Comptes Rendus Physique* 19, 233–243.

Shekhtman LM, et al. (2014) Robustness of a network formed of spatially embedded networks. *Physical Review E* 90, 012809.

Simberloff D, Von Holle B (1999) Positive interactions of nonindigenous species: Invasional meltdown? *Biological Invasions* 1, 21–32.

Soininen J, et al. (2007) The distance decay of similarity in ecological communities. *Ecography* 30, 3–12.

Steidinger BS, et al. (2019) Climatic controls of decomposition drive the global biogeography of forest-tree symbioses. *Nature* 569, 404–408.

Stohlgren TJ, Barnett DT, Kartesz JT (2003) The rich get richer: Patterns of plant invasions in the United States. *Frontiers in Ecology and the Environment* 1, 11–14.

Stone L (2018) The feasibility and stability of large complex biological networks: a random matrix approach. *Scientific Reports* 8, 8246.

Stotz GC, et al. (2020) Not a melting pot: Plant species aggregate in their non-native range. *Global Ecology and Biogeography* 29, 482–490.

Tilman D (1994) Competition and biodiversity in spatially structured habitats. *Ecology* 75, 2–16.

Tilman D (2004) Niche tradeoffs, neutrality, and community structure: A stochastic theory of resource competition, invasion, and community assembly. *Proceedings of the National Academy of Sciences of the United States of America* 101, 10854–10861.

Traveset A, et al. (2013) Invaders of pollination networks in the Galápagos Islands: Emergence of novel communities. *Proceedings of the Royal Society B: Biological Sciences* 280, 20123040.

Vaknin D, et al. (2017) Spreading of localized attacks in spatial multiplex networks. *New Journal of Physics* 19, 073037.

Van Kleunen M, et al. (2010) Are invaders different? A conceptual framework of comparative approaches for assessing determinants of invasiveness. *Ecology Letters* 13, 947–958.

Van Wilgen BW, et al. (2011) National-scale strategic approaches for managing introduced plants: Insights from Australian acacias in South Africa. *Diversity and Distributions* 17, 1060–1075.

Vasseur DA, Fox JW (2009) Phase-locking and environmental fluctuations generate synchrony in a predator–prey community. *Nature* 460, 1007–1010.

Vicente JR, et al. (2019) Different environmental drivers of alien tree invasion affect different life-stages and operate at different spatial scales. *Forest Ecology and Management* 433, 263–275.

Vilà M, et al. (2011) Ecological impacts of invasive alien plants: A meta-analysis of their effects on species, communities and ecosystems. *Ecology Letters* 14, 702–708.

Vizentin-Bugoni J, et al. (2019) Structure, spatial dynamics, and stability of novel seed dispersal mutualistic networks in Hawai'i. *Science* 364, 78–82.

Vizentin-Bugoni J, et al. (2021) Ecological correlates of species' roles in highly invaded seed dispersal networks. *Proceedings of the National Academy of Sciences USA* 118, e2009532118.

Wickman J, et al. (2020) How geographic productivity patterns affect food-web evolution. *Journal of Theoretical Biology* 506, 110374.

Willig MR, Kaufman DM, Stevens RD (2003) Latitudinal gradients of biodiversity: Pattern, process, scale and synthesis. *Annual Review of Ecology, Evolution, and Systematics* 34, 273–309.

Wilson JRU, et al. (2009) Something in the way you move: Dispersal pathways affect invasion success. *Trends in Ecology & Evolution* 24, 136–144.

Wilson RJ, et al. (2004) Spatial patterns in species distributions reveal biodiversity change. *Nature* 432, 393–396.

Wright DH (1991) Correlations between incidence and abundance are expected by chance. *Journal of Biogeography* 18, 463–466.

Wright S (1943) Isolation by distance. *Genetics* 28, 114–138.

Zhao Z, et al. (2019) The failure of success: cyclic recurrences of a globally invasive pest. *Ecological Applications* 29, ee01991.

7 · *Rethinking Invasibility*

> So in war, the way is to avoid what is strong, and strike at what is weak.
>
> Sun Tzu, *The Art of War*

7.1 Invasion Science 2.0

Humanity's rise is rapidly moulding the structure and functioning of the biosphere over the surface of our planet, while human-mediated translocations of organisms – an inevitable consequence of this rise – is driving further transformation (Pyšek et al. 2020b). Drawing inspiration and concepts from population ecology, Invasion Science 1.0 (see Chapter 1) has explored the myriad ways a focal alien species can negotiate geographical, ecological and environmental barriers to establish and potentially invade in new novel environments. Coordinated efforts have been made to classify introduction pathways (Hulme et al. 2008; Wilson et al. 2009); forecast invasion risks (Kumschick and Richardson 2013) and impacts (Jeschke et al. 2014); model invasive spread (Hui and Richardson 2017); unify invasion frameworks (Wilson et al. 2020); and prescribe management strategies such as early detection and rapid response to prevent, contain and eradicate problematic species (Wilson et al. 2017). However, the phenomenon of biological invasions involves all types of organisms, ecosystems and a wide range of contexts and framings; this has given rise to a plethora of invasion hypotheses and theories that seek to explain and ultimately predict aspects of invasion dynamics and the expected outcome of specific management actions (Jeschke and Heger 2018). Most invasion hypotheses are relevant in specific contexts and often fail when faced with the reality of contextual complexity. This has led to a wave of syntheses that have attempted to classify invasion cases and hypotheses based largely on three aspects – invasive traits, site characteristics and invasion pathways (Pyšek et al. 2020a). To embrace considerations that arise when attempting to merge

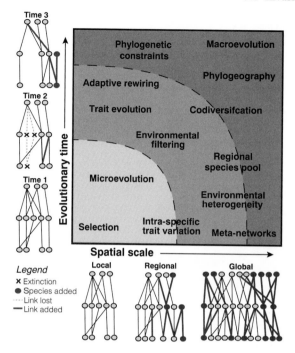

Figure 7.1 A conceptual framework for drivers of network structure across evolutionary and spatial scales. From Segar et al. (2020), reprinted with permission.

insights from all these perspectives, a paradigm shift began emerging at the turn of the millennium, together with the rise of network science. It embraces the complexity of biotic interaction networks (Figure 7.1; e.g., Segar et al. 2020), the trait paradigm in community ecology (Figure 7.2; e.g., McGill et al. 2006; Salguero-Gómez et al. 2018), and considers how functional traits of species dictate their ecology and roles in networks (Figure 7.3; e.g., Mello et al. 2019). This new lens for drawing together threads pertaining to all facets of biological invasions (Invasion Science 2.0) seeks to elucidate the structure and function of an ecological network facing biological invasions. This book has laid out a road map of signposts, hazard warnings and shortcuts for the journey to Invasion Science 2.0, framing and classifying research topics and offering tentative solutions and travel advisories.

The intension of this hitchhiker's guide is to expose the knowledge terrain and stimulate coordinated research to address key challenges facing the new paradigm that underpins Invasion Science 2.0. In doing so, we have taken several routes. Chapter 2 explored the nature of the

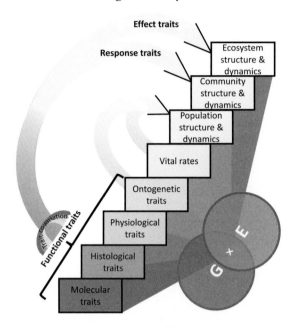

Figure 7.2 Two of the most widely used approaches for examining drivers and consequences in trait variation are the trait-based approach and the demographic approach. In the trait-based approach, molecular (e.g., oxidative stress), histological (e.g., bone/wood density), physiological (e.g., photosynthetic rate) and ontogenetic traits (e.g., adult height) are used to explore trait–trait covariation (e.g., Chave et al. 2009; Díaz et al. 2016; Wright et al. 2004). These so-called functional traits are also used to upscale to describe the structure and dynamics of communities (e.g., McGill et al. 2006) and ecosystems (e.g., Gross et al. 2017) using response traits and effect traits. In this up-scaling, typically the demographic compartment (vital rates and populations) is not considered. The demographic approach examines how vital rates (e.g., survival, development, reproduction) scale with ontogenetic characteristics of individuals (e.g., age, size, development) to inform on population structure and dynamics (e.g., Caswell 2001). Both trait-based and demographic approaches share similarities in the questions they target (e.g., Genetics × Environment interactions; Barks et al. 2018; Vasseur et al. 2018), and the recent macroecological patterns of trait covariation they have reported (Díaz et al. 2016; Salguero-Gómez et al. 2016). From Salguero-Gómez et al. (2018), under CC-BY Licence.

building blocks of an ecological network – biotic interactions. We defined biotic interactions as the reciprocal exploitation between interacting species, as the result of repeated games between individuals of the same or different species, natives or aliens. As games, the outcomes of biotic interactions depend not only on the strategies of any given focal species but also on the strategies of other opponents involved in the

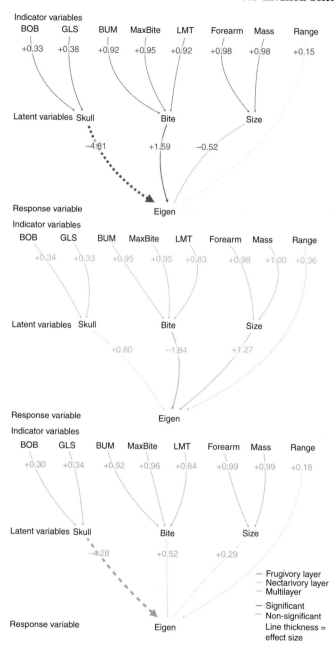

Figure 7.3 Influence of organismal traits on centrality. The relative importance of a bat species in regard to the structure of the network (measured as its eigenvector centrality, Eigen) was indirectly determined by a combination of organismal traits

interactions. Interaction strengths can be defined according to the benefit and cost (thus payoff) from such games, while interaction kernels depict how the interaction strength of an interaction changes with each player's traits (game strategies). The coexistence of two interacting species requires fulfilling the condition for mutual invasibility; that is, between species A and B, species A can invade a community where B resides at the population equilibrium, and similarly B can invade a community where A is at equilibrium. Invasion success, thus, requires only one side of mutual invasibility, without the necessity for species coexistence, and thus makes species extirpation possible. Moreover, as the functional force behind interaction strength, the traits of a species can greatly affect its demographic performance and fitness in an ecological network; they thus experience strong selection that drives evolutionary dynamics. An introduced species partakes in this relentless dance in the invaded ecological network with partners by establishing novel interactions through interaction switching and co-evolution.

When multiple species are assembled in an ecological network, network structures emerge in this multiplayer game through self-organisation via ecological fitting and co-evolution. Chapter 3 explained how network structures can emerge during community succession and assembly, especially from the constant influx of alien propagules in many novel ecosystems. We synthesised information, from the scant literature, on the expected network structures (e.g., connectance, nestedness and modularity) for three types of ecological networks (competitive communities, mutualistic networks and antagonistic networks), and anticipated

---

Figure 7.3 (cont.) (indicator variables). Those traits were used to calculate the latent variables: skull morphology (Skull), bite force (Bite) and body size (Size). Those latent variables, together with a single indicator variable (geographic range size, Range), directly determined the eigenvector centrality. In other words, a bat species was more important to the structure of the frugivory layer (blue lines) when it had a strong bite force and a small body size. In the nectarivory layer, larger bats with weak bite force were the most important. Finally, in the multilayer structure, the most important roles were played by bats that bite more forcefully. Numbers on lines represent effect sizes (standardized path coefficients calculated in the latent variable analysis), and line thickness is drawn proportionally to effect size. Significance was estimated only for the main variables (skull, bite, size and range). Indicator variable names: BOB, breadth of braincase; GLS, greatest length of skull; BUM, breadth across upper molars; MaxBite, maximum bite force; LMT, length of maxillary tooth row; Forearm, forearm length; Mass, body mass. From Mello et al. (2019), reprinted with permission.

changes in network structures affected by biological invasions. Although consensus is yet to be reached, there seem to be neither universal structures in ecological networks nor uniformed structural changes that emerge as a result of biological invasions. Contextual contingency is the rule. In depicting the structural formation in ecological networks we introduced a number of structural emergence models based on randomly picking species along selected trait/niche spectra, and dynamic network models based on co-evolution of adaptive traits and ecological fitting via interaction switching. These models allow us to explore how alien species can be integrated into the ecological networks in recipient ecosystems, and the ripple effects it imposes at both the ecological and evolutionary time scales. Although the structures and patterns of invaded ecological networks can be diverse, a limited number of underlying processes are implicated. It is the interplay of this limited set of processes, especially those involved in trait-mediated biotic interactions, that allow us to make sense of the structural emergence in an ecological network and its responses to biological invasions.

Chapter 4 argues that the key to understanding the diverse emerged network structures from community assembly and biological invasions lies in dissecting and unpacking the multifaceted meaning of system stability. In other words, network structures are the manifesto and consequence of network functioning (stability being the crux of system functioning). When invaded ecosystems are considered as an open adaptive system (OAS), the exhibited network structures should reflect the dynamic behaviour of the regime of the OAS (or the loss and reorganisation of its regime) when challenged by perturbations, including biotic incursions. This reorganisation of system regimes has been crystallised as *panarchy*. Furthermore, as an open system, we argue that a more explicit measure of stability (or instability in this case) is the temporal turnover of network compositions and interactions from invasions and resulting extinctions. Importantly, the susceptibility of an ecological network to invasions is related to the loss of system stability, or the propensity to move away from a system's standing regime. Biological invasions, involving the continuous bombardment of alien propagules, can push the recipient ecological network out of its current basin of attraction towards alternative regimes. We followed a few leads that suggest that invaded ecological networks often experience constant collapse and turnover at marginal instability, as the alternative regime is yet to be formed. At this critical point of marginal instability, the well-known relationship between system complexity and stability, dubbed the stability criterion, is constantly being violated, and is consequently decoupled. At this criticality, the network becomes highly flexible in handling incoming perturbations. Network structures emerge spontaneously at the marginal instability from perturbation and biological invasions, while network compositions and interactions undergo persistent transitions and turnover. A well-shaped whirlpool tentatively emerges from the constant flows of species.

With the loss of stability, we could inform invasion management using a weather vane that can forecast the transient population dynamics of all resident and alien species. In Chapter 5, we first discussed the floundering of early warning signals that have been proposed in attempting to forecast imminent regime shift in complex systems. The reason, we think, is that the identity and composition of ecological networks are in constant transition under biological invasions; such behaviour is reflected in the persistent temporal turnover of both species and interactions in the ecological network. Instead of forecasting potential regime shifts, it is more practical to predict the direction and magnitude of change – hence our proposed weather vane for dynamic forecasting in invaded ecological networks. The weather vane of an ecological network is defined as the major axis of the network interaction matrix. We also showed that rare species, especially those that lack self-regulation, play critical roles in the stability of ecological networks. The rarity gradient and the weather vane could be driven by different mechanisms in different networks. Consequently, we argue that identifying this rarity gradient and its relationship with the weather vane is key to explaining how an initially rare invader can potentially establish and invade an ecological network. Effectively, invasion dynamics follow the procedure of climbing this rarity gradient, from being rare to becoming common, while some resident species are evicted from the community along the gradient. This highlights the need to explore the gradient of commonness and rarity in ecological networks, and its relationship with the weather vane.

An ecological network is not spatially isolated, but is embedded in a larger regional meta-network. Chapter 6 discusses how 'alien biomes' emerge; how network structure and stability change with spatial scales; and how dispersal of propagules between local networks and beyond the boundary of regional networks affects the stability and transient dynamics of meta-networks. We meld these topics with meta-networks using the phenomenon of network percolation at critical transition. Each species can fine-tune its spatially explicit dispersal rate at the percolation threshold to maximise the benefit and minimise the cost of being connected. This highlights two key drivers of the structure and stability of meta-networks: biotic interactions and dispersal. Chapters 2 to 6 provide a travel guide on the road to Invasion Science 2.0. In particular, we discussed five interlinked excursions on ecological networks facing biological invasions, including network interactions, structures, stability, dynamics and spatial scaling. The aim of this final chapter is to traverse the terrain of Invasion Science 2.0 using the breadcrumb trail assembled in the previous chapters. We seek to derive some universal features of invasion performance in an invaded ecological network and explore how network invasibility can be captured by invasion fitness over the trait space.

7.2 Eco-evolutionary Dynamics of an Open Adaptive Network

An invaded ecological network possesses two key features: open systems and adaptive components. The system experiences constant fluxes of native and alien propagules entering and exiting the system, and its dynamics are increasingly influenced by human activities. This can be effectively formulated as dispersal; although influxes are often independent from the population dynamics in the target ecosystem they can nonetheless have a major effect on the eco-evolutionary dynamics of that system. Being adaptive means that the residing species do not simply retain their prevailing strategies during this play, but change adaptively, fine-tuning the strategies through natural selection to improve each player's own payoff (and thus fitness). This can be effectively formulated as the evolution of adaptive traits that can result in co-evolution and adaptive interaction switching. As explained in Chapter 2, this grand game is ultimately played at the individual level, while the population- and species-level payoff and fitness, on which we often focus, are only aggregated quantities. The open adaptive network resulting from this multiplayer eco-evolutionary game is not system-level optimisation – optimality for one species is often suboptimal to another, which drives relentless co-evolution. We now discuss the eco-evolutionary dynamics of an open adaptive network under biological invasions that embraces these two features (Hui et al. 2021). The model paves the way for us to revisit the concepts of trait-mediated invasiveness and invasibility in ecological networks in the following sections.

Following the trait paradigm in community ecology, we can consider traits simply as a player's strategies in this eco-evolutionary game. Importantly, these traits help to define and differentiate the functional role of a species in an ecological network. To this end, not only these typical life-history traits but also any features, such as preference, plasticity, phenology and phylogeny, as well as abundances, can be considered as traits as long as they can affect a species' payoff. As outlined in Chapter 2, how these traits can affect the interaction strength defines interaction kernels. Together with trait-dependent influx propagules, these traits dictate a species' demographic perform-ance and thus eco-evolutionary dynamics in the ecological network. We now trace a few key steps of formulating an open adaptive eco-logical network with trait-mediated dispersal and interactions (for detailed discussion, see Hui et al. 2021). Assume that there are S distinct resident species in a network, with species i characterised by its popula-tion size (n_i), trait value (x_i), influx rate of propagules (γ_i) from outside the network with trait value (z_i). From Chapter 2 and 3, we have shown that, without propagule influxes, the per-capita population change rate of species i, f_i, is simply a function of the abundances of and interaction strengths from all resident species (see Section 2.2), $f_i = r_i + \sum_{j=1}^{S} \alpha_{ij} n_j$, where r_i is the intrinsic rate of increase (i.e., per-capita growth rate in an empty network), and α_{ij} ($= \partial f_i/\partial n_j$ at zero abundances) the per-capita interaction impact of species j on the intrin-sic rate of species i. Following the trait paradigm, all demographic components are trait dependent. In particular, the intrinsic rate of increase of species i can be considered a function of its trait, $r_i = r(x_i)$. The per-capita interaction strength follows the typical interaction kernel (Eq. (2.12)), $\alpha_{ij} = \alpha(x_i, x_j) = \pm \exp\left(-(x_i - x_j - \mu)^2/2\sigma^2\right)$. That is, individuals of two different species but with identical traits experi-ence equal demographic performance, $f_i = f_j$ if $x_i = x_j$; however, this does not mean that two individuals with equal performance need to have the same trait.

The constant influx of propagules opens up this ecological network. In particular, over a short period (τ), let there be a number of $\gamma_i \tau$ propagules of species i introduced into this network. When including propagule pressure from external sources as a constant inflow of individuals at rate γ_i, we could estimate the magnitude of population change after this short period as $n_i(t + \tau) - n_i(t) = \gamma_i(t)\tau + n_i(t)f_i(t)\tau$. This then leads to the following differential equation (by letting $\tau \to 0$) which describes the population dynamics of species i (Hui et al. 2021),

$$\dot{n}_i = \gamma_i + n_i \left(r(x_i) + \sum_{j=1}^{S} \alpha(x_i, x_j) n_j \right). \tag{7.1}$$

When the network is isolated without the influx of propagules (i.e. $\gamma_i = 0$), the population dynamics depend entirely on its trait value in relation to the distribution of traits of other species. In contrast, when there is an influx of propagules into this open network ($\gamma_i > 0$), propagule pressure, unsurprisingly, affects the population dynamics of a species.

As a species' traits can affect its demographic performance and fitness, it experiences selection forces as a result of such trait-dependent fitness. To express the evolutionary dynamics of the trait, following Section 2.6, we first estimate the per-capita growth rate of rare mutants, f_i'. Assuming incremental evolution that the mutant inherits the ecological function of its source population but with minute differences, we can have a similar but not identical per-capita growth rate of mutants, $f_i' = r(x_i') + \sum_{j=1}^{S} \alpha(x_i', x_j)n_j$, from which the selection gradient can be defined (Section 2.6), $s_i \equiv \partial f_i'/\partial x_i'\big|_{x_i}$, while the dynamics of trait evolution is proportional to the selection gradient (Dieckmann and Law 1996; Geritz et al. 1998), $\dot{x}_i = v_i s_i$, where v_i reflects a compound factor of trait variability of species i in the network and is chosen to be a small positive number in practice to scale the pace of trait evolutionary dynamics relative to ecological dynamics.

Propagule influx can alter such trait evolution driven by biotic interactions in an open ecological network. As mentioned previously, with an influx of propagules the amount of population change of species i from time t to $t + \tau$ equals the amount of influx propagules $\gamma_i(t)\tau$ plus the amount of population change within the network, $n_i(t + \tau) - n_i(t) = \gamma_i(t)\tau + n_i(t)f_i(t)\tau$. We can therefore derive the following trait evolutionary dynamics under persistent propagule influxes (Hui et al. 2021),

$$\dot{x}_i = \frac{\gamma_i}{n_i}(z_i - x_i) + v_i s_i. \tag{7.2}$$

Trait evolution is driven by two components. The first term on the right-hand side depicts how the difference between the trait of influx propagules and the resident trait, $(z_i - x_i)$, can steer trait dynamics, while the evolutionary force from this component diminishes either when the influx propagules have the same trait as the resident population ($z_i = x_i$) or when the influx rate relative to the resident population size drops. This latter scenario could materialize either by halting the influx ($\gamma_i \to 0$) or with the natural increase of the resident population size. This first term therefore describes the effect of propagule pressure on trait evolution in an open ecological network. Although propagule pressure can have a lasting effect on population dynamics, its role in steering trait evolution declines with the natural increase of the resident population size. The second component is related to the selection gradient (s_i) that is jointly determined by the sensitivity of the intrinsic rate of increase to the trait change $(\partial r_i / \partial x_i)$ and the sensitivity of the total interaction pressure experienced by an average individual of species i to its trait change $\left(\sum_{j=1}^{S} n_j \left(\partial \alpha(x, x_j) / \partial x \big|_{x_i} \right) \right)$. Therefore, Eq. (7.1) and Eq. (7.2) depict the population and trait dynamics of a resident species in an open adaptive network.

The performance of an alien species depends on both the propagule pressure and the biotic interactions experienced as a function of the non-native species traits relative to the traits of resident species. The anticipated demographic performance, or the invasiveness, of the non-native species can be defined as its per-capita population growth rate $f_A = \dot{n}_A / n_A$. According to the population dynamics in Eq. 7.1, we can have the following equation for a newly introduced alien species,

$$\dot{n}_A = \gamma_A + n_A \left(r_A(x_A) + \alpha(x_A, x_A) n_A + \sum_{j=1}^{S} \alpha(x_A, x_j) n_j \right). \quad (7.3)$$

The demographic performance of a non-native species ($\dot{n}_A / n_A$) can be partitioned into four components (schematically shown in Figure 7.4), propagule pressure, intrinsic rate of increase, intraspecific density dependence and trait-mediated interspecific interaction pressure (total per-capita interaction strength from other resident species in the network). In particular, during the introduction phase, the umbrella effect of propagule pressure dominates the demographic performance, γ_A / n_A. If there is a constant influx of propagules (the yellow surface for $\gamma_A > 0$ in Figure 7.1), the non-native species can always grow from a zero initial population size ($n_A = 0$), with the initial population dynamics solely determined by propagule pressure, $\dot{n}_A = \gamma_A$. However, this does not guarantee successful establishment and invasion, as the roles of other components kick in as population size increases, while the effect of propagule pressure recedes. As part of the post-introduction stages, the second component in Eq. (7.3) reflects the intrinsic rate of growth, r_A. Large positive values of r_A contribute to faster population growth, whereas negative values indicate unsuitable abiotic environments or the presence of a critical Allee effect. The third component in Eq. (7.3) reflects density-dependent self-regulation and can be ignored when

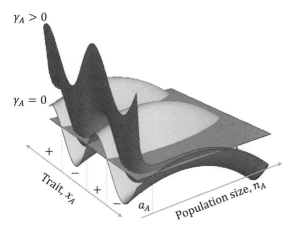

Figure 7.4 A detailed landscape of invasion fitness, illustrating the purpose of the concepts encompassed in the generic model proposed. Invasiveness of a non-native species ($\dot{n}_A/n_A$, vertical axis) is determined by its trait (x_A) and population size (n_A) as the horizontal plane, as well as propagule pressure (γ_A). The blue surface represents invasiveness for one-off introduction ($\gamma_A = 0$), while the yellow surface represents invasiveness with a constant rate of propagule influx ($\gamma_A > 0$). Green plane represents zero fitness (for reference purpose). For one-off introduction ($\gamma_A = 0$), invasion fitness (blue surface) divides the trait axis into a number of positive ($+$) and negative ($-$) performance pockets when the initial population size n_A is small. A non-native species possessing a trait value located in the positive pockets in the trait space will successfully invade, while other traits fail. For a particular non-native trait, its invasiveness also depends on the initial population size: for instance, when $n_A < a_A$ the non-native population could suffer from positive density dependence (Allee effect) and fail to establish, but can establish when $n_A > a_A$; ultimately, demographic and invasiveness will be constrained by negative density dependence (thus the blue and yellow curves eventually bend downwards when the population size n_A becomes too large (e.g., exceeding the carrying capacity)). From Hui et al. (2021), under CC-BY Licence. *A black and white version of this figure will appear in some formats. For the colour version, please refer to the plate section.*

considering the initial invasion performance and trait dynamics (when the population size is small).

The last component in the demographic performance of a non-native species (Eq. (7.3)) captures the role of interspecific biotic interactions from other resident species in the ecological network. It is the sum of interaction strength from all individuals of other resident species, where the interaction strength depends on the position of the non-native trait in the trait space relative to the resident traits. Through the influence of trait-mediated demographic fitness, interspecific interactions can constrain or expand the fundamental niche of an incoming non-native species in the trait space, forming pockets of positive or negative trait-dependent invasiveness (along the trait axis in Figure 7.4). These pockets of positive performance are opportunistic empty niches waiting to be invaded. Consequently, the invasibility of an ecological network is encapsulated by the empty niches in an ecosystem that remain under-exploited by resident species at a given time. Should an introduced species possess a trait within such an empty niche, it can establish even without the continuous influx of propagules. This means that the empty niches in an ecological network can be defined as pockets of trait space with positive demographic performance ($\dot{n}_A/n_A > 0$), typically under zero propagule influx ($\gamma_A = 0$) and negligible initial propagule pressure ($n_A \to 0$). The total width of empty niches over the entire feasible range of trait values, or the total area or volume of empty niches over the entire feasible range of 2D or 3D trait space, thus defines the invasibility of an ecological network, while the invasiveness of a particular invader (with its trait given) is defined by the corresponding height on the demographic performance surface (Figure 7.4; Hui et al. 2016, 2021).

The evolutionary dynamics of the non-native trait, $\dot{x}_A$, can also be formulated according to Eq. 7.2,

$$\dot{x}_A = \frac{\gamma_A}{n_A}(z_A - x_A) + v_A\left(\frac{\partial r_A}{\partial x_A} + n_A\frac{\partial \alpha(x,x_A)}{\partial x}\bigg|_{x_A} + \sum_{j=1}^{S}\left(n_j\frac{\partial \alpha(x,x_j)}{\partial x}\bigg|_{x_A}\right)\right).$$

$$(7.4)$$

Note that as the trait of influx propagules is the trait of the initial non-native population, the first term on the right side of Eq. (7.4), representing propagule pressure, does not contribute to the initial trait dynamics of the non-native species. It only begins steering trait dynamics once the trait difference emerges, although its influence wanes with increasing population size. Propagule pressure and the selection gradient (second term on the right hand side) may work synergistically to speed up trait evolution when the two forces have the same sign. However, they can also cancel each other to slow down trait evolution. The four equations listed in this section therefore capture the eco-evolutionary dynamics of all involved species in an open adaptive network invaded by non-native species. In the next two sections, we elaborate on the implications of network invasibility as defined previously and then use four examples to illustrate what kind of trait strategies can emerge as a robust winner of the multi-player game in open ecological networks.

7.3 Network Invasibility

Within the discussions on the role of functional traits in determining invasiveness in the literature (e.g., Catford et al. 2018; Enders et al. 2020), the prominent hypotheses in invasion science are centred on the concept of opportunistic ecological niches (e.g., Simberloff 1981; Herbold and Moyle 1986; Shea and Chesson 2002). An empty niche is defined by the specific absence of a species along particular gradients in the resource space (Hutchinson 1957; Holt 2009). Ecological niches in communities have been argued to be largely unsaturated and thus open to invasion (Simberloff 1981; Walker and Valentine 1984; Rohde 2005). The presence of unsaturated niches has been hypothesised to explain the lack of biotic resistance to some biological invasions and the lack of impact in some invaded communities (Mack et al. 2000; Sax et al. 2007). In a fitness landscape, empty niches are represented by pockets of positive fitness in the trait space that are 'waiting' to be filled through invasion or incremental trait evolution (Figure 7.4). Non-native species with traits that match these empty niches can establish in the newly invaded environment without the necessity of intensively competing with and affecting native species. The concept of invasiveness and invasibility can thus be measured by the shape and quantity of such empty niches in an ecological network (Lonsdale 1999; Shea and Chesson 2002; Hui et al. 2016). With the concept of the landscape of invasion fitness, these two concepts – niches and traits – are therefore closely mirrored, as in our model.

Once a non-native species has engaged with resident species in co-evolving dynamics, the network of biotic interactions can drastically adjust its fundamental niche to form the realised niche accessible to the non-native species. Co-evolution in the context of invasion could impose evolutionary barriers that could eventually prohibit trait radiation among resident species of a community. Evolutionary barriers are mechanisms that prohibit directional evolution of resident species with certain traits. In other words, rare mutants with traits similar to those of resident species cannot establish and replace the resident trait. This would constrain the distribution of functional traits in the community and potentially create empty niches that can only be filled by the invaders with *large* trait differences. For instance, co-evolution via facilitative interactions can generate positive reinforcing feedbacks (e.g., mutualism). Such reinforcing feedbacks can lead to a lock-in of trait evolution in a sub-optimal state in terms of functional trait distribution in resident species. This can happen when selfish mutualists benefit excessively from the interactions, thereby creating evolutionary barriers that thwart further radiation of traits. Such empty niches set up by evolutionary barriers can also give rise to priority effects: the traits and sequence/history of inva-sion can greatly affect how the fitness landscape – and therefore invasibility – of an ecological network unfolds (Minoarivelo and Hui 2018). Over ecological timescales that are relevant to invasion manage-ment, co-evolution can therefore have a great effect on the demograph-ics and invasiveness of both resident and non-native species (Saul and Jeschke 2015; Le Roux et al. 2017).

Coupled with rapid environmental changes, co-evolution of entan-gled biotic interactions can drive changes in trait-mediated and density-dependent interaction strengths (Thompson 2013), creating both empty niches and invasion barriers that are dynamic at both ecological and evolutionary timescales. As highlighted in the open adaptive network (Figure 7.4; Hui et al. 2021), such empty opportunistic niches with positive invasion fitness are enclosed by valleys of negative invasion fitness in the trait space (Figure 3.30; Hui et al. 2016). Similar to empty niches, ecological and evolutionary barriers are also constructed through trait-mediated biotic interactions that can drastically bend and reshape the invasion fitness landscape. Negative (antagonistic) interactions, such as those involving competitors and predators, can constrain the funda-mental niche and form ecological barriers to invasion with respect to specific non-native traits. In contrast, positive (mutualistic) interactions can expand the fundamental niches through the provision of mutualistic benefits into otherwise unsuitable niche space (Rodriguez-Cabal et al. 2012; Stachowicz 2012; Afkhami et al. 2014), thereby unravelling eco-logical barriers to invasion.

Invasibility defines the vulnerability and susceptibility of recipient eco-systems to invasion (Lonsdale 1999). In an open adaptive network, invasi-bility is related to community assembly processes and a result of the loss of system stability (see Chapter 3 and 4). In the trait space, invasibility can be measured as the size of empty niches (pockets of positive invasion fitness) that depends on how species are packed in the trait space in terms of trait density and dispersion. Both deterministic niche-based and stochastic neu-tral processes affect community assembly (Hubbell 2001; Tilman 2004), while network structure and architecture, as well as complex interactions (e.g., intransitive and higher-order ones), further affect species coexistence and thus network assembly (May 1974; McCann 2000; Allesina and Tang 2012). As shown earlier, dispersal can further cause the umbrella effect of propagule pressure (Simberloff 2009), affecting species performance and thus network stability. Spatial interdependence between local networks further affects a local network's stability and invasibility (Chapter 6). As a species' abundance is a proxy of its temporal performance in an ecological network, how these assembly processes and network structures are correlated with the rarity gradient and the weather vane of network dynamics provide further details for the staircase and doorway of network instability and invasibility (Chapter 5). We now move on to ask a few questions that can be typically addressed with trait-mediated interactions in ecological networks: (i) where the strategic locations in the trait space lie that correspond to these pockets of empty niches of an ecological network; and (ii) what trait dispersion of the recipient ecological network brings the most resistance or susceptibility to invasion.

7.4 Central-to-Reap, Edge-to-Elude

Centrality is a key metric of network location that can differentiate a well-connected hub from peripheral largely isolated nodes in a network (Chapter 3). When a network is fully parameterised by its interaction or adjacency matrix, many centrality metrics can be used to assess each node's network position, which often reflects the node's function and role in the network. In an ecological network, centrality allows us to identify those species that function as network hubs. With interaction strength formulated by interaction kernels, centrality can reveal the trait centroid and periphery in the trait space according to the trait dispersion of resident species. That is, with interaction kernels (interaction strength as a function of functional traits), we could identify the strategic trait position of an ecological network. Here we introduce a specific centrality metric proposed for open adaptive networks with trait-mediated interactions formulated in Section 7.2. Within a functional guild where species are engaging in competitive interactions that reach the maximum between species with the same trait value, we can define the following *centrality* of an alien species with trait x_A in the same functional guild (Figure 7.5; Hui et al. 2021),

$$C_A^{\{P\}} = 1 / \sum_{j=1}^{S} d_{Aj}^2 w_j, \qquad (7.5)$$

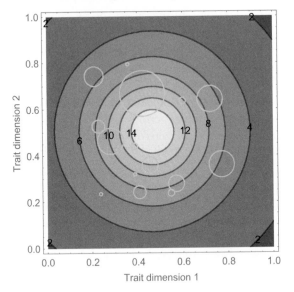

Figure 7.5 An illustration of centrality over two-dimensional trait space. White circles indicate trait positions of twenty species generated randomly (white circle centres), with the corresponding abundances (proportional to the radius) randomly generated from a geometric distribution with a mean of 5. From Hui et al. (2021), under CC-BY Licence.

where $d_{Aj} = x_A - x_j$ is the trait distance between the alien species and resident species j; $J_P = \sum n_j$ the total number of individuals of all resident species (i.e., the community size) and $w_j = n_j/J_P$ the relative abundance of species j. In other words, the centrality of the non-native species within the competitive functional guild is the inverse of the sum of weighted squares of the trait distance between resident species and the non-native species, with the relative abundances of resident species as weights. This definition reflects the overall pressure of biotic interactions experienced by the alien species. This centrality metric is related to, but is distinct from, the typical harmonic centrality in network science (Marchiori and Latora 2000). A variant of the denominator has been proposed as an index of biotic novelty (Schittko et al. 2020); this centrality therefore also captures functional familiarity of an invader to its recipient community. We now use the trait centrality metric to assess the necessary condition of elevated invasion performance in different types of ecological networks, including a flat competitive community, a bipartite mutualistic network, a bipartite antagonistic network and a food web.

Using the typical form of interaction kernel in Eq. (2.12), the trait-mediated interaction strength between two competitive species can be formulated as $\alpha(x_i, x_j) = -\exp(-(x_i - x_j)^2/2\sigma_P^2)$, with σ_P the width of the competition kernel. This kernel is symmetric and the strongest for intraspecific competition ($\alpha(x_i, x_i) = -1$). According to Eq. (7.1) and Eq. (7.2), we can thus formulate the ecological and evolutionary dynamics of all resident species before invasion. According to Eq. (7.3) and Eq. (7.4), we can formulate a non-native species invading an ecological community with S_p number of resident species (Figure 7.6A). For simplicity, we consider a small one-off introduction ($\gamma_A = 0$ and $n_A \to 0$) and therefore derive the following inequality to ensure the demographic performance $\dot{n}_A/n_A$ is positive,

$$C_A^{\{P\}} < \frac{J_P}{J_P - r_A} \frac{1}{2\sigma_P^2} \approx \frac{1}{2\sigma_P^2}. \tag{7.6}$$

This suggests that for a non-native species to be able to invade a competitive community, the centrality of its trait must be lower than a threshold set by the competition kernel. Invasion is more likely to succeed in communities with more specialised competitors (i.e., a narrower kernel); that is, such communities beget a high level of invasibility. Importantly, the requirement of low centrality suggests that the non-native species needs to possess a trait lying at the periphery of the traits of resident species in the community. In other words, to become invasive, the trait of the non-native species needs to be distinct from those of resident species to elude intense interspecific competition. This conforms to recent macroecological evidence derived from exploring the distributions of traits of vascular plants when divided into natives, archaeophytes and neophytes (Figure 7.7; Divíšek et al. 2018). So, to become invasive, under the scenario of horizontal resource competition, the trait of the non-native species needs to be distinct from those of resident species to elude intense interspecific competition.

Bipartite mutualistic networks are ubiquitous in nature (Bronstein 2015); they include pollination networks, seed dispersal networks and below-ground networks of plant–microbe symbiosis (e.g., Steidinger et al. 2019). The symmetric trait-mediated strength of mutualistic interaction can be formulated as $\alpha(x_i, y_j) = \exp(-(x_i - y_j)^2/2\sigma_M^2)$, where x_i and y_j are traits of two species from separate guilds that are engaging in assortative interactions, with σ_M the width of the interaction kernel of

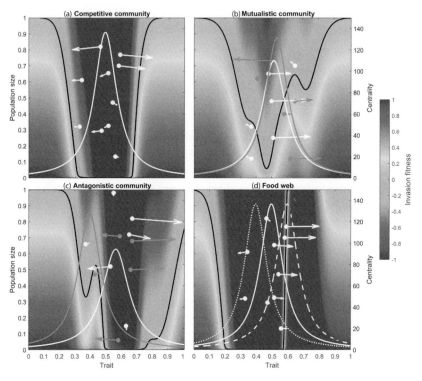

Figure 7.6 Four examples of eco-evolutionary dynamics on demographic fitness landscapes (see Figure 7.1). (a) A competitive community; (b) a bipartite mutualistic network between animals and plants with within-guild competition, facing the introduction of a non-native plant species; (c) a bipartite antagonistic network with within-guild competition, facing the introduction of a non-native resource species; (d) a food web with interspecific competition. White/grey dots represent trait and population size of resident species (white for plants and grey for animals in (b); white for resources and grey for consumers in (c)). Arrows represent the joint ecological dynamics (projection along the population size vertical axis) and evolutionary dynamics (projection along the trait horizontal axis) of resident species. For (b) and (c), grey dots and arrows indicate the other functional guild relative to the non-native species. The background colour represents the demographic performance of an incoming non-native ($\dot{n}_A / n_A$), with black contour lines representing zero invasion fitness. White bell-shaped lines represent within-guild non-native trait centrality (measured by Eq. (7.5)), while grey bell-shaped lines represent centrality for exploiting non-native mutualism (b) and non-native resource (c). In (d) dotted line represents centrality for consuming the non-native resource, while dashed line represents the centrality for the non-native to consume resident resource species. From Hui et al. (2021), under CC-BY Licence. *A black and white version of this figure will appear in some formats. For the colour version, please refer to the plate section.*

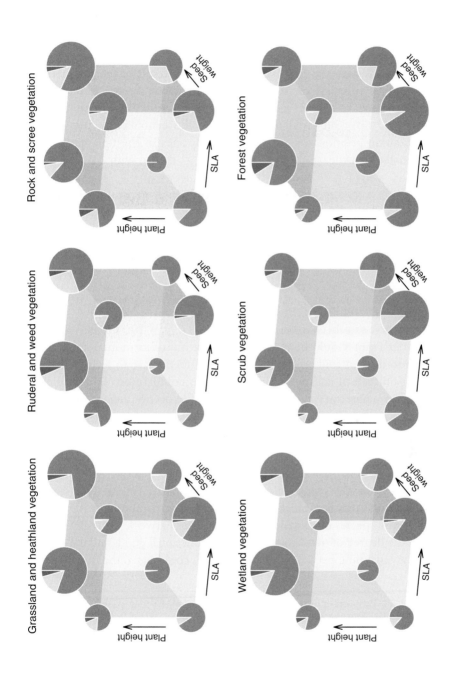

mutualism. For an alien species, let there be a number of S_P resident species within the same guild (e.g., flowering plants) that are assumed to engage in resource competition (see previous paragraph), and let there be a number of S_M species in the other guild engaging in mutualistic interactions with the non-native species. As a result of the symmetry of mutualistic interactions, we only need to formulate the demographic performance of a non-native species from its own functional guild (Figure 7.6B). Similarly, we can define the centrality of the optimal mutualistic position for the non-native species in the trait space of its mutualistic partner guild as $C_A^{\{M\}} = 1/\sum_{k=1}^{S_M} d_{Ak}^2 w_k$, with $d_{Ak} = x_A - y_k$ and $w_k = n_k/J_M$. Following the same line, let $\rho_{P|M} = J_P/J_M$ be the ratio of community size between the focal guild and the mutualistic partner guild, and we have the following inequality for positive demographic performance (Hui et al. 2021),

$$\frac{\rho_{P|M}/\left(2\sigma_P^2\right)}{C_A^{\{P\}}} + 1 > \frac{1/\left(2\sigma_M^2\right)}{C_A^{\{M\}}} + \rho_{P|M}. \tag{7.7}$$

To get a rough picture of this inequality, if $\rho_{P|M} = 1$ and $\sigma_P^2 = \sigma_M^2$, the inequality can be shortened to $C_A^{\{M\}} > C_A^{\{P\}}$; this implies that the invader's trait should be closer to the centroid of its mutualistic partners than to the centroid of its within-guild competitors so to reap the benefit from mutualism and elude the harm from competition. For the simple scenario where species within the same guild do not engage in resource competition, the inequality becomes $C_A^{\{M\}} > 1/\left(2\sigma_M^2\right)$. We can thus conclude that a successful invader in a mutualistic network needs to possess traits towards the centroid of the traits of its mutualistic partners, and in contrast, towards the periphery of the traits of those resident species from the same guild, consistent with the results from Bastolla et al. (2009) that low pressure from interspecific competition is necessary to ensure species coexistence in a mutualistic network. This also implies that the degree of trait dispersion within functional guilds and the level of trait overlapping between functional guilds could signal the invasibility of

Figure 7.7 Proportion of species occupying octants of the trait space in each habitat. Blue, yellow and red colours represent, in turn, native, naturalized non-invasive and invasive species. Octants were defined with respect to the centroid of the native species group in the trait space, i.e., by above-average or below-average SLA, plant height and seed weight of native species. The bigger the pie, the larger number of species occupies the region. From Divíšek et al. (2018), under CC-BY Licence.

a mutualistic network. This is consistent with our condition to elude competition pressure via peripheral trait positioning for invasion success. Moreover, the nested structure that characterises many mutualistic networks has been shown to play a key role in reducing competition (Bastolla et al. 2009). It is, however, debatable whether the nested structure can also affect the success of an invasion into such nested mutualistic networks (Figure 3.33; Minoarivelo and Hui 2016; Valdovinos et al. 2018). Although the nested structure can minimise the negative pressure from competition, it does not necessarily facilitate the establishment of an invader. To become a successful invader, a species needs to possess not only traits that are more similar to the centroid of the traits of its mutualistic partners but also traits that occur towards the periphery of the trait space of resident species from the same guild.

Bipartite antagonistic networks are also ubiquitous in nature (e.g., Morris et al. 2014; Nuwagaba et al. 2015), and include predator–prey networks and host–parasite networks. We use subscript C and R to denote the consumer and resource guilds, respectively. The trait-mediated interaction strength of resource consumption can be formulated as $\alpha(y_C, x_R) = \exp\left(-(y_C - x_R - \mu_G)^2 / (2\sigma_G^2)\right)$, where x_R and y_C are traits of the resource species and the consumer species from two different guilds, with $\mu_G > 0$ the optimal consumer to resource trait gap and σ_G the width of the resource consumption kernel. We further assume that there are competitive interactions between species within the same guild. Consequently, if the non-native species is part of the resource guild, we have the eco-evolutionary dynamics in Figure 7.6C. Let $C_{A+\mu_G}^{\{C\}} = 1/\sum_{k=1}^{S_C} d_{Ak}^2 w_k$, with $d_{Ak} = y_k - (x_A + \mu_G)$ and $w_k = n_k/J_C$, be the centrality of the maximum consumption position in the consumer trait space to consume the non-native resource species, and $C_A^{\{R\}}$ the centrality of the non-native species within the resource guild; we thus have the following inequality for an alien resource species to have positive demographic performance (Hui et al. 2021),

$$\frac{\rho_{R|C}/(2\sigma_R^2)}{C_A^{\{R\}}} + \frac{1/(2\sigma_G^2)}{C_{A+\mu_G}^{\{C\}}} > \rho_{R|C} + 1. \tag{7.8}$$

Again, if $\rho_{R|C} = 1$ and $\sigma_R^2 = \sigma_G^2$, the inequality can be shortened to $1/C_A^{\{R\}} + 1/C_{A+\mu_G}^{\{C\}} > \sigma_G^2$; the non-native resource species will have a better demographic performance when its optimal consumer position (with trait $x_A + \mu_G$) has a lower centrality. Without competition within the guild, we simply have $C_{A+\mu_G}^{\{C\}} < 1/\left(2\sigma_G^2\right)$. This means that if the non-native resource can only be effectively consumed by resident species at the periphery of the consumer trait space, the non-native resource will experience low levels of consumption from the resident consumers and thus have a higher chance of establishing and invading.

If the non-native species is part of the consumer guild, we can define $C_{A-\mu_G}^{\{R\}} = 1/\sum_{k=1}^{S_R} d_{Ak}^2 w_k$, with $d_{Ak} = \left(\gamma_A - \mu_G\right) - x_k$ and $w_k = n_k/J_R$ is the centrality of the non-native consumer's optimal resource position in the resource trait space, and $C_A^{\{C\}}$ is the centrality of the non-native species in the consumer trait space. We have the following inequality to ensure the non-native consumer a positive demographic performance (Hui et al. 2021),

$$\frac{\rho_{C|R}/\left(2\sigma_C^2\right)}{C_A^{\{C\}}} + 1 > \frac{1/\left(2\sigma_G^2\right)}{C_{A-\mu_G}^{\{R\}}} + \rho_{C|R}. \tag{7.9}$$

This bears similarity to the inequality of Eqn. (7.8). Following similar arguments, if $\rho_{C|R} = 1$ and $\sigma_C^2 = \sigma_G^2$, the inequality can be shortened to $C_{A-\mu_G}^{\{R\}} > C_A^{\{C\}}$; that is, the optimal resource position of the non-native species needs to be more central in the resource trait space than its position in the competitive consumer trait space. If we ignore within-guild competition, the inequality becomes $C_{A-\mu_G}^{\{R\}} > 1/\left(2\sigma_G^2\right)$; that is, the position of the non-native species' optimal resource should be closer to the centroid in the resource trait space, to ensure the invasion success of the non-native consumer. To become a successful invader, when the non-native is a resource species (e.g., prey), it should possess traits that ensure the traits of its optimal consumer locating at the periphery of the consumer trait space (to reduce consumption rates by residents); when the non-native is consuming resident resources, the position of the non-native species' optimal resource should be close to the centroid of resident resource trait space (to maximize consumption rates of the invader).

The eco-evolutionary dynamics of an alien species invading a food web are illustrated in Figure 7.6D. Following a similar procedure as above, we can derive the following condition for an alien species to possess positive demographic performance in a food web (Hui et al. 2021),

$$\frac{1}{2\sigma_P^2}\frac{1}{C_A} + \frac{1}{2\sigma_G^2}\left(\frac{1}{C_{A+\mu_G}} - \frac{1}{C_{A-\mu_G}}\right) > 1. \tag{7.10}$$

Assuming no competition, the inequality becomes $C_{A+\mu_G} < 1/\left(2\sigma_G^2 + 1/C_{A-\mu_G}\right)$, in addition to $C_{A+\mu_G} < C_{A-\mu_G}$. This means that the invader's optimal consumer position $(x_A + \mu_G)$ should be towards the trait periphery while its optimal resource position $(x_A - \mu_G)$ should be central in the trait space. The invader's optimal consumer position should lean towards the trait periphery while its optimal resource position should be central in the trait space. To ensure elevated performance, an alien species with a low trophic level needs to occupy trait periphery in the food web, while an alien species with a high trophic level needs to be in the trait centroid.

A cohesive trait strategy emerges: to be successful, invaders need to position their traits relative to the trait distributions of resident species from different functional guilds. They must also mitigate negative interactions by occupying peripheral trait positions and increase positive interactions by seeking central trait positions. This trait strategy of *central-to-reap, edge-to-elude*, highlights the leverage trait position in an ecological network for a non-native species to achieve elevated invasiveness. A similar strategy in repeated games, known as *win-stay, lose-shift*, describes heuristic learning of a player of its opponent by sticking to the same strategy that yielded a positive payoff in the last round of play but shifting to an alternative strategy if it resulted in a loss (see Chapter 2). This strategy has been confirmed as the most robust winning strategy in game theory (Nowak and Sigmund 1993). Evolution via natural selection, as discussed in Chapter 3, can also be effectively implemented as *fit-survive, unfit-eliminate* (Wallace 1889). This describes, again, the heuristic learning of a species to fine-tune its traits over evolutionary timescales. We also discussed the *good-stay, bad disperse* strategy of species movement when exploring heterogeneous resource landscapes (Chapter 6); a similar non-spatial recent experience-driven foraging strategy, *flexible when hungry, picky when stomach full*, also reflects the heuristic learning and decision making in an adaptive forager that can quickly reach optimal energy intake rate when facing a novel resource landscape (Zhang and Hui 2014). All these clues suggest that these heuristic learning strategies reflect a type of robust 'winning' strategy that can ensure an intruder outcompeting other players in the multiplayer game. In particular, the *central-to-reap, edge-to-elude* trait strategy connects a species' network position to its functional performance.

The *central-to-reap, edge-to-elude* strategy, of course, can be further elaborated by considering more realistic interaction kernels. For instance, within the hull of traits of the resident species in Figure 7.6 (with more specific non-linear interaction kernels), an introduced species with a trait similar to a resident species may have a greater probability of successful establishment as a result of the presence of required niches to ensure its survival. However, this same introduced species may suffer severely from biotic resistance, ultimately limiting its invasiveness (Divíšek et al. 2018). In contrast, an introduced species with a trait sitting between the traits of two resident species faces an uncertain outcome: either there is an empty niche to allow invasion or no niche available for invasion (Hui et al. 2016). However, it should be noted that excessive elaborations could increase the intrinsic system complexity, creating computational irreducibility and actually lowering the realised predictability (Beckage et al. 2011). Given the contextual complexity of any invasion event (Pyšek et al. 2020a), a fine balance of system elaboration that can clearly contain system uncertainty should be preferred (Latombe et al. 2019).

7.5 Management in a VUCA World

A special feature of an open adaptive system, when compared to a typical complex adaptive system, is that the components of the former, and their interactions, are fluid and are in constant transition and turnover. As discussed throughout this book, ecological networks facing constant bombardment by alien propagules are open adaptive systems; the transient membership of their resident species and the dynamic nature of interactions form and explain emerged structures, functions and dynamics of ecological networks. In a typical complex adaptive system, such as the neural network of a brain, the functioning of the system does not rely on the births and deaths of neurons but on the *action potentials* of neurons (pulses of electrical activity) to transmit and process information through the neural network (Gerstner et al. 1997). Neural networks and ecological networks are similar in some respects. Recent syntheses in neurobiology suggest that neural networks can self-organize to near criticality, which tentatively explains brain activities such as epilepsy and neonatal hypoxia (Beggs and Plenz 2003; Wilting and Priesemann 2019; Zimmern 2020), as well as the level of consciousness (Demertzi et al. 2019). As we have shown, an ecological network under constant biological invasions is also self-organized at its marginal instability (criticality). Before an ecosystem reaches criticality, its ecological networks can tolerate and accommodate a growing assemblage resulting from persistent influxes of propagules, but only until criticality is reached, at which point an avalanche of species extinctions and invasion failures is triggered. Self-organized criticality is thus a robust attractor of system evolution in open adaptive networks (Bak et al. 1988). The features of this criticality include diverse contextually dependent network structures and dynamics, while the two forces responsible for the fluid nature of an ecological network – influxes of alien and regional propagules and co-evolution within the network – push the system to operate near marginal instability. To this end, we can simply replace the slowly accumulating grains of sand in the sand-pile model (Figure 4.23) with species of different functional shapes (traits). The system will self-organize towards marginal instability, spontaneously exhibiting many structural features at criticality such as the power law, including species abundance distribution, node-degree distribution and dispersal kernel, as well as spatial synchrony.

The functional level of our brain neural network, as Sigmund Freud proposed, can be divided into unconscious, preconscious and conscious. A recent review suggested four types of system consciousness (Morin 2006): unconsciousness – being non-responsive to self and environment; consciousness – focusing attention on the environment and processing incoming external stimuli; self-awareness – focusing attention on self and processing private and public self-information; meta-self-awareness – being aware that one is self-aware. Adopting this classification, ecological networks can 'experience', 'digest' and 'respond' to environmental perturbations and biotic intrusions by fine-tuning and rotating their constituent components and interactions. Ecological networks, being perturbed by biological invasions at marginal instability, can also operate at different levels of organization depending crucially on the introduction and influx rate of alien propagules. It resembles the laminar and turbulent flow of tap water, the flow patterns being hypersensitive to the speed and pressure of water flow. Human-mediated biological invasions and reshuffling of biotic interactions are happening at a fast rate and at great magnitude, leading to novel network structures and ecosystem functions that spontaneously emerge at criticality. Following this train of thought, ecological networks could be considered as possessing 'consciousness', while invasion and conservation management impose 'self-awareness' on a targeted ecosystem. How does such informed invasion management address the challenge facing a 'conscious' invaded ecosystem that is open, adaptive and deeply embedded in a rapidly changing socioeconomic system? The feasibility and desirability of management targets and practices requires deep consideration from the 'consciousness' of invasion and conservation management when targeting such highly open, adaptive and novel ecosystems (Hobbs et al. 2014).

Let us apply the concept of VUCA (Volatility, Uncertainty, Complexity and Ambiguity) in dissecting the challenges facing current invasion science along the spectra of systems knowledge [how much we know about the system] and predictability [how well we can predict the system's response to intervention] (see Figure 1.24). We find that Invasion Science 1.0 has progressed steadily beyond the zone of Ambiguity regarding invasion phenomena. Invasion Science 2.0 is now edging into the Complexity block to quantify key drivers and their interplay (the theme of this book) and into the Uncertainty block to address known unknowns (as evident from recent collective attempts to fill data shortages on historical and current trends of different taxa, regions and pathways). To elucidate Complexity and minimise Uncertainty pertinent to managing invaded ecosystems, we need to map out the interaction matrix of embedded ecological networks. This can be achieved by estimating interaction strengths between any two or more species from comprehensive datasets of their functional traits, or by using new molecular techniques (e.g., generating large network structures from next-generation sequencing based on co-occurrence patterns of co-exclusions and co-associations; Figure 7.8). This necessitates a stronger focus on traits of species (including their origin) and functional roles of species. In response to the criticism of invasion science as being obsessed with profiling species according to their origin (Davis et al. 2009), we think it is necessary to use origin as a proxy for the potential functional role in an ecological network, especially when information on species traits is scarce. In the absence of trait data for mapping quantitative networks, semi-quantitative or even qualitative networks based on the knowledge of local experts and drawing on published sources would suffice for a rapid assessment (e.g., Figure 4.15).

Ecological networks interact with socioeconomic networks to affect the perception and management actions towards specific components (e.g., alien species) and the entirety (e.g., nature as culture) of the hyper-network (The QUINTESSENCE Consortium 2016; Kull et al. 2018). Removing alien species from such, often human-dominated, systems is increasingly impractical, and management is in some (but not all) cases more effectively directed at managing features of such novel ecosystems to ensure the sustainable delivery of certain functions or services (Richardson 2015). Management is a prioritisation art, while competing agendas wax and wane. Climate change, economy and employment, disease and human wellbeing, globalisation and food security are all issues interlinked with ecosystem management; these issues are increasingly

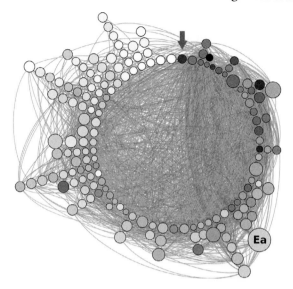

Figure 7.8 Microbial association network of the leaves of an oak tree (*Quercus robur* L.) susceptible to the foliar fungal pathogen *Erysiphe alphitoides* (Ea). Each node represents a microbial taxon (either bacterial or fungal) and each link represents a significant correlation between their abundances. Dark and light grey links indicate co-exclusions and co-associations, respectively. The arrow indicates the node with the highest degree (i.e., the highest total number of links). Degree decreases clockwise, with nodes stacked on the same line having the same degree. The size of the nodes is inversely proportional to the sum of the correlation coefficients: larger nodes have more numbers and/or stronger negative associations. Darker nodes have higher betweenness centrality (calculated on the absolute values of associations), suggesting that they are topological keystone taxa. *E. alphitoides* is predominantly connected to the network through strong negative links (co-exclusions) but is not a good candidate for topological keystone species. Legend from Bohan et al. (2017); figure reproduced with permission from Vacher et al. (2016).

competing for resources and attention from the public and policy makers. How should we better allocate management resources to prevent the influx of particular species, or preserve some critically endangered species and interactions? Feasibility and costs are key factors to consider when formulating effective and sustainable management actions. Managing open adaptive ecosystems is not simply about comparing which management targets are better. Managers must increasingly face the stark realisation that perhaps none of the management targets on the table will be achieved. Many debates relating to options for managing invasions are, thus, about comparing desired, but largely untenable, targets. Based on the principle of VUCA, a conservation framework comprising four interrelated phases for managing the vulnerability and risks in ecosystems subjected to human influence has been proposed (Figure 7.9). This is designed for implementing adaptive management strategies aimed at reducing, not eliminating, human impacts (Ibisch and Hobson 2014; Schick et al. 2017).

Although invasion science and the trajectory of biological invasions are yet to encounter volatility as conceptualized in the VUCA model, we are seeing early warning signals. The escalating scale of biological invasions, and the increasing complexity of synergies between invasions and other facets of global change that generate multifaceted influences on

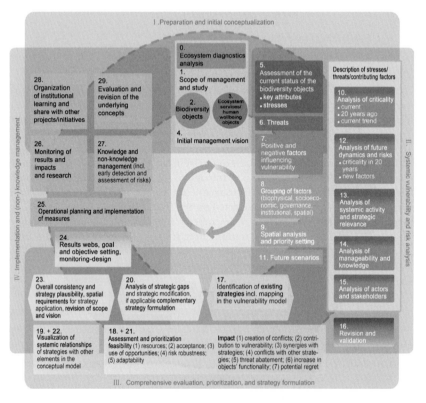

Figure 7.9 The steps of the management of vulnerability and risks at conservation method. Phase I: Preparation and initial conceptualization; phase II: Systemic risk and vulnerability analysis; phase III: Comprehensive evaluation, prioritization and strategy formulation and phase IV: Implementation and (non-) knowledge management. From Schick et al. (2017), under CC-BY Licence.

ecosystems are creating truly daunting problems for managers. Restoring ecosystems to some 'pristine' or 'historical' condition or pre-invasion status is, in most cases, impractical and infeasible. Invasion management aimed simply at removing invasive species often exacerbates rather than reduces problems. Measures aimed at preventing the introduction of high-risk alien species and at early detection and, where feasible, eradication of such species before they spread and cause damage are crucial. However, the roles of alien species in human-dominated ecosystems need to be viewed pragmatically.

Various sub-fields of invasion science have developed strong momentum and promise to deliver important new insights on the dynamics of invasions and on ways to manage them. For example, much remains to be understood about pathways of biological invasions and how these could potentially be managed. To this end, Essl et al. (2015) advocate more attention to the following techniques and analyses in 'pathway science': spatio-temporal changes in pathways and other covariates of invasions; network analysis of pathways; identifying future changes of pathways: horizon scanning; and geographic profiling. Recommendations on key gaps to be filled have also been made on a diversity of other sub-fields of invasion science, e.g., on methods for engaging with stakeholders on the management of non-native species (Novoa et al. 2018); developing guidelines for the sustainable use of certain useful non-native species to prevent and mitigate negative impacts (Brundu et al. 2020); and provision of a standardized protocol for quantifying impacts (Kumschick et al. 2015). We are definitely not suggesting a deviation from such priorities – they clearly reflect the need to diversify and to send the roots and tentacles of invasion science into many other fields to discover new approaches and to formulate solutions.

We do suggest, however, that the toolbox of invasion science needs a rather radical make-over to better equip researchers to tackle the increasingly wicked problems associated with biological invasions and other facets of global change. This book has highlighted the potential benefits to be gained by viewing invaded ecosystems as open adaptive networks, with critical transitions and turnover, with resident species learning heuristically and fine-tuning their niches and roles in a multiplayer eco-evolutionary game. It defines an agenda for Invasion Science 2.0 by providing new framings and a classification of research topics and by offering tentative solutions to vexing problems. A cursory examination of the glossary and the reference list of this book will make it clear that the terrain to be traversed to seize the opportunities that this approach offers will be unfamiliar to most researchers working on biological invasions. The book has provided a catalogue of options to guide the re-stocking of the toolbox for dealing with open adaptive networks to explore innovative eco-evolutionary games, the core–peripheral role of functional traits, a weather vane for dynamic forecasting, commonness–rarity gradients, and many analytic and numerical tools. We hope that you will share our excitement and enthusiasm for the journey that lies ahead.

We end with T. H. Huxley's (1869) rendering of J. W. von Goethe's Aphorisms on Nature,

Nature! We are surrounded and embraced by her: powerless to separate ourselves from her, and powerless to penetrate beyond her. Without asking, or warning, she snatches us up into her circling dance, and whirls us on until we are tired, and drop from her arms. She is ever shaping new forms: what is, has never yet been; what has been, comes not again. Everything is new, and yet nought but the old. . . . She is always building up and destroying; but her workshop is inaccessible. . . . Each of her works has an essence of its own; each of her phenomena a special characterisation: and yet their diversity is in unity. . . . Her mechanism has few springs–but they never wear out, are always active and manifold. The spectacle of Nature is always new, for she is always renewing the spectators. Life is her most exquisite invention; and death is her expert contrivance to get plenty of life. . . . She is complete, but never finished. As she works now, so can she always work. Everyone sees her in his own fashion. She hides under a thousand names and phrases, and is always the same. She has brought me here and will also lead me away. I trust her. She may scold me, but she will not hate her work. It was not I who spoke of her. No! What is false and what is true, she has spoken it all. The fault, the merit, is all hers.

References

Afkhami ME, McIntyre PJ, Strausss SY (2014) Mutualist-mediated effects on species' range limits across large geographic scales. *Ecology Letters* 17, 1265–1273.

Allesina S, Tang S (2012) Stability criteria for complex ecosystems. *Nature* 483, 205–208.

Bak P, Tang C, Wiesenfeld K (1988) Self-organized criticality. *Physical Review A* 38, 364.

Barks PM, et al. (2018) Among-strain consistency in the pace and shape of senescence in duckweed. *Journal of Ecology* 106, 2132–2145.

Bastolla U, et al. (2009) The architecture of mutualistic networks minimizes competition and increases biodiversity. *Nature* 458, 1018–1020.

Beckage B, Gross LJ, Kauffman S (2011) The limits to prediction in ecological systems. *Ecosphere* 2, 1–12.

Beggs JM, Plenz D (2003) Neuronal avalanches in neocortical circuits. *Journal of Neuroscience* 23, 11167–11177.

Bohan DA, et al. (2017) Next-generation global biomonitoring: Large-scale, automated reconstruction of ecological networks. *Trends in Ecology & Evolution* 32, 477–487.

Bronstein J (ed.) (2015) *Mutualism*. Oxford: Oxford University Press.

Brundu G, et al. (2020). Global guidelines for the sustainable use of non-native trees to prevent tree invasions and mitigate their negative impacts. *NeoBiota* 6, 65–116.

Caswell H (2001) *Construction, Analysis, and Interpretation*. Sunderland: Sinauer.

Catford JA, Bode M, Tilman D (2018) Introduced species that overcome life history tradeoffs can cause native extinctions. *Nature Communications* 9, 2131.

Chave J, et al. (2009) Towards a worldwide wood economics spectrum. *Ecology Letters* 12, 351–366.

Davis MA, et al. (2009) Don't judge species on their origins. *Nature* 474, 153–154.

Demertzi A, et al. (2019) Human consciousness is supported by dynamic complex patterns of brain signal coordination. *Science Advances* 5, eaat7603.

Díaz S, et al. (2015) A Rosetta stone for nature's benefits to people. *PLoS Biology* 13, e1002040.

Díaz S, et al. (2016) The global spectrum of plant form and function. *Nature* 529, 167–171.

Dieckmann U, Law R (1996) The dynamical theory of coevolution: A derivation from stochastic ecological processes. *Journal of Mathematical Biology* 34, 579–612.

Divišek J, et al. (2018) Similarity of introduced plant species to native ones facilitates naturalization, but difference enhance invasion success. *Nature Communications* 9, 4631.

Enders M, et al. (2020) A conceptual map of invasion biology: Integrating hypotheses into a consensus network. *Global Ecology and Biogeography* 29, 978–991.

Essl F, et al. (2015) Crossing frontiers in tackling pathways of biological invasions. *BioScience* 65, 769–782.

Geritz SAH, et al. (1998) Evolutionarily singular strategies and the adaptive growth and branching of the evolutionary tree. *Evolutionary Ecology* 12, 35–57.

Gerstner W, et al. (1997) Neural codes: Firing rates and beyond. *Proceedings of the National Academy of Sciences USA* 94, 12740–12741.

Gross N, et al. (2017) Functional trait diversity maximizes ecosystem multifunctionality. *Nature Ecology and Evolution* 1, 0132.

Herbold B, Moyle PB (1986) Introduced species and vacant niches. *The American Naturalist* 128, 751–760.

Hobbs RJ, et al. (2014) Managing the whole landscape: Historical, hybrid, and novel ecosystems. *Frontiers in Ecology and the Environment* 12, 557–564.

Holt RD (2009) Bringing the Hutchinsonian niche into the 21st century: Ecological and evolutionary perspectives. *Proceedings of the National Academy of Sciences USA* 106, 19659–19665.

Hubbell SP (2001) *The Unified Neutral Theory of Biodiversity and Biogeography*. Princeton: Princeton University Press.

Hui C, Richardson DM (2017) *Invasion Dynamics*. Oxford: Oxford University Press.

Hui C, et al. (2016) Defining invasiveness and invasibility in ecological networks. *Biological Invasions* 18, 971–983.

Hui C, et al. (2021) Trait positions for elevated invasiveness in adaptive ecological networks. *Biological Invasions* 23, 1965–1985.

Hulme PE, et al. (2008) Grasping at the routes of biological invasions: A framework for integrating pathways into policy. *Journal of Applied Ecology* 45, 403–414.

Hutchinson GE (1957) Concluding remarks. *Population Studies: Animal Ecology and Demography. Cold Spring Harbor Symposium on Quantitative Biology* 22, 415–457.

Huxley TH (1869) Nature: Aphorisms by Goethe. *Nature* 1, 9–11.

Ibisch PL, Hobson PR (2014) *MARISCO: Adaptive Management of Vulnerability and Risk at Conservation Sites. A Guidebook for Risk-robust, Adaptive and Ecosystem-based Conservation of Biodiversity.* Eberswalde: Centre for Econics and Ecosystem Management.

Jeschke J, Heger T (eds.)(2018) *Invasion Biology: Hypotheses and Evidence.* Wallingford: CAB International.

Jeschke JM, et al. (2014). Defining the impact of non-native species: Resolving disparity through greater clarity. *Conservation Biology* 28, 1188–1194.

Kull CA, et al. (2018) Using the 'regime shift' concept in addressing social-ecological change. *Geographical Research* 56, 26–41.

Kumschick S, Richardson DM (2013) Species-based risk assessments for biological invasions: Advances and challenges. *Diversity and Distributions* 19, 1095–1105.

Kumschick, S et al. (2015). Ecological impacts of alien species: Quantification, scope, caveats and recommendations. *BioScience* 65, 55–63.

Latombe G, et al. (2019) A four-component classification of uncertainties in biological invasions: Implications for management. *Ecosphere* 10, e02669.

Le Roux JJ, et al. (2017) Co-introduction versus ecological fitting pathways to the establishment of effective mutualisms during biological invasions. *New Phytologist* 215, 1354–1360.

Lonsdale WM (1999) Global patterns of plant invasions and the concept of invasibility. *Ecology* 80, 1522–1536.

Mack RN, et al. (2000) Biotic Invasions: Causes, epidemiology, global consequences, and control. *Ecological Applications* 10, 689–710.

Marchiori M, Latora V (2000) Harmony in the small-world. *Physica A: Statistical Mechanics and its Applications* 285, 539–546.

May RM (1974) Biological Populations with nonoverlapping generations: Stable points, stable cycles, and chaos. *Science* 186, 645–647.

McCann KS (2000) The diversity-stability debate. *Nature* 405, 228–233.

McGill BJ, et al. (2006) Rebuilding community ecology from functional traits. *Trends in Ecology & Evolution* 21, 178–185.

Mello MAR, et al. (2019) Insights into the assembly rules of a continent-wide multilayer network. *Nature Ecology & Evolution* 3, 1525–1532.

Minoarivelo HO, Hui C (2016) Invading a mutualistic network: To be or not to be similar. *Ecology and Evolution* 6, 4981–4996.

Minoarivelo HO, Hui C (2018) Alternative assembly processes from trait-mediated co-evolution in mutualistic communities. *Journal of Theoretical Biology* 454, 146–153.

Morin A (2006) Levels of consciousness and self-awareness: A comparison and integration of various neurocognitive views. *Consciousness and Cognition* 15, 358–371.

Morris RJ, et al. (2014) Antagonistic interaction networks are structured independently of latitude and host guild. *Ecology Letters* 17, 340–349.

Novoa A, et al. (2018). A framework for engaging stakeholders on the management of alien species. *Journal of Environmental Management* 205, 286–297.

Nowak M, Sigmund K (1993) A strategy of win-stay, lose-shift that outperforms tit-for-tat in the Prisoner's Dilemma game. *Nature* 364, 56–58.

Nuwagaba S, Zhang F, Hui C (2015) A hybrid behavioural rule of adaptation and drift explains the emergent architecture of antagonistic networks. *Proceedings of the Royal Society B: Biological Sciences* 282, 20150320.

Pyšek P, et al. (2020a). Macroecological Framework for Invasive Aliens (MAFIA): Disentangling large-scale context dependency in biological invasions. *NeoBiota* 62, 407–461.

Pyšek P, et al. (2020b) Scientists' warning on invasive alien species. *Biological Reviews* 95, 1511–1534.

The QUINTESSENCE Consortium (2016) Networking our way to better ecosystem service provision. *Trends in Ecology & Evolution* 31, 105–115.

Richardson DM (2015) Conservation is complicated, and all approaches need to be on the table. Dispatches from the front line suggest an important role for a novel ecosystem approach to conservation. Ensia Available online at https://ensia.com/voices/conservation-is-complicated-and-all-approaches-need-to-be-on-the-table/ (accessed 22 February 2021).

Rodriguez-Cabel MA, et al. (2012) Disruption of ant-seed dispersal mutualisms by the invasive Asian needle ant (*Pachycondyla chinensis*). *Biological Invasions* 14, 557–565.

Rohde K (2005) *Nonequilibrium Ecology*. Cambridge: Cambridge University Press.

Salguero-Gómez R, et al. (2016) Fast-slow continuum and reproductive strategies structure plant life-history variation worldwide. *Proceedings of the National Academy of Sciences USA* 113, 230–235.

Salguero-Gómez R, et al. (2018) Delivering the promises of trait-based approaches to the needs of demographic approaches, and vice versa. *Functional Ecology* 32, 1424–1435.

Saul W, Jeschke JM (2015) Eco-evolutionary experience in novel species interactions. *Ecology Letters* 18, 236–245.

Sax DF, et al. (2007) Ecological and evolutionary insights from species invasions. *Trends in Ecology & Evolution* 22, 465–471.

Schick A, et al. (2017) Conservation and sustainable development in a VUCA world: The need for a systemic and ecosystem-based approach. *Ecosystem Health and Sustainability* 3, e01267.

Schittko C, et al. (2020) A multidimensional framework for measuring biotic novelty: How novel is a community? *Global Change Biology* 26, 4401–4417.

Segar ST, et al. (2020) The role of evolution in shaping ecological networks. *Trends in Ecology & Evolution* 35, 454–466.

Shea K, Chesson P (2002) Community ecology theory as a framework for biological invasions. *Trends in Ecology & Evolution* 17, 170–176.

Simberloff D (1981) Community effects of introduced species. In Nitecki MH (ed.) *Biotic Crises in Ecological and Evolutionary Time*, pp. 53–81. New York: Academic Press

Simberloff D (2009) The role of propagule pressure in biological invasions. *Annual Review of Ecology, Evolution, and Systematics* 40, 81–102.

Stachowicz J (2012) Niche expansion by positive interactions: Realizing the fundamentals. A comment on Rodriguez-Cabal et al. *Ideas in Ecology and Evolution* 5, 42–43.

Steidinger BS, et al. (2019) Climatic controls of decomposition drive the global biogeography of forest-tree symbioses. *Nature* 569, 404–408.

Thompson JN (2013) *Relentless Evolution*. Chicago: University of Chicago Press.

Tilman D (2004) Niche tradeoffs, neutrality, and community structure: A stochastic theory of resource competition, invasion, and community assembly. *Proceedings of the National Academy of Sciences USA* 101, 10854–10861.

Vacher C, et al. (2016) Learning ecological networks from next-generation sequencing data. *Advances in Ecological Research* 54, 1–39.

Valdovinos FS, et al. (2018) Species traits and network structure predict the success and impacts of pollinator invasions. *Nature Communications* 9, 2153.

Vasseur F, et al. (2018) Adaptive diversification of growth allometry in the plant *Arabidopsis thaliana*. *Proceedings of the National Academy of Sciences USA* 115, 3416–3421.

Walker TD, Valentine JW (1984) Equilibrium models of evolutionary species diversity and the number of empty niches. *The American Naturalist* 124, 887–899.

Wallace AR (1889) *Darwinism*. London: MacMillan.

Wilson JR, Panetta FD, Lindgren C (2017) *Detecting and Responding to Alien Plant Incursions*. Cambridge: Cambridge University Press.

Wilson JRU, et al. (2009) Something in the way you move: Dispersal pathways affect invasion success. *Trends in Ecology & Evolution* 24, 136–144.

Wilson JRU, et al. (2020) Frameworks used in invasion science: Progress and prospects. *NeoBiota* 62, 1–30.

Wilting J, Priesemann V (2019) 25 years of criticality in neuroscience: Established results, open controversies, novel concepts. *Current Opinion in Neurobiology* 58, 105–111.

Wright IJ, et al. (2004) The worldwide leaf economics spectrum. *Nature* 428, 821–827.

Zhang F, Hui C (2014) Recent experience-driven behaviour optimizes foraging. *Animal Behaviour* 88, 13–19.

Zimmern V (2020) Why brain criticality is clinically relevant: A scoping review. *Frontiers in Neural Circuits* 14, 54.

Glossary

Adaptive dynamics – (also known as evolutionary invasion analysis) A set of modelling techniques that trace the dynamics of trait evolution as propelled by the **selection gradient** according to **invasion fitness**.

Adaptive foraging – The heuristic behaviour of a forager whereby it optimises its energy intake rate through trial-and-error, modifying its diet according to resource availability and profitability, often in a novel or changing environment.

Adaptive management – An iterative management practice that constantly assesses the context and performance and revises the management target and strategy, especially when facing high complexity and uncertainty.

Adjacency matrix – A matrix with its entries indicating the adjacency (presence or absence) of links between nodes corresponding to the column and the row.

Alien biome – (also known as alien species assemblage zone) A biogeographical cluster of co-occurring **alien species**.

Alien species – (synonyms: adventive-, exotic-, foreign-, introduced-, non-indigenous-, non-native species) Species whose presence in a region is attributable to human actions that enabled them to overcome fundamental biogeographical barriers (i.e., human-mediated extra-range dispersal).

Antagonistic network – An **ecological network** that only considers antagonistic interactions between species, often between species of two separate guilds (e.g., a host–parasite **network**).

Anthropocene – The geological era during which human activity has imposed substantial influence on climate and the environment.

Bifurcation – The qualitative change of dynamic behaviour in the system regime from a minute smooth change in a system parameter.

Biological invasions – (synonyms: bioinvasions, biotic invasions, species invasions) The phenomenon of and suite of processes involved in determining: (1) the transport of organisms through human activity (intentionally or accidentally, via **Introduction pathways**) to areas outside the potential range of those organisms as defined by their natural dispersal mechanisms and biogeographical barriers; and (2) the fate of such organisms in their new ranges, including their ability to survive, establish, reproduce, disperse, spread, proliferate, interact with **resident** biota and exert influence in many ways on and in invaded ecosystems.

Biotic interactions – Relationships established between at least two organisms of one or more species. The outcome of such relationships may result in the involved organisms benefitting, being harmed or being unaffected, depending on the environmental context in which the interaction takes place.

Biotic resistance – A notion, derived from **limiting similarity theory**, which argues that species-rich communities can withstand and even resist **biological invasions** (but see **invasion paradox**).

Bipartite network – A type of complex **network** whose nodes are divided into two guilds, with typically only cross-guild links considered.

Bistability – The state of a system that possesses two (or more) basins of attraction and can operate at alternative regimes depending solely on the initial values of state variables.

Centrality – A measure describing a node's relative position in **network** topology; a node's centrality reflects its importance.

Central-to-reap, edge-to-elude – A heuristic learning trait and strategic position, in the trait space, for introduced species to establish and invade a trait-mediated **ecological network**.

Co-evolution – Trait evolution through two (or more) species engaging in **biotic interactions** that reciprocally affect each other's fitness, posing frequency-dependent selection on each other.

Community assembly – (synonym: community succession) The formation and succession of species assemblages in a community, often driven by disturbance, and increasingly mediated by **biological invasions** (see also **community disassembly**).

Complex adaptive system (CAS) – A decentralized system consisting of many diverse and autonomous components or parts (agents) which are interrelated, interdependent and linked through many interconnections, and which behave as a unified whole by learning from experience and adjusting (not just reacting) to changes in the environment.

Complexity–stability relationship – The relationship between system complexity (as measured by both the number of and connections between components) and stability (measured primarily by system **resilience** and **persistence**). The debate on this relationship is rooted in the need for inference of system function from system structure, or inferring processes from observed patterns. The relationship that complexity leads to the loss of system stability has, arguably, led to the modular design of many **complex systems** (e.g., aircraft; to reduce system complexity so to ensure stability).

Complex systems – Systems whose behaviour is intrinsically difficult to model due to the dependencies, competitions, relationships or other types of interactions between their parts or between a given system and its environment.

Complexity theory – A set of concepts that attempts to explain complex phenomena not explainable by traditional (mechanistic) theories. It integrates ideas derived from chaos theory, cognitive psychology, computer science, evolutionary biology, general systems theory, fuzzy logic, information theory and other fields to deal with the natural and artificial systems as they are, and not by simplifying them (breaking them down into their constituent parts). It recognizes that complex behaviour emerges from a few simple rules, and that all **complex systems** are **networks** of many interdependent parts which interact according to those rules.

Connectance – The number of realised links in a **network** divided by the number of feasible links.

Critical slowing down (CSD) – see **early warning signals**.

Darwin's naturalization hypothesis – The notion that **alien species** with close native relatives in their introduced range may have reduced chances of establishment and **invasion**; based on ideas formulated by Charles Darwin in Chapter 3 of *The Origin of Species*, drawing on ideas from Alphonse de Candolle.

Dispersal kernel – The probability of an individual moving between two locations, often as a decaying function of the distance between the locations. Typical forms include negative exponential (e.g., Brownian motion) or power law (e.g., Lévy flight).

Dynamical system – A set of (often) differential equations with each depicting the dynamics of a state variable (a node) as a function of its interactions with other components in a system.

Early warning signals (EWS) – Detectable system behaviours that appear when the system approaches the tipping point of its current regime. A popular EWS is the **critical slowing down** of system dynamics, which is often accompanied by elevated variations.

Ecological fitting – The process whereby organisms establish novel associations with other species or display **interaction switching**, especially when colonizing a novel environment or experiencing environmental change, to preserve the supply of essential resources, often as a result of the suites of traits that they carry at the time for resource acquisition.

Ecological network – (synonym: ecological interaction **network**) A web of co-evolving and co-fitting interactions among species in an **ecosystem**.

Ecosystem – A biological community of interacting organisms and their physical environment.

Eigenvector – Vectors that do not change their directions after matrix transformation; eigenvalues depict whether its corresponding eigenvector has been extended or shortened.

Empty-niche hypothesis – The view that **ecological networks** with specialized interactions could hamper the effect of **co-evolution,** leaving unexploited niches from incremental evolution, thus creating opportunities for **alien species** to establish and exploit such empty niches through **ecological fitting**.

Evolution – Trait dynamics of a multiplayer game, with each trait following the rule of natural selection (heuristically, fit-survive, unfit-eliminate).

Evolution of cooperation – Theories, e.g., inclusive fitness and reciprocity, that explain the invasion and establishment of cooperative strategies in a community with selfish defective individuals.

Evolutionarily stable strategy (ESS) – A strategy which, if adopted by a population in a given environment, cannot be invaded by any alternative strategy that is initially rare.

Food web – An **ecological network** with its resident species only engaging in feeding and trophic interactions (normally resulting in energy transferring from one species to another).

Forecasting conundrum – The low predictability, and sometimes impossibility, of forecasting the future dynamics and behaviour of **complex adaptive systems** even though the past dynamics and retrospective behaviours are highly structured.

Game theory – The study of strategic interactions among rational decision-makers based on the benefit and cost of each player's strategy.

Good-stay, bad-disperse – A dispersal strategy whereby individuals choose to stay or leave a habitat based on their experience of habitat quality.

Heuristic learning – A generic learning strategy through trial and error.

Higher-order interaction – Interactions often involving individuals from more than three species.

Hill number – A unified measure of biodiversity that encompasses species richness, the exponential of the Shannon index and the reciprocal of the Simpson index as special cases.

Impact – The description or quantification of how an **alien species** affects the physical, chemical and biological environment.

Integrodifference equation – A type of spatial modelling technique with the unique feature that the movement of individuals between locations can be implemented with any specific or realistic **dispersal kernel**.

Interaction kernel – The formulation of **interaction strength** as a function of the traits of interacting species, often as a function of their trait complementarity and degree of trait matching. In assortative interactions, **interaction strength** typically declines with increasing trait dissimilarity.

Interaction matrix – A matrix with its entries recording the interaction strength between corresponding species represented by the specific column and row.

Interaction rewiring – see **interaction switching**

Interaction strength – The demographic impact of a species on another, normally measured as the sensitivity of the population change rate of the receiving species to changes in the population size of the acting species. It is often measured using interaction strength proxies that only consider the sensitivity of one specific demographic component of the receiving species.

Interaction switching – (synonym: species rewiring) The phenomenon whereby species can switch their interaction partners in an **ecological network**, either randomly or adaptively. Interaction switching can happen at different temporal scales, from rapid changes in behaviours to slow changes that coincide with phylogeographical events.

Intransitive interactions – A number of directed interactions forming a loop that normally involves at least three species and lacks a definite winner from the interactions (e.g., rock–paper–scissors), as opposed to transitive interactions that have a clear hierarchic structure in outcome of the interactions.

Introduction – Movement of a species, intentionally or accidentally, as a result of human activity, from an area where it is **native** to a region outside that range ('**introduced**' is synonymous with **alien**). The act of introduction (inoculation of propagules) may or may not lead to **invasion**.

Introduction-naturalization-invasion (INI) continuum – A conceptualization of the progression of stages and phases in the status of an **alien** organism in a new environment which posits that the organism must negotiate a series of barriers. The extent to which a species is able to negotiate sequential barriers (mediated by **propagule pressure** and **residence time** and other factors) determines the organism's status as an **alien species**, i.e., whether it will become **naturalized** (established) and/or **invasive** (Richardson et al. 2000 *Diversity Distribution* 6, 93–107).

Introduction pathway – The processes that result in the **introduction** of **alien** species from one geographical location to another.

Invasibility – The properties of a community, habitat, ecosystem or **network** that determine its vulnerability to **invasion**. Early studies applied the concept deterministically (particular systems were deemed either invasible or not), but invasibility is more appropriately considered probabilistically, and the degree of invasibility may change markedly over time because, for instance, of changes in biotic or abiotic features.

Invasion debt – A concept that posits that even if **introductions** cease (and/or other drivers of **invasion** are relaxed, e.g., **propagule pressure** is reduced), new invasions will continue to emerge and already-invasive species will continue to spread and cause potentially greater impacts, since large numbers of alien species are already present, many of them in a lag phase (Essl et al. 2010 *Proceedings of the National Acadamy of Sciences USA* 108, 203–207; Rouget et al. 2016 *Diversity Distribution* 22, 445–456).

Invasion dynamics – Changes in population demography, niches, spatiotemporal patterns and genetic/phenotypic makeup during the **introduction** and spread of **alien species**, and the ecological, evolutionary and socio-economic responses of recipient **social-ecological systems** (Hui and Richardson 2017 *Invasion Dynamics*. Oxford: Oxford University Press).

Invasion ecology – The study of the causes and consequences of the **introduction** of organisms to areas outside their native range as governed by their dispersal mechanisms and biogeographical barriers. The field deals with all aspects relating to the **introduction** of organisms; their ability to establish, **naturalize** and **invade** in the target region; their interactions with **resident organisms** in their new location; and the consideration of costs and benefits of their presence and abundance with reference to human value systems. This term is often used interchangeably with 'invasion biology'; see also **invasion science**.

Invasion fitness – The per-capita population growth rate or fitness of a particular trait when its population size is negligibly small.

Invasion frameworks – A way of organising concepts relating to biological invasions that can be easily communicated to allow for shared understanding or that can be implemented to allow for generalisations useful for research, policy or management (Wilson et al. 2020 *NeoBiota* 62, 1–30).

Invasion paradox – The positive correlation between **native** and **alien species** at landscape and regional scales that seemingly defies predictions from the **biotic resistance hypothesis**.

Invasion science – A term used to describe the full spectrum of fields of enquiry that address issues pertaining to **alien species** and **biological invasions**. The field embraces **invasion ecology**, but increasingly involves non-biological lines of enquiry, including economics, ethics, sociology and inter- and transdisciplinary studies (see also **Invasion Science 1.0**; **Invasion Science 2.0**).

Invasion Science 1.0 – The field of **invasion science** as defined by agendas that prevailed mainly in the period 1980 to 2010, whereby synergies were sought mainly with population ecology, initially, and macroecology, more recently, to understand **biological invasions** and to guide management.

Invasion Science 2.0 (the theme of this book) – The field of **invasion science** that seeks to move beyond the state of paradox that exists in the field with low predictability despite significant knowledge and data availability. It emphasises the openness and adaptability of invaded **ecological networks**, and seeks synergies from **network** and systems sciences.

Invasion syndrome – A combination of pathways, alien species traits and characteristics of the recipient ecosystem which collectively result in predictable dynamics and impacts, and that can be managed effectively using specific policy and management actions (Novoa et al. 2020 *Biological Invasions* 22, 1801–1820).

Invasional meltdown – A phenomenon whereby **alien species** facilitate one another's establishment, spread and **impacts** (Simberloff and von Holle 1999 *Biological Invasions* 1, 21–32).

Invasive species – **Alien species** that sustain self-replacing populations over several life cycles, produce reproductive offspring, often in very large numbers at considerable distances from the parent and/or site of **introduction**, and have the potential to spread over long distances. Invasive species are a subset of **naturalized species**; not all **naturalized species** become invasive. This definition explicitly excludes any connotation of **impact**, and is based exclusively on ecological and biogeographical criteria (Wilson et al. 2009 *Trends in Ecology & Evolution* 24, 586).

Invasiveness – The features of an **alien** organism, such as its life-history traits and modes of reproduction that define its capacity to invade, i.e., to overcome various barriers to **invasion**. The level of invasiveness of a species can change over time as a result of, for example, changes in genetic diversity through hybridization, introgression or the continued arrival of new propagules of the same species that is already established in a region, but from new and different (meta)populations, such that genetic diversity may increase. This latter concept is important in management strategies, which sometimes assume that less concern needs to be paid to the continued **introduction** of species (the continued arrival of propagules, whether accidental or intentional) that are *already* well-established in a region, overlooking the critical potential for elevated **invasiveness** over time.

Island biogeography – The study of species composition and species richness on islands (or insular habitats), aimed at establishing and explaining the factors that affect species diversity of a specific community.

Jacobian matrix – The linear approximation of a dynamical system at a particular point (often at one equilibrium), captured in a matrix.

Limiting similarity theory – A concept in theoretical ecology and community ecology that proposes the existence of a maximum level of niche overlap between two given species that will allow continued coexistence.

Link-species scaling (law) – A law that states that species interact with a constant number of species independently of species richness.

Marginal instability – The condition when a system has just lost the stability and resilience of its current regime. It is different from a typical tipping point because the alternative regime, as required to define a tipping point, has yet to form in the system, or will not form as a result of persistent system transitions and turnover. **Complex**

adaptive systems do not often evolve towards maximum stability, but only marginal stability; system stability is not the target for system evolution (Nuwagaba et al. 2015 *Proceedings of the Royal Society B* 282,20150320).

Metacommunity – Multiple communities maintained at the regional level, with migrations of propagules between local communities and from regional species pool constantly invading local communities.

Meta-network – An extension of the concept of **metacommunity** where each local community consists of a local **ecological network**; propagules can be exchanged by dispersal of individuals between different local **networks**.

Meta-population – Local populations, often in ephemeral patches, connected via dispersal of individuals and persist through the recolonization of empty patches.

Modifiable Areal Unit Problem (MAUP) – A statistical biasing effect that arises when samples in a given area are used to represent information such as density in a given area. The area defined by an analyst is often arbitrary, which means that measurement (e.g., of density) could be deceptive because that measure could have widely different results depending on the shape and scale chosen for analysis.

Modularity – The measure of the structure of **networks**. It indicates the strength of division of a **network** into modules (also called groups, clusters or communities). **Networks** with high modularity have dense connections between the nodes within modules but sparse connections between nodes in different modules.

Moran effect – The effect of spatially autocorrelated environmental forcing on species distributions.

Multilayer network – An **ecological network** that comprises different types of biotic interactions.

Mutualistic network – An **ecological network** in which one class of nodes represents one type of species (e.g., plants) and the other class represents another type of species (e.g., pollinators), while links connecting nodes of the two different classes represent the mutualistic interactions (e.g., pollination).

Native species – One that originated in a given area without human involvement or that arrived there without intentional or unintentional intervention of humans from an area in which it is native (compare with **alien species**).

Naturalized species – (synonym: established species) Alien species that sustain self-replacing populations over considerable periods without direct intervention by people (or despite human intervention) by recruitment from propagules capable of independent growth (the concept is mainly applied in plant invasion ecology).

Nestedness – A measure of **network** asymmetry where highly specialised species only interact with a subset of interacting partners of generalised species.

Network – a representation of biotic interactions in an ecosystem in which species (nodes) are connected by pairwise interactions.

Network architecture – (synonyms: network structure, network topology) The structure of the **adjacency matrix** of a **network**, depicting the **interaction strength** between any two nodes in a **network**. Metrics, such as nestedness, modularity and node-degree distribution, are used to depict the matrix/**network** structure.

Network assembly – (related to network transition, network dynamics and network turnover) A depiction of the dynamics of both the changes in species composition and the interactions between residing species.

Network compartmentalisation – The process whereby network nodes form motifs or modules with the **network** exhibiting a high level of **modularity**.

Network complexity – The number of nodes and links in a **network**.

Network emergence – Processes that give rise to the formation of particular **network architecture** and functions, often from an initially randomly assembled **network**.

Network invasibility – The areal size of positive **invasion fitness** in the feasible trait space for an **ecological network** with trait-mediated interactions. **Network** invasibility is equivalent to system instability when **interaction strength** is not specified as trait dependent.

Network percolation – A theory that describes the phase transition of connectivity in a **network**. With the increase of the number of randomly placed links, a **network** can change from many scattered clusters to suddenly forming a single connected cluster; the number of links needed is called the percolation threshold.

Network persistence – The **network** that results via ecological fitting is often only a subset of a randomly assembled initial **network**. The proportion of what remains defines persistence.

Network regime – The particular **network** structure when the **network** has converged to a steady state. It does not necessarily require each node to reach its equilibrium but the overall structure maintains the same, especially for open adaptive networks.

Network scaling – The change of **network** structure and regime when assessed over an increasing physical area or an increasing sampling effort (area and duration).

Network size – The number of nodes in a **network**. It is species richness when each species is represented by a single node (not always the case).

Novel ecosystems – **Ecosystems** comprising species that occur in combinations and relative abundances that have not occurred previously at a given location or biome. Such ecosystems result from either the degradation or **invasion** of natural ecosystems (those dominated by **native species**) or the abandonment of intensively managed systems (Hobbs et al. 2006 *Global Ecology and Biogeography* 15, 1-7).

Open adaptive system (OAS) – (Synonym: open adaptive network) – A special but arguably typical type of **complex adaptive system** (CAS) that emphasises both adaptiveness of its components (and their interactions), as is the case in a typical CAS, and the feature of openness (often captured as a constant influx of alien and native propagules in an **ecological network**).

Open network emergence – The emergence of **network** structure and regime that is predominantly driven by the influx of propagules, without neglecting the role of co-evolution within the **network**.

Optimal foraging theory – A theory stating that natural selection favours animals whose behavioural strategies maximize their net energy intake per unit time spent foraging.

Panarchy – A conceptual framework that depicts the process of the formation, collapse, transformation and reorganisation of regimes in **complex systems**.

Priority effect – The order and timing of species colonization during **community assembly** can affect community structures across scales, with early pioneer species playing important roles in deciding the trajectory of community succession.

Prisoner's Dilemma – A paradox in decision analysis in which two players acting in their own self-interests do not produce the optimal outcome. The typical prisoner's dilemma is set up so that both players choose to protect themselves at the expense of the other participant, even though mutual cooperation can lead to higher payoff to each.

Propagule pressure – A concept that encompasses variation in the quantity, quality, composition and rate of supply of **alien** organisms resulting from the transport conditions and pathways between source and recipient regions (Lockwood et al. 2005 *Trends in Ecology & Evolution* 5, 223–228).

Rapid evolution – Changes in heritable trait distribution or allele frequency within a population over a few generations.

Regime shift – Large, abrupt and persistent changes in the structure and function of an **ecosystem**, as a result of either **bifurcation** from changing ecosystem processes or jumping basins of attraction due to a large disturbance.

Reid's paradox (of Rapid Plant Migration) – The failure of ordinary reaction–diffusion equations, which produce a Gaussian tail to the dispersal curve, to reconstruct certain range expansions, notably the postglacial rate of spread of trees. Only by invoking long-distance dispersal can such expansions be accurately reconstructed.

Residence time – The time since the **introduction** of a species to a region.

Resident species (biota/organisms) – Species that are present in an **ecosystem**, community, habitat or region at the time of **introduction** of an **alien species**. The pool of resident species includes both **native species** and **alien species** introduced previously.

Resilience – Capacity of a system to maintain structure, functioning and feedbacks despite shocks and perturbations.

Robustness – The ability of an **ecosystem** to withstand species loss, often measured as the number of species removed so that the number of secondary extinctions could reach 50 per cent of the original species richness.

Scale resonance – Patterns often explained strongly by processes operating at similar spatiotemporal scales.

Selection gradient – The slope/gradient in the landscape of **invasion fitness** in the trait space; can be calculated as the partial derivate of **invasion fitness** with respect to mutant traits.

Self-organised criticality (SOC) – A process of open system evolution towards a critical point where system structures and functions, which often follow power laws, emerge spontaneously. In invaded **ecological networks**, **network** structures can be achieved by self-organisation via **ecological fitting** and **biological invasions**.

Singularity – The trait strategy where selection gradient diminishes.

Social-ecological system – An ecological system intricately linked with and affected by one or more social systems.

Spatial network – The spatial realisation and arrangement of an **ecological network**; see also **meta-networks**.

Spatial scaling – The phenomenon of changing structures in species distribution and **ecological networks** as one changes the spatial scale of survey.

Spatial sorting – The phenomenon whereby strong dispersers accumulate at the advancing range front and potentially also through assortative mating at the range front. This leads to acceleration of range expansion.

Spatial synchrony – The covariance of time series of two geographically distinct populations could be explained by shared environmental forcing (**Moran effect**), common natural enemies, and migration (**Meta-populations**).

Species rewiring – see **Interaction switching**

Species turnover – Compositional changes of species assemblages across sites (or time); pairwise turnover is typically measured as beta diversity and multi-site turnover as **zeta diversity**.

Spectrum – The distribution of eigenvalues of a matrix.

Stability criterion – An equation of inequality that depicts the condition of a stable **network**, often including the features of **interaction strength**, **network size**, complexity and connectance.

Steady state – A state that often exists in open systems whereby a static structure is reached (e.g., a constant rate of heat dissipation).

Strategic positions – Positions in a **network** where addition or removal of nodes (and interactions) at that place could cause profound changes in the **network** structure and function. With trait-mediated interactions, this can be visualised in trait space.

Structural emergence models – A set of mathematical models that can explain the emergent **network** structures from implementing a number of often trait-mediated processes.

System stability – The size and shape of the current basin of attraction of the current regime and its ability to withstand perturbations.

Tipping point – The specific state variable values that divide the basins of attraction of two regimes in the phase plane.

Tragedy of the Commons – A problem in economics occurring when individuals neglect the wellbeing of society in pursuing personal gain, leading to over-consumption and ultimately depletion of the common resource, to everybody's detriment.

Trait – (synonym: strategy) Any measurable characteristic of an individual. Functional traits are those that affect performance and, ultimately, fitness.

Trait dispersion – The trait distribution of resident species of a community in the trait space.

Transient dynamics – System dynamics that have not reached an equilibrium, as opposed to asymptotic dynamics.

Unified framework for biological invasions – A construct that reconciles and integrates key features of the most commonly used invasion frameworks into a single conceptual model that is applicable to all **biological invasions**. It combines previous stage-based and barrier models, and provides a terminology and categorisation for populations at different points in the invasion process (Blackburn et al. 2011 *Trends in Ecology & Evolution* 26, 333–339).

VUCA – Volatility, uncertainty, complexity and ambiguity. A framework proposed by the US Army War College to handle complex and uncertain situations.

Weather vane – The major axis or the eigenvector of the leading eigenvalue of a dynamic **network**'s **Jacobian matrix**, revealing the transient dynamics of each node in an open adaptive **network** at **marginal instability** resulting from invasions (Hui and Richardson 2019 *Trends in Ecology & Evolution* 31, 121–131).

Wicked problems – Management problems where the cause-and-effect relationships between components, be they logistical components or stakeholders involved in management, are unordered and thus have solutions that are not obvious and require collaboration among stakeholders to determine appropriate actions.

Win–stay, lose–shift – A generalised tit-for-fat reciprocal game strategy that has been shown to be the robust winner in game contests.

Zeta diversity – The extension of the concept of pairwise similarity to multi-site similarity. It provides a common currency to express all incidence-based biodiversity and **network** measures, and can be used to test hypotheses that differentiate turnovers from common versus rare species in metacommunities, and from generalised versus specialised interactions in **meta-networks** (Hui and McGeoch 2014 *The American Naturalist* 184, 684–694).

Index

Note: Page numbers in bold refer to definitions